Proceedings
of the
STEKLOV INSTITUTE
OF MATHEMATICS

1985, ISSUE 3

Algebraic Geometry and Its Applications

Editor-in-chief of Trudy Mat. Inst. Steklov
S. M. Nikol'skiĭ

Assistant to the Editor-in-chief of Trudy Mat. Inst. Steklov
E. A. Volkov

Translation of
ТРУДЫ
ордена Ленина
МАТЕМАТИЧЕСКОГО ИНСТИТУТА
имени В. А. СТЕКЛОВА

Том 165 (1984)

АКАДЕМИЯ НАУК

СОЮЗА СОВЕТСКИХ СОЦИАЛИСТИЧЕСКИХ РЕСПУБЛИК

ТРУДЫ

ордена Ленина

МАТЕМАТИЧЕСКОГО ИНСТИТУТА

имени В. А. СТЕКЛОВА

CLXV

АЛГЕБРАИЧЕСКАЯ ГЕОМЕТРИЯ И ЕЕ ПРИЛОЖЕНИЯ

издательство ''наука''
МОСКВА 1984

Library of Congress Cataloging-in-Publication Data

Algebraicheskaia geometriia i ee prilozheniia.
English.
Algebraic geometry and its applications.
(Proceedings of the Steklov Institute of Mathematics, 0081-5438; 1985, issue 3)
Translation of: Algebraicheskaia geometriia i ee prilozheniia.
Number in Russian series statement: t. 165 (1984)
1. Geometry, Algebraic—Addresses, essays, lectures. I. Nikol'skiĭ, S. M. (Sergei Mikhaĭlovich), 1905– . II. Volkov, E. A. III. Title. IV. Series: Trudy Ordena Lenina Matematicheskogo instituta imeni V. A. Steklova. English; 1985, issue 3.
QA1.A413 1985, issue 3 510s [516.3'5] 85-28762
[QA564]
ISBN 0-8218-3092-9

PROCEEDINGS OF THE STEKLOV INSTITUTE OF MATHEMATICS IN THE ACADEMY OF SCIENCES OF THE USSR

TABLE OF CONTENTS

ABSTRACT. This collection consists of original work in algebraic geometry and its applications to K-theory, arithmetic, differential equations and theoretical physics.

1980 *Mathematics Subject Classification*. Primary 01A70, 05B05, 05B35, 06B15, 10C15, 10D20, 12A80, 12B10, 12B22, 12B25, 12B35, 12E15, 13C15, 14E05, 14F05, 14J25, 16A39, 16A65, 17B20, 17B25, 17B40, 17E65, 18E10, 18F25, 20-02, 20E07, 20G35, 42C10, 53C20, 53C80, 58F05, 81E10; Secondary 10C02, 10D05, 10D40, 12A25, 12A65, 12B22, 13C13, 14C10, 14E05, 14F05, 14J05, 14J25, 14K30, 14M20, 17D06, 17D08, 32J25, 35Q20, 52A10, 53C55, 83E15, 94B25.

RUSSIAN TABLE OF CONTENTS*

*The American Mathematical Society scheme for transliteration of Cyrillic may be found at the end of index issues of *Mathematical Reviews.*

Correspondence between *Trudy Mat. Inst. Steklov.* and *Proc. Steklov Inst. Math.*

Russian imprint	Vol.	English imprint
1966	74	1967
1965	75	1967
1965	76	1967
1965	77	1967
1965	78	1967
1965	79	1966
1965	80	1968
1966	81	1968
1966	82	1967
1965	83	1967
1965	84	1968
1966	85	1967
1965	86	1967
1966	87	1967
1967	88	1969
1967	89	1968
1967	90	1969
1967	91	1969
1966	92	1968
1967	93	1970
1968	94	1969
1968	95	1971
1968	96	1970
1968	97	1969
1968	98	1971
1967	99	1968
1971	100	1974
1972	101	1975
1967	102	1970
1968	103	1970
1968	104	1971
1969	105	1971
1969	106	1972

Russian imprint	Vol.	English imprint
1971	107	1976
1968	108	1971
1971	109	1974
1970	110	1972
1970	111	1972
1971	112	1973
1970	113	1972
1970	114	1974
1971	115	1974
1971	116	1973
1972	117	1974
1972	118	1976
1973	119	1976
1974	120	1976
1972	121	1974
1973	122	1975
1973	123	1975
1976	124	1978, Issue 3
1973	125	1975
1973	126	1975
1975	127	1977
1972	128	1974
1973	129	1976
1978	130	1979, Issue 4
1974	131	1975
1973	132	1975
1973	133	1977
1975	134	1977
1975	135	1978, Issue 1
1975	136	1978, Issue 2
1976	137	1978, Issue 4
1975	138	1977

Russian		English	imprint
imprint	Vol.	Year	Issue
1980	139	1982	1
1976	140	1979	1
1976	141	1979	2
1976	142	1979	3
1977	143	1980	1
1979	144	1980	2
1980	145	1981	1
1978	146	1980	3
1980	147	1981	2
1978	148	1980	4
1978	149	1981	3
1979	150	1981	4
1980	151	1982	2
1980	152	1982	3
1981	153	1982	4
1983	154	1984	4
1981	155	1983	1
1980	156	1983	2
1981	157	1983	3
1981	158	1983	4
1983	159	1984	2
1982	160	1984	1
1983	161	1984	3
1983	162	1985	1
1984	163	1985	4
1983	164	1985	2
1984	165	1985	3

Труды Матем. Инст. Стеклов.
165(1984)

Proceedings of the Steklov
Institute of Mathematics
1985, Issue 3

FROM THE EDITORIAL BOARD

The present collection contains papers in algebraic geometry and its applications to K-theory, arithmetic, differential equations, and theoretical physics. These papers break up naturally into several cycles.

The first cycle involves work in which a local class field theory is constructed, the higher Milnor K-functors are computed in terms of the cohomology of the general linear group, the torsion in the group K_2 of a local field is computed, the multiplicative structure of division algebras of arbitrary finite index over number fields is investigated, and perfect ideals in commutative rings are studied.

Papers which deal with the connections between different aspects of the theory of quadratic forms and algebraic geometry make up the second cycle. There the classification of K3 surfaces with a finite automorphism group is completed, the question of analytic sections in bundles on a conic is investigated, the birational automorphism group is computed for certain rational surfaces, and a connection is established between Hecke operators and representations of binary quadratic forms. Also, an application of the theory of even unimodular lattices to coding theory is given, and the groups of lattice automorphisms compatible with the multiplicative decomposition of a simple Lie algebra of type G_2 are found.

The next cycle consists of papers on algebraic groups and representations. Here a description is given of the lattice of intermediate subgroups of the full linear group, and the indecomposable representations of finitely presented transcendental lattices over division rings are studied.

The fourth and last cycle contains papers on applications of algebraic geometry to differential equations and physics. In these articles analytic Poisson brackets are investigated, new solutions of the Yang-Mills equation are constructed, and the indictrix of a four-dimensional manifold is studied.

On the whole, the papers in this volume are concentrated around contemporary problems in algebraic geometry, one of the topical areas of mathematics, whose development in our country is closely tied to the name of the Lenin Prize Laureate and Corresponding Member of the Soviet Academy of Sciences, Professor I. R. Shafarevich. Over the last quarter century a strong school of algebraic geometry has been created in our land, and a great deal of credit for this is certainly due to Igor Rostislavovich Shafarevich, who turned 60 in June 1983.

Translated by H. H. McFADEN

Труды Матем. Инст. Стеклов.
165(1984)

Proceedings of the Steklov
Institute of Mathematics
1985, Issue 3

HECKE OPERATORS AND REPRESENTATIONS OF BINARY QUADRATIC FORMS

UDC 511.342 + 517.54

A. N. ANDRIANOV

ABSTRACT. The author finds explicit expressions for the representation matrices of all regular Hecke operators on the spaces of degree 1 and 2 theta-series of integral positive definite binary quadratic forms in terms of the Eichler matrices of these forms. He finds all of the eigenvectors of the Eichler matrices and all of the eigenfunctions of the Hecke operators in this situation. A technique is given for applying Hecke operators to study the number of integral representations of binary forms by binary forms.

Bibliography: 10 titles.

§1. Introduction

Why are Hecke operators needed for the study of integral representations of quadratic forms by quadratic forms? In the classical situation of representation of numbers by quadratic forms, at least if the forms have an even number of variables, the answer is simple: the Fourier coefficients of the modular forms and, in particular, the number of integral representations of numbers by quadratic forms can be expressed explicitly in terms of the eigenvalues of Hecke operators acting on suitable spaces (see [1] and [2]), and this, because of the connections between Hecke operators and algebraic geometry and representation theory, enables one to give a virtually complete solution to the Diophantine problem under consideration. In the situation of representation of forms by forms, one also has Hecke operators (they always exist), their connections with the representation theory of Lie groups are no less promising, and there seems to be no reason to doubt the existence of connections with algebraic geometry, although this still remains for the future. One is less encouraged, however, by the prospects for applying information about the Hecke operators to study the number of representations of forms by forms or, more generally, the Fourier coefficients of Siegel modular forms. The computations that have been done show that the Hecke operators are not sufficient to be able to express the Fourier coefficients of the eigenfunctions in terms of their eigenvalues; the most that one can obtain is certain relations between the eigenvalues and the Fourier coefficients (see [3]–[5]). It appears that the path leading directly from classical analogies is not very hopeful. But it is possible that other techniques of applying Hecke operators which are not constrained by classical analogy are potentially of deeper significance for the representation of forms by forms. The one way to clarify this is to turn to the example of Diophantine equations.

1980 *Mathematics Subject Classification*. Primary 10C15, 10D20; Secondary 10C02, 10D40, 10D05.

0081-5438/85 $1.00 + $.25 per page

In the present paper we examine the role of Hecke operators in the simplest nonclassical case—the problem of integral representations of binary quadratic forms by one another. The results of the paper are stated in §3. Perhaps these formulas will suggest which questions to raise and how to approach them, at least in the situation of integral representations of quadratic forms by forms in the same number of variables.

General notation. $\mathbf{Z}$ and $\mathbf{C}$ are, respectively, the ring of rational integers and the field of complex numbers.

$M_\rho(\mathbf{A})$ is the ring of square matrices of order ρ over the ring $\mathbf{A}$. If $M, N \in M_\rho(\mathbf{A})$, then we set

$$N[M] = MN{}^tM \quad \text{and} \quad e\{M\} = \exp(\pi i \sigma(M)),$$

where ${}^tM$ is the transpose of M and $\sigma(M)$ is the trace of M.

If $a, b \in \mathbf{Z}$, we let (a, b) denote their greatest common divisor; $a|b$ means that a divides b.

§2. Definitions

2.1. Binary quadratic forms. Let

$$Q(x, y) = ax^2 + bxy + cy^2$$

be a binary quadratic form. Q is called an *integral* form if $a, b, c \in \mathbf{Z}$. In this case the greatest common divisor $\delta = \delta(0)$ of the numbers a, b, c is called the *divisor of the form* 0. A form is called *primitive* if its divisor is 1. The matrix

$$Q = \begin{Vmatrix} 2a & b \\ b & 2c \end{Vmatrix}$$

is called the *matrix of the form* Q, and the number $D(Q) = b^2 - 4ac$ is called its *discriminant*.

We let $\mathfrak{N}$ and $\mathfrak{N}^+$, respectively, denote the set of matrices of all semidefinite (positive definite) integral binary quadratic forms. The sets

$$\{Q\} = \{Q[u];\ u \in U\} \quad \text{and} \quad \{Q\}_+ = \{Q[v];\ v \in V\},$$

where

$$U = \mathrm{GL}_2(\mathbf{Z}) \quad \text{and} \quad V = \mathrm{SL}_2(\mathbf{Z}),$$

are called, respectively, the *class* and the *proper class* of the matrix $Q \in \mathfrak{N}$.

There is a natural bijection between the set $H_+(D)$ of all proper classes of primitive matrices $Q \in \mathfrak{N}$ of discriminant $D(Q) = D < 0$ and the group of classes of similar modules of the field $K = \mathbf{Q}(\sqrt{D})$ having as a ring of multipliers the subring of discriminant D of the ring of integers of this field (see, for example, [6], Chapter II, §7). Thus, the set $H_+(D)$ can be regarded as an abelian group, and classes in $H_+(D)$ can be multiplied together (Gaussian composition of quadratic forms) and multiplied by suitable modules of the field K.

If $Q, Q' \in \mathfrak{N}$, we shall let $r(Q, Q')$ denote the number of integral representations of the form Q' by the form Q, i.e., the number of solutions of the equation $Q[X] = Q'$ in integral 2×2 matrices X; we let $r^*(Q, Q')$ denote the number of reduced representations, i.e., for which the entries in X are relatively prime; and we let $e(Q) = r(Q, Q)$ denote the number of units of the form Q.

For every $D < 0$ we arbitrarily fix an ordered set of representatives

$$R(D) = (Q_1, \dots, Q_h) \tag{2.1}$$

of all classes $\{Q\}$ for primitive $Q \in \mathfrak{N}$ of discriminant D. If $R(D) = (Q_1, \dots, Q_h)$ and $R(D') = (Q'_1, \dots, Q'_{h'})$, then the $h \times h'$ matrices of the form

$$\begin{aligned} E_{D,D'}(m) &= \left(e(Q_i)^{-1} r\left(Q_i, mQ'_j\right)\right), \\ E^*_{D,D'}(m) &= \left(e(Q_i)^{-1} r^*\left(Q_i, mQ'_j\right)\right) \end{aligned} \tag{2.2}$$

will be called, respectively, the *Eichler matrices* and the *reduced Eichler matrices*; we shall write $E_{D,D}(m) = E_D(m)$ and $E^*_{D,D}(m) = E^*_D(m)$. It is clear that

$$E_D(m) = \sum_{d^2|m} E^*_D(m/d^2), \qquad E^*_D(m) = \sum_{d^2|m} E_D(m/d^2)\mu(d), \tag{2.3}$$

where μ is the Möbius function.

2.2. Modular forms and Hecke operators. Let $\rho, q \geqslant 1$. We set

$$M^\rho_q(m) = \left\{ M = \begin{Vmatrix} A & B \\ C & D \end{Vmatrix} \in M_{2\rho}(\mathbf{Z});\ J_\rho[M] = mJ_\rho,\ C \equiv 0 \pmod q \right\},$$

where A, B, C, D are blocks of size $\rho \times \rho$,

$$J_\rho = \begin{pmatrix} 0 & E_\rho \\ -E_\rho & 0 \end{pmatrix},$$

and E_ρ is the identity matrix of order ρ.

If F is a function on the *Siegel upper half-plane of degree* ρ,

$$H_\rho = \left\{ Z = X + iY \in M_\rho(C);\ {}^tZ = Z, Y > 0 \right\}$$

and if

$$M = \begin{Vmatrix} A & B \\ C & D \end{Vmatrix} \in M^\rho_q(m) \quad \text{for } m = 1, 2, \dots,$$

then for any integer k and Dirichlet character χ modulo q, we set

$$F|_{k,\chi}M = m^{k\rho - \rho(\rho+1)/2}\chi(\det A)\det(CZ + D)^{-k} F\left((AZ + B)(CZ + D)^{-1}\right).$$

Then $F|_{k,\chi}M$ is also a function on H_ρ, and

$$F|_{k,\chi} M|_{k,\chi} M_1 = F|_{k,\chi} MM_1. \tag{2.4}$$

The space $\mathfrak{M}^\rho_k(q, \chi)$ of *modular forms of degree* ρ, *weight* k *and character* χ *for the group* $\Gamma^\rho_0(q) = M^\rho_q(1)$ is, by definition, the set of all functions F which are holomorphic on the space H_ρ and which for each $M \in \Gamma^\rho_0(q)$ satisfy the functional equation $F|_{k,\chi}M = F$ and, if $\rho = 1$, are also regular at all cusps of $\Gamma^1_0(q)$. It easily follows from the definition of modular forms and (2.4) that for every $m \geqslant 1$, $(m, q) = 1$, the *Hecke operator*

$$T^\rho(m)\colon\ F \to F|_{k,\chi}T^\rho(m) = \sum_{M \in \Gamma^\rho_0(q)\backslash M^\rho_q(m)} F|_{k,\chi}M$$

on the space $\mathfrak{M}^\rho_q(q, \chi)$ does not depend on the choice of left coset representatives of the set $M^\rho_q(m)$ modulo the group $\Gamma^\rho_0(q)$, and it takes this space to itself.

2.3. Theta-series of binary forms. To every matrix $Q \in \mathfrak{N}^+$ (or to the corresponding binary form) we associate the generating functions

$$\theta^2(Z,Q) = \sum_{Q' \in \mathfrak{N}} r(Q,Q')e\{Q'Z\} \qquad (Z \in H_2)$$

and

$$\theta^1(Z,Q) = \sum_{m=0}^{\infty} r(Q,m)e\{mZ\} \qquad (Z \in H_1),$$

where $r(Q,m)$ is the number of solutions of the equation $Q(x,y) = m$ in integers x and y; these functions are called the *theta-series of the form* 0 *of degree* 2 and *degree* 1, respectively. If Q is primitive, then

$$\theta^\rho(Z,Q) \in \mathfrak{M}_1^\rho(-D(Q), \chi_{D(Q)}) \qquad (\rho = 1,2),$$

where $D(Q)$ is the discriminant of Q and $\chi_{D(Q)}$ is the real Dirichlet character modulo $-D(Q)$, which can be written as follows in terms of the Jacobi symbol:

$$\chi_{D(Q)}(d) = (\operatorname{sgn} d)\left(\frac{D(Q)}{|d|}\right)$$

(see [2] and [7]).

§3. Formulas

Suppose that $D < 0$ and $R(D) = \{Q_1, \dots, Q_h\}$ is a set of representatives as in (2.1) of all classes $\{Q\}$ for primitive $Q \in \mathfrak{N}$ of discriminant D. By Corollary 4.3.1 of Theorem 4.3.1 in [5], the space spanned by the theta-series

$$\theta_1^\rho = \theta^\rho(Z,Q_1), \dots, \theta_h^\rho = \theta^\rho(Z,Q_h), \quad \text{where } \rho = 1,2,$$

is invariant under any Hecke operator $T^\rho(m)$ on the space $\mathfrak{M}_1^\rho(-D,\chi_D)$. Consequently, there exists a matrix $T_D^\rho(m)$ of order h with constant coefficients such that

$$\begin{Vmatrix} e_1^{-1}\theta_1^\rho | T^\rho(m) \\ \vdots \\ e_h^{-1}\theta_h^\rho | T^\rho(m) \end{Vmatrix} = T_D^\rho(m) \begin{Vmatrix} e_1^{-1}\theta_1^\rho \\ \vdots \\ e_h^{-1}\theta_h^\rho \end{Vmatrix} \qquad (\rho = 1,2), \tag{3.1}$$

where $e_i = e(Q_i)$ is the number of units of the form Q_i.

THEOREM 1. *Suppose that* $m, n \geqslant 1$, $D, D' < 0$ *and* $(m, nDD') = 1$. *Then*

$$T_D^2(m)E_{D,D'}(n) = \sum_{m=d_1d_2} d_2^{-1}\chi_D(d_2)E_{D,D'}(d_1 n)E_{D'}^*(d_2), \tag{3.2}$$

where E *and* E^* *are the Eichler matrices* (2.2). *In particular,*

$$T_D^2(m) = \sum_{m=d_1d_2} d_2^{-1}\chi_D(d_2)E_D(d_1 n)E_D^*(d_2). \tag{3.3}$$

THEOREM 2. *Suppose that* $m \geqslant 1$, $D < 0$ *and* $(m, D) = 1$. *Then*

$$T_D^1(m) = E_D(m). \tag{3.4}$$

From (3.4) and Hecke's [1] rules for multiplying the operators $T^1(m)$ we immediately obtain

COROLLARY. *If* $m, m_1 \geqslant 1$, $D < 0$ *and* $(mm_1, D) = 1$, *then*

$$E_D(m)E_D(m_1) = \sum_{d|(m,m_1)} \chi_D(d)E_D(mm_1/d^2). \tag{3.5}$$

We let $\widehat{H_+(D)}$ denote the character group for the group $H_+(D)$ of proper classes of primitive binary quadratic forms of discriminant $D < 0$. To every character $\chi \in \widehat{H_+(D)}$ we associate a function X_χ on the classes $\{Q\}$ of primitive forms of discriminant D by setting

$$X_\chi(Q) = \chi(Q) + \chi(\overline{Q}), \quad \text{where } \overline{Q} = Q\left[\begin{Vmatrix} 1 & 0 \\ 0 & -1 \end{Vmatrix}\right]. \tag{3.6}$$

Since $\{Q\} = \{Q\}_+ \cup \{\overline{Q}\}_+$ and the classes $\{Q\}_+$ and $\{\overline{Q}\}_+$ are inverse to one another under composition, it follows that $X_\chi(Q)$ does not depend on the choice of representative $Q \in \{Q\}$ and $X_\chi = X_{\overline{\chi}}$, where $\overline{\chi} = \chi^{-1}$ is the conjugate character.

THEOREM 3. *To each character $\chi \in \widehat{H_+(D)}$, where $D < 0$, associate the vector*

$$X_\chi = X_{\chi,D} = \left(X_\chi(Q_1), \dots, X_\chi(Q_h)\right), \quad \textit{where } \{Q_1, \dots, Q_h\} = R(D).$$

Then X_χ is an eigenvector of every Eichler matrix $E_D(m)$ for $(m, D) = 1$:

$$X_\chi E_D(m) = \lambda(m, \chi) X_\chi. \tag{3.7}$$

COROLLARY 1. *The functions*

$$\theta^\rho(Z, \chi) = \sum_{i=1}^{h} X_\chi(Q_i) e(Q_i)^{-1} \theta^\rho(Z, Q_i), \quad \textit{where } \rho = 1, 2,$$

are eigenfunctions of all of the Hecke operators $T^\rho(m)$, where $(m, D) = 1$.

REMARK. Using the linear independence of the functions $\theta^\rho(Z, Q_1), \dots, \theta^\rho(Z, Q_h)$ (see [8]) and the orthogonality relations for characters, we easily find that the functions of the form $\theta^\rho(Z, \chi)$ which correspond to a largest possible subset of pairwise nonconjugate characters in $\widehat{H_+(D)}$ make up a basis for the linear span of the theta-series $\theta^\rho(Z, Q_1), \dots, \theta^\rho(Z, Q_h)$.

COROLLARY 2. *The eigenvalues $\lambda(m, \chi)$ in (3.7) are given by*

$$\lambda(m, \chi) = \sum_{i=1}^{h} X_\chi(Q_i) e(Q_i)^{-1} r(Q_i, m), \tag{3.8}$$

where $r(Q_i, m)$ is the number of integral representations of m by the quadratic form Q_i.

COROLLARY 3. *The following formula holds for the entries $e(Q_i)^{-1} r(Q_i, mQ_j)$ in the Eichler matrix $E_D(m)$, where $(m, D) = 1$:*

$$e(Q_i)^{-1} r(Q_i, mQ_j) = e(Q_i \cdot Q_j)^{-1} r(Q_i \cdot Q_j, m) + e\left(Q_i \cdot \overline{Q}_j\right)^{-1} r\left(Q_i \cdot \overline{Q}_j, m\right), \tag{3.9}$$

where the product is understood in the sense of Gaussian composition and $\overline{Q}_j$ is defined in (3.6).

§4. Hecke operators and Eichler matrices

The computation of the action of the operators $T^2(m)$ on the Fourier coefficients of modular forms is based on an expression for these operators as a convolution of certain elementary operators, which, however, do not stay within the original space of modular forms. Thus, in order to justify these computations, we must first extend both the set of operators under consideration and our original space of functions. For more details about this, see [5], §§3.1 and 3.2.

We let $\mathfrak{M}_0$ denote the set of all functions F holomorphic on the upper half-plane H_2 which for any matrix

$$\begin{Vmatrix} A & B \\ 0 & D \end{Vmatrix} \in \Gamma_0^2 = \left\{ \begin{Vmatrix} A & B \\ 0 & D \end{Vmatrix} \in M_1^2(1) \right\}$$

satisfy the functional equation

$$F((AZ+B)D^{-1}) = F(Z) \qquad (Z \in H_2).$$

Every function F in $\mathfrak{M}_0$ has a Fourier expansion of the form

$$F(Z) = \sum_{Q \in \mathfrak{N}} f(Q) e\{QZ\}, \tag{4.1}$$

which converges absolutely and uniformly on any subset of H_2 of the form $\{Z = X + iY \in H_2;\ Y \geqslant cE_2\}$, where $c > 0$ (the Koecher effect), and the Fourier coefficients $f(Q)$ satisfy the relations

$$f(uQ{}^t u) = f(Q) \qquad (u \in U,\ Q \in \mathfrak{N}).$$

Conversely, any Fourier series (4.1) with the above properties gives a function in $\mathfrak{M}_0$.

For any matrix $M \in \bigcup_{m,(m,q)=1} M_q^2(m)$ of the form $\|{}^{A\ B}_{0\ D}\|$ the double coset $\Gamma_0^2 M \Gamma_0^2$ contains only finitely many left (right) cosets modulo the group Γ_0^2. Thus, for any function $F \in \mathfrak{M}_0$, integer k and Dirichlet character χ, it makes sense to consider the sum

$$F|T(M) = F|_{k,\chi}T(M) = \sum_{M_i \in \Gamma_0^2 \backslash \Gamma_0^2 M \Gamma_0^2} F|_{k,\chi} M_i \tag{4.2}$$

(see §2.2). If k and χ satisfy the condition $\chi(-1) = (-1)^k$, then $F|T(M)$ does not depend on the choice of the set of representatives M_i, and it is also an element of the space $\mathfrak{M}_0$. We shall use the term *Hecke operator on* $\mathfrak{M}_0$ to refer to the operators $T(M)$, and also to linear combinations with complex coefficients of products of these operators. In general, Hecke operators on $\mathfrak{M}_0$ do not commute.

If k and χ satisfy the condition $\chi(-1) = (-1)^k$, then obviously

$$\mathfrak{M}_k^2(q,\chi) \subset \mathfrak{M}_0.$$

Furthermore, it is well known that each left coset $\Gamma_0^2(q)M$, where $M \in M_q^2(m)$, contains a representative in the set

$$M_0^2(m) = \left\{ \begin{Vmatrix} A & B \\ 0 & D \end{Vmatrix} \in M_q^2(m) \right\}.$$

Thus, the definition of the Hecke operators $T^2(m)$ on the space $\mathfrak{M}_k^2(q,\chi)$ can be written in the form

$$F|T^2(m) = \sum_{M \in \Gamma_0^2 \backslash M_0^2(m)} F|_{k,\chi} M = \sum_{\Gamma_0^2 M T_0^2 \subset M_0^2(m)} F|_{k,\chi} T(M'), \tag{4.3}$$

where in the last expression F is regarded as an element of $\mathfrak{M}_0$, $F|T(M')$ is understood in the sense of (4.2), and the summation is taken over all distinct double cosets $\Gamma_0^2 M' \Gamma_0^2$ in the set $M_0^2(m)$. That is, if we define the function $F|T^2(m)$ by (4.3) for an arbitrary $F \in \mathfrak{M}_0$ and m an integer with $(m,q) = 1$, then we may regard $T^2(m)$ as a Hecke operator on the entire space $\mathfrak{M}_0$ whose restriction to the subspace $\mathfrak{M}_k^2(q,\chi)$ coincides with the classical operator $T^2(m)$.

LEMMA 1. *Let k, q and χ be fixed as above, and $(m,q)=1$. Then the Hecke operator $T^2(m)$ on the space $\mathfrak{M}_0$ can be expanded in the form*

$$T^2(m)=\sum_{m=d_1d_2d_3}\Pi_+(d_1)\Pi(d_2)\Pi_-(d_3),$$

where for $d=1,2,\dots$

$$\Pi_+(d)=T\left(\begin{Vmatrix}E_2 & 0\\ 0 & dE_2\end{Vmatrix}\right),\qquad \Pi(d)=T\left(\begin{Vmatrix}\tilde{A}_d & 0\\ 0 & A_d\end{Vmatrix}\right)$$

and

$$\Pi_-(d)=T\left(\begin{Vmatrix}dE_2 & 0\\ 0 & E_2\end{Vmatrix}\right),$$

where

$$A_d=\begin{Vmatrix}1 & 0\\ 0 & d\end{Vmatrix},\qquad \tilde{A}_d=\begin{Vmatrix}d & 0\\ 0 & 1\end{Vmatrix}.$$

PROOF. It is easy to check that a complete set of left coset representatives modulo the group Γ_0^2 in the sets

$$\Gamma_0^2\begin{Vmatrix}E_2 & 0\\ 0 & dE_2\end{Vmatrix}\Gamma_0^2,\quad \Gamma_0^2\begin{Vmatrix}\tilde{A}_d & 0\\ 0 & A_d\end{Vmatrix}\Gamma_0^2,\quad \Gamma_0^2\begin{Vmatrix}dE_2 & 0\\ 0 & E_2\end{Vmatrix}\Gamma_0^2$$

and $M_0^2(m)$ can be taken to be, respectively, the sets

$$\left\{\begin{Vmatrix}E_2 & S\\ 0 & dE_2\end{Vmatrix};\ S=\begin{Vmatrix}s_1 & s_2\\ s_2 & s_3\end{Vmatrix},\ 0\leqslant s_i<d\ (i=1,2,3)\right\},\tag{4.4}$$

$$\left\{\begin{Vmatrix}\tilde{A}_d^t v^{-1} & Bv\\ 0 & A_d v\end{Vmatrix};\ v\in U,\ A_dv\in U\setminus UA_dU,\ B=\begin{pmatrix}0 & 0\\ 0 & b\end{pmatrix},\ 0\leqslant b<d\right\},\tag{4.5}$$

$$\left\{\begin{Vmatrix}dE_2 & 0\\ 0 & E_2\end{Vmatrix}\right\},\tag{4.6}$$

and

$$\left\{\begin{Vmatrix}d_3\tilde{A}_{d_2}^t v^{-1} & Bv\\ 0 & d_1A_{d_2}v\end{Vmatrix};\ d_1d_2d_3=m,\ d_i>0,\ v\in U,\right.$$

$$\left. A_{d_2}v\in U\setminus UA_{d_2}U,\ B=\begin{Vmatrix}b_1 & d_2b_2\\ b_2 & b_3\end{Vmatrix},\ 0\leqslant b_1,\ b_2<d_1,\ 0\leqslant b_3<d_1d_2\right\}.\tag{4.7}$$

By multiplying matrices, we see that the set (4.7) consists of all possible products $M_1M_2M_3$, where M_1, M_2, and M_3 belong, respectively, to the set (4.4) for $d=d_1$, to the set (4.5) for $d=d_2$, and to the set (4.6) for $d=d_3$, where $d_1d_2d_3=m$ and $d_i>0$. The expansion in the lemma now follows from the definition of the Hecke operators and the equalities (2.4).

The lemma is proved.

For each function $F \in \mathfrak{M}_0$ with Fourier expansion (4.1) we set

$$F^*(Z) = \sum_{Q \in \mathfrak{R}^+} f(Q)e\{QZ\}. \tag{4.8}$$

It is clear that

$$F^* \in \mathfrak{M}_0 \quad \text{and} \quad (F|T)^* = F^*|T \tag{4.9}$$

for any Hecke operator T on $\mathfrak{M}_0$. For $D < 0$ and $m \geqslant 1$ we set

$$f_D(m) = (f(mQ_1), \dots, f(mQ_h)),$$

where $(Q_1, \dots, Q_h) = R(D)$ (see §2.1), and

$$\varepsilon_D(Z) = \begin{Vmatrix} \varepsilon(Z, Q_1) \\ \vdots \\ \varepsilon(Z, Q_h) \end{Vmatrix}, \tag{4.10}$$

where for $Q \in \mathfrak{R}^+$

$$\varepsilon(Z, Q) = \sum_{Q' \in \{Q\}} e\{Q'Z\} = e(Q)^{-1} \sum_{u \in U} e\{Q[u]Z\} \in \mathfrak{M}_0.$$

With this notation, the function F^* can be written in the form

$$F^*(Z) = \sum_{\substack{n \geqslant 1, \\ D<0}} \sum_{Q_i \in R(D)} f(nQ_i) \sum_{Q \in \{Q_i\}} e\{nQ_iZ\} = \sum_{\substack{n \geqslant 1, \\ D<0}} f_D(n)\varepsilon_D(nZ). \tag{4.11}$$

We call this expansion the *D-expansion of the function* F^*, and we call the vector $f_D(n)$ the *D-expansion coefficient corresponding to D and n*.

PROPOSITION 1. *Suppose that m, k and q are natural numbers, $(m, q) = 1$, χ is a Dirichlet character modulo q, and $\chi(-1) = (-1)^k$. Suppose that $F \in \mathfrak{M}_0$ and*

$$\left(F|_{k,\chi}T^2(m)\right)^* = \sum_{n \geqslant 1,\, D<0} f_D(n, m)\varepsilon_D(nZ)$$

is the D-expansion of the function $(F|T^2(m))^*$. Then, if $(m, nD) = 1$, it follows that

$$f_D(n, m) = \sum_{m = d_1 d_2} d_2^{k-2}\chi(d_2) f_D(d_1 n) E_D^*(d_2), \tag{4.12}$$

where $f_D(d_1 n)$ are the D-expansion coefficients of the function F^* and $E_D^*(d_2)$ is the reduced Eichler matrix (see §2.1).

PROOF. Using the sets of coset representatives (4.4)–(4.6) and the definition of the Hecke operators on $\mathfrak{M}_0$, we easily find that

$$G = \sum_{Q \in \mathfrak{R}} g(Q)e\{QZ\} \in \mathfrak{M}_0$$

satisfy the following formulas for $d \geqslant 1$, $(d, q) = 1$:

$$G|\Pi_+(d) = \sum_{Q \in \mathfrak{R}} d(Q)e\{QZ\},$$

$$G|\Pi(d) = d^{k-2}\chi(d) \sum_{\substack{Q \in \mathfrak{R},\, M \in U \backslash UA_d U, \\ dQ[M^{-1}] \in \mathfrak{R}}} g(Q)e\{dQ[M^{-1}]Z\}$$

$$= d^{k-2}\chi(d) \sum_{Q \in \mathfrak{R}} \left(\sum_{\substack{M \in U \backslash UA_d U, \\ Q[M] \in d\mathfrak{R}}} g(d^{-1}Q[M]) \right) e\{QZ\}.$$

If $Q \in \mathfrak{N}$ and M satisfies the conditions in the inner sum, then the matrix $d^{-1}Q[M]$ is contained in $\mathfrak{N}$ and its discriminant is equal to the discriminant of Q. Consequently, if $D(Q) \neq 0$, then this matrix is equivalent to some matrix Q_i in the set $S(D(Q))$ of representatives of all classes $\{Q'\} \in \mathfrak{N}^+$ with $D(Q') = D(Q)$. It is easy to see that

$$\begin{gathered}\text{the number of } M \in U \setminus UA_dU \text{ for which } d^{-1}Q[M] \in \{Q_i\} \\ \text{is equal to } e(Q_i)^{-1}r^*(Q, dQ_i) = e(Q_i)^{-1}r^*(Q_i, dQ).\end{gathered} \tag{4.13}$$

Thus,

$$(G|\Pi(d))^* = d^{k-2}\chi(d) \sum_{Q \in \mathfrak{N}^+} \left(\sum_{Q_i \in S(D(Q))} g(Q_i)e(Q_i)^{-1}r^*(Q_i, dQ) \right) e\{QZ\}.$$

Finally,

$$G|\Pi_-(d) = d^{2k-3}\chi(d^2) \sum_{Q \in \mathfrak{N}} g(Q)e\{dQZ\}.$$

Applying Lemma 1 and these formulas, we obtain

$$\begin{aligned}(F|T(m))^* = &\sum_{m=d_1d_2d_3} d_2^{k-2}d_3^{2k-3}\chi(d_2d_3^2) \\ &\times \sum_{D<0} \sum_{Q \in S(D)} \left(\sum_{Q' \in S(D)} f(d_1Q')e(Q')^{-1}r^*(Q', d_2Q) \right) \varepsilon(Z, d_3Q).\end{aligned} \tag{4.14}$$

If $d_3Q = nQ_j$, where $Q_j \in R(D)$ and $(m, nD) = 1$, then $d_3 = 1$, and for $Q' \in S(D)$ the number $r^*(Q', d_2Q)$ can only be nonzero if Q is equivalent to a matrix of the form nQ_i, where $Q_i \in R(D)$, and in this case it is equal to $r^*(Q_i, d_2Q_j)$. Thus, the coefficient of $\varepsilon(Z, nQ_i)$ in the sum (4.14), i.e., the j-component of the vector $f_D(n, m)$, in the case under consideration is equal to

$$\sum_{m=d_1d_2} d^{k-2}\chi(d_2) \sum_{Q_i \in R(D)} f(d_1nQ_i)e(Q_i)^{-1}r^*(Q_i, d_2Q_j).$$

The proposition is proved.

§5. Action of the Hecke operators on theta-series

Proof of Theorem 1. From (3.1) it obviously follows that

$$\left\| \begin{matrix} e_1^{-1}(\theta_1^2|T^2(m))^* \\ \vdots \\ e_h^{-1}(\theta_h^2|T^2(m))^* \end{matrix} \right\| = T_D^2(m) \left\| \begin{matrix} e_1^{-1}(\theta_1^2)^* \\ \vdots \\ e_h^{-1}(\theta_h^2)^* \end{matrix} \right\|, \tag{5.1}$$

where F^* denotes the function (4.8) for any $F \in \mathfrak{M}_0$.

From the definition of the D-expansion it is easy to see that the D-expansion coefficient of the function $e(Q_i)^{-1}\theta^2(Z, Q_i)^*$ where $Q_i \in R(D)$, which corresponds to some $D' < 0$ and $n \geqslant 1$, is the ith row of the Eichler matrix $E_{D,D'}(n)$. Thus, we can write

$$\left\| \begin{matrix} e_1^{-1}(\theta_1^2)^* \\ \vdots \\ e_h^{-1}(\theta_h^2)^* \end{matrix} \right\| = \sum_{\substack{D'<0, \\ n \geqslant 1}} E_{D,D'}(n)\varepsilon_{D'}(nZ),$$

where the $\varepsilon_{D'}$ are the vectors (4.10). Similarly, if $f^i_{D'}(n, m)$ denotes the D-expansion coefficient of the function $e_i^{-1}(\theta_i^2|T^2(m))^*$ which corresponds to D' and n, and if $F_{D'}(n, m)$ is the matrix with rows $f^1_{D'}(n, m), \ldots, f^h_{D'}(n, m)$, then the left side of (5.1) can be written in the form

$$\sum_{D'<0,\, n\geqslant 1} F_{D'}(n, m)\varepsilon_{D'}(nZ).$$

Substituting these expansions in (5.1) and equating corresponding coefficients (this can be done, because the functions $\varepsilon(Z, Q)$ corresponding to pairwise inequivalent Q are linearly independent), we obtain the equalities

$$F_{D'}(n, m) = T^2_D(m)E_{D,D'}(n),$$

which hold for any $D' < 0$, $n \geqslant 1$ and m prime to D. If we also have $(m, nD') = 1$, then, by Proposition 1, each row of the matrix $F_{D'}(n, m)$ is equal to the corresponding row of the matrix

$$\sum_{m=d_1d_2} d_2^{-1}\chi_D(d_2)E_{D,D'}(d_1n)E^*_{D'}(d_2).$$

Equating these expressions for the matrix $F_{D'}(n, m)$, we obtain (3.2). Then (3.3) follows from (3.2) if we set $n = 1$ and $D' = D$.

Theorem 1 is proved.

To prove Theorem 2 we shall need some information about the *Siegel operator* Φ on the spaces $\mathfrak{M}^2_k(q, \chi)$. If $F \in \mathfrak{M}^2_k(q, \chi)$ and $Z \in H_1$, then it follows from the properties of modular forms that the limit

$$(\Phi F)(Z) = \lim_{\lambda\to+\infty} F\left(\begin{Vmatrix} Z & 0 \\ 0 & i\lambda \end{Vmatrix}\right)$$

exists and $\Phi F \in \mathfrak{M}^1_k(q, \chi)$.

PROPOSITION 2. *Suppose that $m \geqslant 1$ and $(m, q) = 1$. Then any form $F \in \mathfrak{M}^2_k(q, \chi)$ satisfies the relation*

$$\Phi(F|_{k,\chi}T^2(m)) = (\Phi F)|_{k,\chi}(\Phi T^2)(m),$$

where

$$(\Phi T^2)(m) = \sum_{m=d_1d_2} d_2^{k-2}\chi(d_2)T^1(d_1)T^1(d_2)^*, \tag{5.2}$$

$$T^1(d)^* = \sum_{\delta^2|d} \mu(\delta)T^1(d/\delta^2),$$

and μ is the Möbius function.

PROOF. In the case $q = 1$ equivalent relations were found by Maass ([9], Theorem 20). The general case can be treated analogously. However, it is simpler to make use of the general theory of Hecke operators in the form explained, for example, in [3]–[5].

In the first place, since $T^\rho(m)T^\rho(m_1) = T^\rho(mm_1)$ if $(m, m_1) = 1$, it follows that one need only prove (5.2) for $m = p^\delta$, where p is a prime. Using the set of coset representatives (4.7) and the definition of the map $\Omega = \Omega^2_p$ in §1.2 of [5], we easily find that

$$\sum_{\delta=0}^{\infty} \Omega(T^2(p^\delta))t^\delta$$
$$= (1 - p^{-1}x_0^2x_1x_2t^2)\{(1 - x_0t)(1 - x_0x_1t)(1 - x_0x_2t)(1 - x_0x_1x_2t)\}^{-1}.$$

Here $T^2(p^\delta)$ is regarded as an element of the abstract Hecke ring, and this relation is understood in the sense of equality of formal series in t. Similarly (only more simply), if we use the standard triangular set of representatives for $T^1(p^\delta)$, we find that

$$\sum_{\delta=0}^{\infty} \Omega(T^1(p^\delta))t^\delta = \{(1 - x_0t)(1 - x_0x_1t)\}^{-1}.$$

Passing to the representations $T \to |_{k,\chi}T$, we can use the relations (2.22) and (3.2.15) and Proposition 3.3.1 of [5] to derive the following relation from these equalities:

$$\sum_{\delta=0}^{\infty} (\Phi T^2)(p^\delta)t^\delta$$
$$= \left(1 - (p^{k-2}\chi(p)t)^2\right) \sum_{\delta=0}^{\infty} T^1(p^\delta)t^\delta \cdot \sum_{\delta=0}^{\infty} T^1(p^\delta)(p^{k-2}\chi(p)t)^\delta.$$

Equating the coefficients of t^δ on both sides of this relation, we obtain (5.2) for $m = p^\delta$.

The proposition is proved.

Proof of Theorem 2. If we apply the Siegel operator Φ to both sides of (3.1) with $\rho = 2$ and take into account that $\Phi\theta_i^2 = \theta_i^1$, then, by (5.2) and (3.1) with $\rho = 1$, we obtain

$$\sum_{m=d_1d_2} d_2^{-1}\chi_D(d_2)T_D^1(d_1)T_D^1(d_2)^* \begin{Vmatrix} e_1^{-1}\theta_1^1 \\ \vdots \\ e_h^{-1}\theta_h^1 \end{Vmatrix} = T_D^2(m) \begin{Vmatrix} e_1^{-1}\theta_1^1 \\ \vdots \\ e_h^{-1}\theta_h^1 \end{Vmatrix},$$

from which, using (3.3) and the linear independence of $\theta_1^1, \dots, \theta_h^1$ (see [8]), we obtain

$$\sum_{m=d_1d_2} d_2^{-1}\chi_D(d_2)T_D^1(d_1)T_D^1(d_2)^* = \sum_{m=d_1d_2} d_2^{-1}\chi_D(d_2)E_D(d_1)E_D^*(d_2).$$

The equalities (3.4) are easily deduced from these relations using induction on the number of divisors of m.

The theorem is proved.

§6. Eigenvectors of the Eichler matrices

Proof of Theorem 3. From (3.5) it follows that each matrix $E_D(m)$ is a polynomial with scalar coefficients in the matrices $E_D(p)$, where p runs through the prime divisors of m. Thus, it suffices to prove the theorem for the case $m = p$. For prime p, obviously $E_D(p) = E_D^*(p)$. If $j = 1, \dots, h$, then according to (4.13) the jth component of the vector $X_\chi E_D^*(p)$ is equal to

$$\sum_{Q_i \in R(D)} X_\chi(Q_i)e(Q_i)^{-1}r^*(Q_i, pQ_j) = \sum_{\substack{M \in U\backslash UA_pU \\ p^{-1}Q_j[M] \in \mathfrak{R}}} X_\chi\left(p^{-1}Q_j[M]\right).$$

By Theorem 2.3.1 of [3], the last sum depends on the type of splitting of the ideal $p\mathfrak{O}_f$ into prime ideals of the ring $\mathfrak{O}_f$: if $p\mathfrak{O}_f$ is the product of two conjugate prime ideals $\mathfrak{P}$ and $\overline{\mathfrak{P}}$, then this sum is equal to

$$\begin{aligned} X_\chi(\mathfrak{P}\cdot Q_j) + X_\chi(\overline{\mathfrak{P}}\cdot Q_j) &= \chi(\mathfrak{P}\cdot Q_j) + \chi(\overline{\mathfrak{P}}\cdot \overline{Q}_j) + \chi(\overline{\mathfrak{P}}Q_j) + \chi(\mathfrak{P}\overline{Q}_j) \\ &= (\chi(\mathfrak{P}) + \chi(\overline{\mathfrak{P}}))X_\chi(Q_j), \end{aligned}$$

while if the ideal $p\mathfrak{O}_f$ remains prime, then the sum is zero, since there are no matrices M satisfying the conditions in the sum. In any case, the jth coordinate of the vector $X_\chi E_D^*(p)$ can be written in the form $\lambda(p)X_\chi(Q_j)$, where $\lambda(p)$ depends only on p.

The theorem is proved.

From (3.3), (3.4) and Theorem 3 it follows that the vector X_χ is an eigenvector for all of the matrices $T_D^\rho(m)$, where $\rho = 1, 2$ and $(m, D) = 1$. Corollary 1 follows from this fact and the definition of these matrices.

By Theorem 2, $\lambda(m,\chi)$ is the eigenvalue of the Hecke operator $T^1(m)$ corresponding to the eigenfunction $\theta^1(Z,\chi)$. It is easy to see that the coefficient of $e\{Z\}$ in the Fourier expansion of this function is equal to 1. But then the classical theory of Hecke operators (see [1] or [2]) shows that the coefficient of $e\{mZ\}$ in the Fourier expansion of the function $\theta^1(Z,\chi)$ is equal to $\lambda(m,\chi)$. This proves Corollary 2.

From (3.7) and (3.8) it follows that for any matrix $Q_j \in R(D)$ and any character $\chi \in \widehat{H_+(D)}$

$$\sum_{i=1}^{h} X_\chi(Q_i)e(Q_i)^{-1}r(Q_i, mQ_j) = \sum_{i=1}^{h} X_\chi(Q_i)e(Q_i)^{-1}r(Q_i, m)X_\chi(Q_j).$$

Since

$$\begin{aligned} X_\chi(Q)X_\chi(Q') &= \chi(Q\cdot Q') + \chi(\overline{Q}\cdot\overline{Q}') + \chi(Q\cdot\overline{Q}') + \chi(\overline{Q}\cdot Q') \\ &= X_\chi(Q\cdot Q') + X_\chi(Q\cdot Q') \end{aligned}$$

and the set $Q_1\cdot Q', \ldots, Q_h\cdot Q'$, where Q' is a fixed primitive form of discriminant D, is also a complete set of representatives of all classes $\{Q\}$ of primitive Q of discriminant D, it follows that the last relation can be rewritten in the form

$$\begin{aligned} \sum_{i=1}^{n} X_\chi(Q_i)e(Q_i)^{-1}r(Q_i, mQ_j) = \sum_{i=1}^{h} X_\chi(Q_i)\Big(&e(Q_i\cdot Q_j)^{-1}r(Q_i\cdot Q_j, m) \\ &+ e(Q_i\cdot\overline{Q}_j)^{-1}r(Q_i\cdot\overline{Q}_j, m)\Big). \end{aligned}$$

From the theory of characters of finite abelian groups it easily follows that the vectors X_χ corresponding to pairwise nonconjugate characters are linearly independent. From this and the last equality Corollary 3 follows.

§7. Conclusion

Formulas (3.2), (3.3) and (3.9) show that the Hecke operators on the spaces of theta-series of binary quadratic forms enable one to express the number $r(Q, mQ')$ of integral representations of the form mQ' by the form Q, where Q' and Q are primitive, in terms of the number $r(Q'', m)$ of integral representations of the number

m by primitive forms Q'' of the same discriminant as Q. We shall not write out these formulas, since they are already known [10] and can be more simply obtained by elementary considerations. But for quadratic forms in a larger number of variables the elementary considerations generally fail to carry over, while the Hecke operators remain.

Bibliography

1. E. Hecke, *Über Modulfunktionen und die Dirichletschen Reihen mit Eulerscher Produktentwicklung*. I, II, Math. Ann. **114** (1937), 1–28, 315–361.

2. Andrew Ogg, *Modular forms and Dirichlet series*, Benjamin, 1969.

3. A. N. Andrianov, *Euler products corresponding to Siegel modular forms of genus* 2, Uspekhi Mat. Nauk **29** (1974), no. 3(177), 43–110; English transl. in Russian Math. Surveys **29** (1974).

4. S. A. Evdokimov, *Euler products for congruence subgroups of the Siegel group of genus* 2, Mat. Sb. **99(141)** (1976), 483–513; English transl. in Math. USSR Sb. **28** (1976).

5. A. N. Andrianov, *The multiplicative arithmetic of Siegel modular forms*, Uspekhi Mat. Nauk **34** (1979), no. 1(205), 67–135; English transl. in Russian Math. Surveys **34** (1979).

6. Z. I. Borevich and I. R. Shafarevich, *Number theory*, 2nd rev. ed., "Nauka", Moscow, 1972; English transl. of 1st ed., "Nauka", Moscow, 1966.

7. A. N. Andrianov and G. N. Maloletkin, *Behavior of theta series of degree N under modular substitutions*, Izv. Akad. Nauk SSSR Ser. Mat. **39** (1975), 243–258; English transl. in Math. USSR Izv. **9** (1975).

8. Yoshiyuki Kitaoka, *Representations of quadratic forms and their applications to Selberg's zeta functions*, Nagoya Math. J. **63** (1976), 153–162.

9. Hans. Maass, *Die Primzahlen in der Theorie der Siegelschen Modulfunktionen*, Math. Ann. **124** (1951), 87–122.

10. Hubert S. Butts and Gordon Pall, *Modules and binary quadratic forms*, Acta Arith. **15** (1968), 23–44.

Translated by N. KOBLITZ

Труды Матем. Инст. Стеклов.
165(1984)

Proceedings of the Steklov
Institute of Mathematics
1985, Issue 3

UDC 512.7

ANALYTIC SECTIONS IN CONIC BUNDLES

F. A. BOGOMOLOV

ABSTRACT. The author proves the existence of an analytic section over an open algebraic variety for any bundle with a projective space as fiber. The notion of weak meromorphic equivalence for compact varieties is introduced and investigated.

Bibliography: 4 titles.

1. In this article we study analytic properties of complex algebraic varieties. We restrict ourselves to noncompact algebraic varieties, because in the projective case analytic equivalence coincides with algebraic equivalence.

We begin with the discussion of a problem proposed to the author by M. Narasimhan and A. Hirschowitz.

Let V be an algebraic variety, let W be an algebraic bundle over V whose fiber is the rational curve P^1, and let $p\colon W \to V$ denote the projection.

Does there exist an algebraic divisor $D \subset V$ such that the restriction of the bundle to $V \setminus D$ is analytically isomorphic to $P(\mathcal{E})$ for an analytic vector bundle $\mathcal{E}$?

The following theorem contains a positive answer to the question:

THEOREM 1. *Let $p\colon W \to V$ be an algebraic family of the projective spaces P^n over V. Then there exists an algebraic divisor $D \subset V$ such that the restriction of the family to $V \setminus D$ is analytically equivalent to $P(\mathcal{E})$ for an analytic vector bundle $\mathcal{E}$.*

We shall only consider the case when V is an affine variety. Let

$$\omega \in H^1(V, O(\mathrm{PGL}(n+1)))$$

be an element corresponding to the bundle W. We have an exact sequence of analytic sheaves of groups on V:

$$0 \to O^* \overset{i}{\to} Q(\mathrm{GL}(n+1)) \overset{r}{\to} O(\mathrm{PGL}(n+1)) \to 0.$$

It yields the exact sequence

$$H^1(V, O(\mathrm{GL}(n+1))) \to H^1(V, O(\mathrm{PGL}(n+1))) \to H^2(V, O^*).$$

The question of whether W is isomorphic to $P(\mathcal{E})$ on V is equivalent to the question of whether ω can be lifted to an element of the group $H^1(V, O(\mathrm{GL}(n+1)))$.

We consider the image $\partial_i(H^1(V, O(\mathrm{PGL}(n+1))))$ in $H^2(V, O^*)$. For this, we fix an inclusion $i_1\colon Z_{n+1} \to C^*$ and the image $i_1(H^2(V, Z_{n+1})) \subset H^2(V, O^*)$.

1980 *Mathematics Subject Classification*. Primary 14F05; Secondary 32J25, 14C10.

LEMMA 1.1. $\partial_i(H^1(V, O(\mathrm{PGL}(n+1)))) \subset i_1(H^2(V, Z_{n+1}))$.

PROOF. The group $\mathrm{GL}(n+1)$ contains the subgroup $\mathrm{SL}(n+1)$ that is projected onto $\mathrm{PGL}(n+1)$ with kernel $Z_{n+1} \subset C^*$. Therefore we get a commutative diagram of exact sequences:

$$\begin{array}{ccccc} H^1(V, O(\mathrm{GL}(n+1))) & \to & H^1(V, O(\mathrm{PGL}(n+1))) & \to & H^2(V, O^*) \\ \uparrow & & \uparrow & & \uparrow \\ H^1(V, O(\mathrm{SL}(n+1))) & & & & H^2(V, Z_{n+1}) \end{array}$$

The lemma follows from the diagram.

LEMMA 1.2. *The image*

$$i_1 H^2(V, Z_{n+1}) \approx H^2(V, Z_{n+1})/H^2(V, Z) \otimes Z_{n+1}.$$

To prove the lemma we consider two exact sequences of analytic sheaves:

1. $1 \to Z_{n+1} \to O^* \overset{q}{\to} O^* \to 1$;
2. $1 \to Z \to O \to O^* \to 1$

and the commutative diagram of maps

$$\begin{array}{ccccccccc} & & 0 & & 0 & & & & \\ & & \downarrow & & \downarrow & & & & \\ 0 & \to & Z & \to & Z & \to & 0 & & \\ & & \downarrow n+1 & & \downarrow & & & & \\ 0 & \to & Z & \to & O & \to & O^* & \to & 0 \\ & & r\downarrow & & \downarrow & & \wr\wr & & \\ 0 & \to & Z_{n+1} & \to & O^* & \overset{q}{\to} & O^* & \to & 0 \\ & & \downarrow & & \downarrow & & & & \\ & & 0 & & 0 & & & & \end{array}$$

By assumption, V is an affine variety; hence $H^1(V, O) = 0$. It follows that $H^i(V, O^*) \approx H^{i+1}(V, Z)$ and the connecting homomorphism ∂_q: $H^i(V, O^*) \to H^{i+1}(V, Z_{n+1})$ coincides with the map $H^{i+1}(V, Z) \to H^{i+1}(V, Z_{n+1})$ obtained from the projection r: $Z \to Z_{n+1}$.

The lemma is proved.

We arrive at the following general cohomological problem.

Let α be an element of $H^i(V, Z_n)$. Is it possible to find an algebraic divisor $D \subset V$ such that the image $i^*\alpha$ under the embedding i: $V \setminus D \to D$ belongs to the subgroup $H^i(V \setminus D, Z) \otimes Z_n$?

Recently, a positive solution to the problem for $H^2(V, Z_n)$ was found by the methods of algebraic K-theory [4].

In this article we give a direct geometric proof of the existence of a divisor D with the required properties for the elements $\partial_i \omega \in H^2(V, Z_{n+1})$ with $n \leqslant 3$.

LEMMA 1.3. *For any* $\alpha \in H^1(V, Z_n)$, *there exists a divisor* D *such that* $i^*\alpha \in H^1(V \setminus D, Z) \otimes Z_n$, *where* i: $V \setminus D \subset V$.

Indeed, α determines a linear bundle F_α over V. It follows from the exact sequence that $\alpha \in H^1(V \setminus D, Z) \otimes Z_n$ if and only if F_α is trivial.

Let s be a section of F_α. Then F_α is trivial on the complement to the divisor of zeros of s. If we denote this divisor by D, we obtain the lemma.

COROLLARY. *The problem has a solution for α in the subring of $H^n(V, Z_m)$ generated by $H^1(V, Z_m)$.*

We now return to the investigation of the bundles $p: W \to V$ with the fiber P^n. The relative tangent sheaf T_p is defined on W. Its restriction to the fiber P^n is isomorphic to $T(P^n)$. We consider the bundle $E = R_p^0(T_p)$ on V. The projectivization of E, $P(E)$, contains $P(\Omega_p^1)$, the projectivization of the relative tangent bundle Ω_p^1 on W, as a subvariety. Indeed $T_p \approx \mathrm{Hom}(\Omega_p^1, O)$. Therefore, the group of sections $H^0(P^n, T)$ defines a linear map $\Omega_p^1|_{P^n} \to H^0(P^n, T)$ on each fiber. We obtain the required embedding i: $P(\Omega_p^1) \to P(E)$ by extending the map to all of V and taking the projectivization. The vector bundle E is trivial on an open affine variety $V' \subset V$. We fix a trivialization of $P(E)$ on V'; that is, a projection of $P(E)$ on the projective space P^N, π: $P(E) \to P^N$, which is an isomorphism on each fiber $P(E)_x$. The composition πi: $P(\Omega_p^i) \to P^N$ gives the characteristic map of the restriction of $P(\Omega_p^1)$ to V' into P^N. The fiber of $P(\Omega_p^1)$ over $x \in V'$, denoted by F, is a homogeneous $\mathrm{PGL}(n + 1)$-variety. Furthermore, $\mathrm{PGL}(n + 1)$ coincides with the group of complex automorphisms of F and the natural projection q: $F \to P^n$ is a $\mathrm{PGL}(n + 1)$-map. We have a homogeneous linear bundle $O(1)$ on F with $R_qO(1) \approx T(P^n)$. Therefore we get a canonical isomorphism $H^0(F, O(1)) \approx H^0(P, T) \approx \mathfrak{g}$, where $\mathfrak{g}$ is the Lie algebra of $\mathrm{PGL}(n + 1)$. Restricting πi to the fiber F_x, $x \in V'$, we see that it coincides with the map determined by the complete linear series $O(1)$ on F_x. Under deformations of $x \in V'$, the map πi_x is deformed by a projective map $g_x \in \mathrm{PGL}(N + 1)$. Therefore the trivialization of $P(E)$ determines the characteristic map

$$\gamma\colon V' \to \mathrm{PGL}(N + 1)/\mathrm{Aut}\, F \approx H_n \approx \mathrm{PGL}(N + 1)/\mathrm{PGL}(N + 1).$$

We denote by F_n the universal bundle on H_n with fiber F. Obviously $\gamma^* F_n \approx P(\Omega_p^1)|_{V'}$. Therefore the general case of the problem is reduced to the similar problem for F_n and H_n.

LEMMA 1.4. *The variety H_n is rational for $n \leqslant 3$.*

PROOF. We realize $\mathrm{PGL}(n + 1)$ as an open subvariety of $P = P(\oplus_1^N \mathfrak{g}_i)$ with $\mathfrak{g}_i$ isomorphic to $\mathfrak{g}$; we fix this isomorphism. The group $\mathrm{PGL}(n + 1)$ acts on $\mathfrak{g}_i$ as the adjoint representation: $g \to xgx^{-1}$, where $g \in \mathfrak{g}_i$ and $x \in \mathrm{PGL}(n + 1)$. It acts on P as on the projectivization of the direct sum $\oplus\, \mathfrak{g}_i$. We consider the natural projection of P onto $\mathfrak{g}_1 + \mathfrak{g}_2$. We shall show that the field of functions on $\mathfrak{g}_1 + \mathfrak{g}_2$ invariant with respect to $\mathrm{PGL}(n + 1)$ is rational. To do this, we consider the subspace Λ of $\mathfrak{g}_1$ of all diagonal matrices with trivial traces. Then we consider the subgroup Λ_Σ of $\mathrm{PGL}(n + 1)$ which is the normalizer of the group $\Lambda^* = \exp \Lambda$. The subgroup Λ_Σ is isomorphic to an extension of Λ^* by the group of permutations $\Sigma(n)$. The subspace $\Lambda + \mathfrak{g}_2 \subset \mathfrak{g}_1 + \mathfrak{g}_2$ is invariant with respect to Λ_Σ, and the $\mathrm{PGL}(n + 1)$-orbit of $\Lambda + \mathfrak{g}_2$ is an open subvariety of $\mathfrak{g}_1 + \mathfrak{g}_2$. In fact, if (x, g) is a pair with a regular element x, then there exists a transformation $y \in \mathrm{PGL}(n + 1)$ mapping $(x, \mathfrak{g})$ into $\Lambda + \mathfrak{g}_2$. Hence the field of $\mathfrak{g}_\Sigma$-invariant functions on $\Lambda + \mathfrak{g}_2$ is isomorphic to the field of PGL($n+1$)-invariant functions on $\mathfrak{g}_1+\mathfrak{g}_2$. We consider the projection of $\Lambda+\mathfrak{g}_2/\Lambda_\Sigma$ in $\Lambda/\Sigma(n)$. Its base is the rational variety $\Lambda/\Sigma(n)$ and its fiber is the quotient space $\mathfrak{g}_2/\Lambda^*$ which is also rational, because it contains an open subvariety isomorphic to a torus $\Pi\mathbf{C}^*$. We now show that $\Lambda+\mathfrak{g}_2/\Lambda_\Sigma$ is rational. We shall extend the functions from the fiber $\mathfrak{g}_2/\Lambda^*$ over a general point $\lambda\in\Lambda/\Sigma(n)$ to the whole variety. The ring of Λ^*-invariant functions on $\mathfrak{g}_2$ is generated by functions of the form $\{a_{ii}\}$, $\{a_{ij}a_{ji}\}$, and $\{a_{ij}a_{jk}a_{ki}\}$ for all collections $\{i\}$, $\{i,j\}$, and $\{i,j,k\}$, where a_{ij} is the (i,j) entry of $g\in\mathfrak{g}_2$. Let $\lambda_1,\dots,\lambda_n\in\Lambda$ be the eigenvalues of the diagonal matrix $\lambda\in\Lambda$. The group $\Sigma(n)$ acts on the set $\{\lambda_1,\dots,\lambda_n\}$ as a permutation group.

We consider the special functions on $\Lambda+\mathfrak{g}_2/\Lambda_\Sigma$:

$$\varphi_1^s=\sum_{\sigma\in\Sigma(n)} a_{\sigma(1)\sigma(1)}\lambda^s_{\sigma(2)}\cdots\lambda^s_{\sigma(n)};\qquad \varphi_2^s=\sum_{\sigma\in\Sigma(n)} a_{\sigma(1)\sigma(2)}a_{\sigma(2)\sigma(1)}\lambda^s_{\sigma(3)}\cdots\lambda^s_{\sigma(n)};$$

$$\varphi_3^s=\sum_{\sigma\in\Sigma(n)} a_{\sigma(1)\sigma(2)}a_{\sigma(2)\sigma(3)}a_{\sigma(3)\sigma(1)}\lambda^s_{\sigma(4)}\cdots\lambda^s_{\sigma(n)}.$$

Obviously they are Λ_Σ-invariant.

For each point $\lambda=(\lambda_1,\dots,\lambda_n)\in\Lambda$, we consider the numbers $\Lambda_{\sigma,m}(\lambda)=\lambda_{\sigma(m)}\cdots\lambda_{\sigma(n)}$, where $m=2,3,4$ and $\sigma\in\Sigma(n)$. We shall say that $\lambda=(\lambda_1,\dots,\lambda_n)$ is a general point provided that $\Lambda_{\sigma,m}(\lambda)\neq\Lambda_{\tau,m}(\lambda)$ when the set $\{\sigma(m),\dots,\sigma(n)\}$ does not coincide with $\{\tau(m),\dots,\tau(n)\}$. The restrictions of the functions φ_m^s to the fiber $\mathfrak{g}_2/\Lambda^*$ over a general point $\lambda\in\Lambda/\Sigma(n)$ generate the space M_m, where M_2 is the linear span of the functions $\{a_{ii}\}$, M_3 is the linear span of $\{a_{ij}a_{ji}\}$, $i\neq j$, and M_4 is the linear span of $\{a_{ij}a_{jk}a_{ki}\}$, $i\neq j\neq k\neq i$. Indeed if $s+1=\dim M_m$, then the transformation from the basis of M_m to the functions φ_m^s is given by a Vandermonde matrix, whose determinant is equal to the discriminant of the numbers $\Lambda_{\sigma,m}(\lambda)$, and it is nontrivial because λ is a general point. Therefore every rational function can be extended from the general point to the Λ_Σ-invariant function of the same degree on $\Lambda+\mathfrak{g}_2$. By writing a rational basis of $\mathbf{C}(\mathfrak{g}_2/\Lambda^*)$ using functions of M_2, M_3, and M_4 and extending them to $(\Lambda+\mathfrak{g}_2)/\Lambda_\Sigma$ using φ_m^s, we obtain a collection of functions that determines a rational basis on the general fiber $\mathfrak{g}_2/\Lambda^*$. By adding a rational basis of the field $\mathbf{C}(\Lambda/\Sigma(n))$ to this basis, we obtain a rational basis of the field $\mathbf{C}((\Lambda+\mathfrak{g}_2)/\Lambda_\Sigma)$ isomorphic to $\mathbf{C}((\mathfrak{g}_1+\mathfrak{g}_2)/\mathrm{PGL}(n+1))$. Since a general point of $\mathfrak{g}_1+\mathfrak{g}_2$ has the trivial stabilizer in PGL($n+1$), the field

$$\mathbf{C}(\mathrm{PGL}(N+1)/PGL(n+1))$$

is isomorphic to the field of functions on a linear bundle over $(\mathfrak{g}_1+\mathfrak{g}_2)/\mathrm{PGL}(n+1)$. Therefore, $H_n=\mathrm{PGL}(N+1)/\mathrm{PGL}(n+1)$ is rational for $n\leqslant 3$.

The lemma is proved.

PROOF OF THEOREM 1. We apply a theorem of S. Bloch on the isomorphism between $K_2(F)/nK_2(F)$ and $H^2(F,\mu_n^{\otimes 2})\approx \mathrm{Br}_n F$, where F is the field of rational functions over $\mathbf{C}$. It implies the existence of a divisor $D'\subset H_n$ such that on its complement $j^*\partial_i\omega$ can be written in the form

$$\Sigma\omega_\alpha\otimes\omega_\beta,\quad \text{where } \omega_\alpha,\ \omega_\beta\in H^1(H_n\setminus D',Z_{n+1}).$$

So there exists D such that on its complement the image of $i^*\partial_i\omega$ in the group

$$H^2(H_n\setminus D,Z_{n+1})/H^2(H_n\setminus D,Z)\otimes Z_{n+1}$$

is trivial. This proves the theorem for $n\leqslant 3$.

We note that this ending, using Bloch's theorem instead of the original explicit construction, was shown to the author by A. N. Parshin.

The argument presented above also proves the result of Merkur'ev and Suslin that the map $K_2(F)/nK_2(F) \to \mathrm{Br}_n(F)$ is an epimorphism in the case when F is a field of rational functions on a complex projective variety.

2. The notions of meromorphic equivalence and Kodaira dimension play the most important role in the classification of compact complex varieties. The results obtained in §1 allow us to justify the introduction of weak meromorphic equivalence, which we shall call W–K-equivalence.

DEFINITION 1. We shall say that two compact varieties V_1 and V_2 are *W–K-isomorphic* provided:

1) there exist an open subvariety $U \subset V_1$ whose closure coincides with V_1 and a holomorphic embedding $f\colon U \to V_2$ such that $\overline{f(U)} = V_2$; and

2) f can be extended to a meromorphic map $\tilde{f}\colon V_1 \setminus X_1 \to V_2 \setminus X_2$, where $X_i \subset V_i$ is a closed analytic subvariety.

The map $\tilde{f}$ satisfying conditions 1) and 2) is called the *W–K-isomorphism.*

REMARK. Since $\tilde{f}$ is meromorphic, the complement $(V_1 \setminus X_1) \setminus U$ is a closed analytic subvariety of $(V_1 \setminus X_1)$.

The W–K-isomorphism $\tilde{f}$ induces an isomorphism between the fields of meromorphic functions on open subvarieties $V_1 \setminus X_1$ and $V_2 \setminus X_2$.

DEFINITION 2. The varieties V_1 and V_2 are called *W–K-equivalent* provided there exists a chain of W–K-isomorphisms connecting them. In other words, if we introduce an additional structure of one-dimensional complex on the set of meromorphic equivalence classes by joining the points corresponding to W–K-isomorphic classes, then the connected components of this complex correspond to the W–K-equivalence classes of varieties.

Comparing this equivalence with the usual meromorphic equivalence, we observe that they coincide on the class of varieties of general type according to the following statement:

LEMMA 2.1. *Let $f\colon V_1 \setminus X_1 \to V_2 \setminus X_2$ be a W–K-isomorphism. Replace the pair (V_α, X_α) by a bimeromorphically equivalent pair (V'_α, X'_α) such that $V'_\alpha \setminus X'_\alpha = V_2 \setminus X_\alpha$ and X'_α is a smooth divisor with normal crossings. Then f induces an isomorphism of spaces*:

$$f^* i_2^* H^0\big(V_2', \big(nK + (n-1)X_2'\big) = i_1^* H^0\big(V_1', \big(nK + (n-1)X_1'\big),$$

where $i_\alpha\colon V'_\alpha \setminus X'_\alpha \subset V'_\alpha$ is an embedding.

PROOF. We shall show that the sections $s \in i^*H^0(V'_\alpha, nK + (n-1)X'_\alpha)$ are distinguished among the sections of the n-fold canonical class by the property of finiteness of the volume $\int \sqrt[n]{s \wedge \bar{s}}$. If $s \in H^0(V_\alpha, nK + (n-1)X'_\alpha)$, then s becomes the nth power of a holomorphic section of the canonical class on a finite cover of the variety V'_α. Hence, $s \in i^*H^0(V'_\alpha, nK + (n-1)X'_\alpha)$ satisfies the finite volume property. Conversely, assume that

$$\int_{V_1 \setminus X_1} \sqrt[n]{s \wedge \bar{s}} < \infty.$$

We write s in the form $\varphi dz_1/dz_n$ in a neighborhood of a point x_0 of the divisor X_1, where φ is a holomorphic function in the variables $z_1, \ldots, z_n$ and the divisor X_1 is of the form $z_1 \cdots z_m = 0$ in the same neighborhood. The function φ has a Laurent expansion in the variables $(z_1, \ldots, z_m)$ of the form

$$\varphi = \sum_{(i_1 \cdots i_m) \geqslant 0} \varphi_{i_1 \cdots i_m} z_1^{-i_1} \cdots z_m^{-i_m} + \varphi_0,$$

where φ_0 is a holomorphic function in $(z_1, \ldots, z_n)$, $\varphi_{i_1 \cdots i_m}$ is a holomorphic function in $(z_{m+1}, \ldots, z_n)$, and

$$\int_{U(x_0)} \sqrt[n]{s \wedge \bar{s}} = \int \sqrt[n]{\varphi\bar{\varphi}}\; dz_1 \cdots dz_n\, d\bar{z}_1 \cdots d\bar{z}_n.$$

Since the functions $z_1^{-i_1} \cdots z_m^{-i_m}$ and $\bar{z}_1^{-j_1} \cdots \bar{z}_m^{-j_m}$ are orthogonal provided $I \neq J$, for any torus $T_{r_1 \cdots r_m}$ the integral $\int_{|z_1| = r_1} \sqrt[n]{\varphi\bar{\varphi}}$ can be estimated from below:

$$\sum \left(\sqrt[n]{\int_T (\varphi_I \bar{\varphi}_I)} \right) r_1^{-i_1/n} \cdots r_m^{-i_m/n}.$$

Thus, the order of growth of the function coincides with $\max[r_1^{-i_1/n} \cdots r_m^{-i_k/n}]$. Hence if $i_k \geqslant n$ for some k, then the local integral

$$\int_{U(x_0)} \sqrt[n]{s \wedge \bar{s}}$$

is unbounded. Therefore, if

$$\int_{V_1 \setminus X_1} \sqrt[n]{s \wedge \bar{s}} < \infty,$$

then s belongs to $i^* H^0(V'_\alpha, nK + (n-1)X'_\alpha)$. Since

$$\int_{V_2 \setminus X_2} \sqrt[n]{s \wedge \bar{s}} < \infty$$

and f is a meromorphic map, we get $\int_{V_1 \setminus X_1} \sqrt[n]{f^*s \wedge f^*\bar{s}} < \infty$, and the lemma is proved. We have repeated an argument of Sakai [1].

COROLLARY. *If the variety V_1 is of general type, then W–K-equivalence coincides with meromorphic equivalence.*

Indeed, let $f\colon V_1 \setminus X_1 \to V_2 \setminus X_2$ be a W–K-isomorphism. It transforms the field of functions on $V_1 \setminus X_1$ generated by $H^0(V'_1, nK + (n-1)X'_1)$ into the similar field on $V_2 \setminus X_2$. However, the former field coincides with the field of the meromorphic functions on V_1, and its image coincides with the field of meromorphic functions on V_2, i.e., f can be extended to define a bimeromorphic isomorphism of the varieties V_i.

W–K-equivalence does not preserve, in general, the Kodaira dimension. To construct an example we shall prove the following lemma:

LEMMA 2.2. *Let $p\colon W \to X$ be an algebraic family of Abelian varieties over the base X. Let $J_p\colon \tilde{W} \to X$ be the corresponding Jacobian family. Then W and $\tilde{W}$ are W–K-isomorphic.*

In order to prove the lemma it is necessary to construct an analytic section over an open affine subvariety $\bar{X} \subset X$. Consider a subvariety $\tilde{X}$ such that the restriction is a smooth bundle. Its fiber by fiber universal covering is an affine bundle over $\tilde{X}$ with fiber $\mathbf{C}^n$ and structure group $A(n, \mathbf{C})$. We have the exact sequence

$$H^1(\overline{X}, O^n) \to H^1(\tilde{X}, A(n, \mathbf{C})) \to H^1(\tilde{X}, \mathrm{GL}(n, \mathbf{C})).$$

Further, the group $H^1(\tilde{X}, O^n)$ is trivial on the affine variety $\tilde{X}$. Hence a $\mathbf{C}^n$-bundle over $\tilde{X}$ has a section. This implies the existence of an analytic section of the bundle $W|\tilde{X}$, i.e., the restrictions $\tilde{W}|\tilde{X}$ and $W|\tilde{X}$ are isomorphic, and W and $\tilde{W}$ are W–K-isomorphic.

An Enriques surface with an elliptic pencil, whose Jacobian surface is rational, provides an example where the Kodaira dimension is not preserved under W–K-equivalence.

We now investigate W–K-equivalence on varieties with the Kodaira dimension $k = -\infty$. In this case, we distinguish the W–K-equivalence class of rational varieties. Theorem 1 implies

LEMMA 2.3. *A projective variety W having a projection onto an algebraic variety Q with a general fiber P^n is W–K-equivalent to $Q \times P^n$.*

COROLLARY. *The existence of a nonrational conic pencil with a rational base* [2] *shows that W–K-equivalence is nontrivial for varieties with $k = -\infty$.*

For these varieties, one can introduce a natural notion or complementary Kodaira dimension as follows:

DEFINITION. For $k(V) = -\infty$, we denote by $dk(V)$ the maximum of the Kodaira dimensions of the quotients of V, i.e., of those X such that there exists a meromorphic epimorphism $f\colon V \to X$, $dk(V) = \sup_X k(X)$.

Obviously $dk(V) \leqslant \dim V$. We denote by $\mathcal{M}^n_m$ the class of varieties of dimension n with $k = -\infty$ and $dk(V) = m$.

The following conjectures justify the introduction of W–K-equivalence.

CONJECTURE 1. *The class $\mathcal{M}^n_{-\infty}$ consists of all varieties that are W–K-equivalent to the rational varieties.*

CONJECTURE 2. *The class $\mathcal{M}^n_0$ consists of all varieties that are stably-equivalent to the rational varieties. In other words, $V \times R$ (with R an open rational variety) has a completion which is W–K-equivalent to a rational variety.*

CONJECTURE 3. *For $k > 0$, the class $\mathcal{M}^n_k$ does not contain varieties W–K-equivalent to rational varieties.*

REMARK. It is possible that Conjecture 1 is valid under the following stronger assumption: the tensor algebra of the variety V, $A(V) = \Sigma H^0(V, \Omega^{1\otimes n})$, is trivial.

A well-known example of a variety of $\mathcal{M}^n_0$, W–K-equivalent to a rational variety, is provided by an affine bundle over an Abelian variety A, of dimension $\dim A$, analytically isomorphic to the product $\Pi\mathbf{C}^*$.

So, the introduction of W–K-equivalence is necessary to divide the natural problems in the meromorphic classification into problems about varieties belonging to the same W–K-equivalence class and to distinct W–K-equivalence classes.

We shall describe various results supporting Conjectures 1, 2, and 3.

LEMMA 2.4. *$\mathcal{M}^n_1$ does not contain varieties W–K-equivalent to unirational varieties.*

Let V_0 be a variety of $\mathcal{M}_1^n$. It has a holomorphic projection p on a curve X of genus greater than one. Suppose V is W–K-equivalent to a unirational variety Q, i.e., there exists a chain of W–K-isomorphisms f_i: $V_i \setminus X_i^1 \to V_{i-1} \setminus X_{i-1}^2$, where $Q = V_n$ and X_i^1, $X_i^2 \subset V_i$. We choose a rational curve P^1 in Q such that the composition of maps $\psi = pf_1 \cdots f_n$ is nontrivial, and we shall prove that ψ, defined over an open subset $U \subset P^1$, can be extended to a meromorphic map $\hat{\psi}$: $P^1 \to X^1$. We can write U in the form $U = P^1 \setminus \bigcup_1^n D_i$, where $D_i \subset \{f_i \cdots f_n(P^1 \setminus \bigcup_1^{i-1} D_j) \cap X_i^1\}$ is the intersection of the image of P^1, under the map into V_i, with the subset X_i^1 to which the map f_i cannot be extended meromorphically. Since X_i' is a closed subvariety, D_i is a discrete subset of $(P^1 \setminus \bigcup_1^{i-1} D_j)$. We construct $\hat{\psi}$ by induction. Assume we get a meromorphic extension of ψ to $(P^1 \setminus \bigcup_{i-1}^n D_j)$. For each point $p \in D_i$, we take a small disc Δ_p containing p whose intersection with $\bigcup_{i+1}^n D_j$ is empty. We obtain the map ψ: $\Delta_p \setminus p \to X$. Since X is hyperbolic, there exists a meromorphic extension of ψ to Δ_p; hence we can extend ψ to $(P^1 \setminus \bigcup_{i+1}^n D_j)$. Since n is finite, we obtain a meromorphic map ψ: $P^1 \to X$, contradicting the assumption $g(X) > 1$. So there is no $P^1 \subset Q$ such that $pf_1 \cdots f_n$ is nontrivial. In particular, Q is not unirational.

The lemma is proved.

REMARK. A curve X of genus greater than one can be realized as a quotient space $(P^1 \setminus \Sigma)/\Gamma$, where Γ is a Schottky group acting on P^1 and Σ is a perfect nowhere dense subset of P^1. It shows that the condition that ψ can be written in the form $pf_1 \cdots f_n$ is essential for the validity of the lemma.

The analogue of Lemma 2.4 is true for any variety V having a nontrivial meromorphic map into a hyperbolic variety. We observe that Lemma 2.4. is also true if we replace W–K-rationality by stable W–K-rationality.

Results of Griffiths and Green [3] and the author on curves on surfaces of general type allow us to extend the lemma to most varieties of $\mathcal{M}_2^n$.

LEMMA 2.5. *An algebraic locally constant family of Abelian varieties with a base X is stably W–K-equivalent to X.*

SKETCH OF PROOF. Let p: $W \to V$ be a family of Abelian varieties as in Lemma 2.2. We consider two locally constant sheaves $Z \subset C$ and the sheaf O on W. The map $Z \to C \to O$ induces the morphism of the direct images with respect to p: $R_1 Z \to R_1 C \to R_1 O$. We choose an open algebraic subvariety $\tilde{X}$ in X such that $R_1 C$ and $R_1 O$ are trivial bundles on $\tilde{X}$. We project $R_1 C/R_1 Z$ into $R_1 O/R_1 Z$. The variety $R_1 O/R_1 Z$ is isomorphic to the Jacobian bundle of the restriction of W to $\tilde{X}$. The fiber of $R_1 C/R_1 Z$ coincides with $\Pi\mathbf{C}^*$. By constructing a rational compactification for each fiber of $M = R_1 C/R_1 Z$, we get that M is W–K-equivalent to $P^n \times W$. It follows from Theorem 1 that M and $P^{2n} \times X$ are W–K-equivalent.

REMARK. It follows from the existence of the compactification of the generalized Jacobians of algebraic curves depending on an algebraic parameter that a suitable compactification of $R_1 C/R_1 Z$ does exist.

The W–K-equivalence establishes a nontrivial correspondence between an algebraic subfield of the field of meromorphic functions on the open variety $V_1 \setminus X_1$ and the field of functions on X_1. However, the problem of more detailed description of this correspondence leads to a problem in algebraic K-theory and is outside the framework of the present article.

The essential part of the article was done during the author's stay at the university in Nice. The author is grateful to the scientists of the university for their support. He is grateful to A. Hirschowitz for helpful discussions and useful critiques, and also to J.-L. Colliot-Thélène and J.-J. Sansuc, who pointed out that Theorem 1 follows from a general result of [4].

Bibliography

1. F. Sakai, *Kodaira dimensions of complements of divisors*, Complex Analysis and Algebraic Geometry (Collection Dedicated to K. Kodaira), W. L. Baily, Jr., and T. Shioda, editors, Iwanami Shoten, Tokyo, and Cambridge Univ. Press, Cambridge, 1977, pp. 239–257.

2. C. Herbert Clemens and Phillip A. Griffiths, *The intermediate Jacobian of the cubic threefold*, Ann. of Math. (2) **95** (1972), 281–356.

3. Mark Green and Phillip Griffiths, *Two applications of algebraic geometry to entire holomorphic mappings*, The Chern Symposium 1979, W. Y. Hsiang et al., editors, Springer-Verlag, 1980, pp. 41–74.

4. A. S. Merkur'ev and A. A. Suslin, *K-cohomology of Severi-Brauer varieties and the norm residue homomorphism*, Izv. Akad. Nauk SSSR Ser. Mat. **46** (1982), 1011–1046; English transl. in Math. USSR Izv. **21** (1983).

Translated by R. TREGER

Труды Матем. Инст. Стеклов.
165(1984)

Proceedings of the Steklov
Institute of Mathematics
1985, Issue 3

THE DISTRIBUTION OF SUBGROUPS IN THE FULL LINEAR GROUP OVER A COMMUTATIVE RING

UDC 512.547

Z. I. BOREVICH AND N. A. VAVILOV

ABSTRACT. For an arbitrary commutative ring R, the authors study the lattice of subgroups of the full linear group $G = \mathrm{GL}(n, R)$ that contain a fixed elementary block-diagonal group $E(\nu)$ of given type ν, subject to the condition that all the diagonal blocks in $E(\nu)$ be of degree $\geqslant 3$. For a D-net σ of ideals in R of degree n, denote the corresponding net subgroup by $G(\sigma)$, its normalizer in G by $N(\sigma)$, and by $E(\sigma)$ the subgroup generated by the elementary transvections in $G(\sigma)$. Main result: For every intermediate subgroup H, $E(\nu) \leqslant H \leqslant G$, there is a uniquely determined D-net σ such that $E(\sigma) \leqslant H \leqslant N(\sigma)$. In addition, $E(\sigma)$ is a normal subgroup of $N(\sigma)$, and $N(\sigma)$ coincides with the subnormalizer of $E(\sigma)$ in G.

Furthermore, it is shown that

$$[E(n, R), \mathrm{GL}'(n, R, \mathfrak{a})] = E(n, R, \mathfrak{a}), \qquad n \geqslant 3$$

(the notation is standard; $\mathfrak{a}$ is an arbitrary ideal of R). On the basis of this formula, a new proof is obtained for Wilson and Golubchik's results describing the normal structure of the group $\mathrm{GL}(n, R)$ for $n \geqslant 3$.

Bibliography: 20 titles.

§1. Introduction

Let Λ be an associative ring with identity, and $G = \mathrm{GL}(n, \Lambda)$ the full linear group of degree n over Λ. In the case where Λ is a semilocal ring (subject to some insignificant restrictions), the problem of describing the "intermediate" subgroups in G that contain the group $D = D(n, \Lambda)$ of diagonal invertible marices was considered in [5] (the case of a field was considered in [3]). Later, in [7] and [8], an analogous question was considered for rings satisfying finiteness conditions on the stable rank; however, the subgroup D of [7] was replaced by the group $D(\nu)$ of block-diagonal matrices of fixed type ν (cf. §2), with diagonal blocks of sufficiently large size. Its subgroup $E(\nu)$ generated by all elementary transvections in $D(\nu)$ was also considered. When the results concerning this sufficiently general case are applied to Dedekind (commutative) rings, they yield the "standard" description of the intermediate subgroups for $E(\nu)$, subject to the condition that all diagonal blocks in $E(\nu)$ have degree $\geqslant 3$.

In the present article, we carry over the main theorem, describing the lattice of intermediate subgroups for $E(\nu)$, from Dedekind rings to arbitrary commutative rings. Here too the description is given in terms of D-nets of σ-ideals in the

1980 *Mathematics Subject Classification*. Primary 20G35, 20E07.

commutative ring R, the net subgroup $G(\sigma)$ in the full linear group $G = \mathrm{GL}(n, R)$, its normalizer $N(\sigma)$, and the elementary net subgroup $E(\sigma)$ (cf. §2). To be precise, we establish, subject to the same premise concerning the sizes of the diagonal blocks (each must have degree $\geqslant 3$), the existence (and uniqueness), for each intermediate subgroup H, $E(\nu) \leqslant H \leqslant G$, of a D-net σ such that

$$E(\sigma) \leqslant H \leqslant N(\sigma). \tag{1}$$

Furthermore, we show that each subgroup $E(\sigma)$ in (1) is normal in $N(\sigma)$ (compare with [8]). This allows us to say, in the terminology of [10] (cf. §11), that $E(\nu)$ is a fan-like (and even paranormal) subgroup of G, and all $E(\sigma)$ appearing in (1) for all H are basic subgroups of the fan for $E(\nu)$.

It turns out (cf. [13]) that the problem of describing the intermediate subgroups for $E(\nu)$ is connected with the normal structure of the full linear group $\mathrm{GL}(n, R)$ of degree $\geqslant 3$. By way of application of the apparatus we have developed, and our results obtained, we will give a new proof of the main theorems of [21] and [14], describing all normal subgroups of $\mathrm{GL}(n, R)$, $n \geqslant 3$.

The following commutator formula plays an important role in our proof:

$$[E(n, R), \mathrm{GL}'(n, R, \mathfrak{a})] = E(n, R, \mathfrak{a}).$$

It is dual, in some sense, to Suslin's commutator formula

$$[E(n, R, \mathfrak{a}), \mathrm{GL}(n, R)] = E(n, R, \mathfrak{a}),$$

proved in [17] (for the notation see §9, and note that $\mathfrak{a}$ is an ideal in R).

A number of our definitions and auxiliary propositions may be of independent interest, and we therefore took care to formulate them in more general terms than is required for our purposes.

We have adopted the following notation in our article: Λ is an associative ring with 1; in case Λ is commutative, we will also denote it by R. Λ^* is the group of invertible elements of Λ. $G = \mathrm{GL}(n, \Lambda)$ is the full linear group over Λ of degree n, and $e = e_n$ the identity matrix of degree n (as a rule the degree will be established by the context). e_{ij} is the matrix having a 1 in position (i, j) and zeros everywhere else. $t_{ij}(\xi)$ is the elementary transvection $e + \xi e_{ij}$, $\xi \in \Lambda$ $(i \neq j)$. $E(n, \Lambda)$ is the subgroup of G generated by all elementary transvections $t_{ij}(\xi)$. $d_r(\varepsilon)$ is the diagonal matrix $e + (\varepsilon - 1)e_{rr}$ $(\varepsilon \in \Lambda^*, 1 \leqslant r \leqslant n)$, and $D = D(n, \Lambda)$ is the group of all diagonal matrices in G. δ_{ij} is the Kronecker symbol ($\delta_{ii} = 1$, and $\delta_{ij} = 0$ for $i \neq j$). $[x, y]$ is the commutator $xyx^{-1}y^{-1}$. $I = I(n)$ is the segment $\{1, \ldots, n\}$ (index set) of natural numbers; given an invertible matrix $a = (a_{ij})$, we will denote the entries of the inverse a^{-1} by a_{ij}, i.e. $a^{-1} = (a'_{ij})$.

§2. Nets and net subgroups

Let Λ be an arbitrary associative ring with identity. We consider, for a natural number n, the square array

$$\sigma = (\sigma_{ij}), \qquad 1 \leqslant i, j \leqslant n,$$

consisting of n^2 two-sided ideals of σ_{ij} of Λ. We will call this array a *net of ideals in* Λ *of degree* n, if

$$\sigma_{ir}\sigma_{rj} \subset \sigma_{ij} \tag{1}$$

for all values of the indices i, j and r.

A net σ is called a *D-net* if all its ideals occuring on the main diagonal are 1, i.e. if $\sigma_{ii} = \Lambda$ for all $i \in I$.

For a given net σ, we denote by $M(\sigma)$ the collection of all matrices $a = (a_{ij})$ in the ring $M(n, \Lambda)$ of matrices of degree n over Λ for which $a_{ij} \in \sigma_{ij}$ for all i and j. In view of (1), $M(\sigma)$ is a subring of $M(n, \Lambda)$. The identity matrix e is contained in $M(\sigma)$ if and only if σ is a D-net. The collection $e + M(\sigma)$ consisting of the matrices $e + a$, where $a \in M(\sigma)$, is a multiplicative system. For a D-net σ, the ring $M(\sigma)$ coincides with $e + M(\sigma)$. The following concept was introduced in [2].

DEFINITION. Let σ be an arbitrary net of ideals in Λ of degree n. The largest subgroup of the full linear group $G = \mathrm{GL}(n, \Lambda)$ contained in the multiplicative system $e + M(\sigma)$ is called the *net subgroup in* $\mathrm{GL}(n, \Lambda)$ *that corresponds to the net* σ, and is denoted by $G(\sigma)$. If σ is a D-net, then $G(\sigma)$ is also called a *D-net subgroup*.

The simplest examples of net subgroups are the group of upper (or lower) invertible triangular matrices and the group of block-diagonal invertible matrices, the blocks being of fixed dimension; in particular, the group $D(n, \Lambda)$ of diagonal matrices is such an example. If $\mathfrak{a}$ is a 2-sided ideal in Λ, we obtain a "constant" net by putting $\sigma_{ij} = \mathfrak{a}$ for all i and j; the corresponding net subgroup is the principal congruence subgroup $\mathrm{GL}(n, \Lambda, \mathfrak{a})$ at level $\mathfrak{a}$ (cf. [1], Chapter V, §1). In particular, the trivial group, containing only the identity, and the full linear group G are net subgroups. The corresponding nets are the zero net (consisting entirely of zero ideals) and the identity net (all ideals coincide with Λ).

According to our definition, an invertible matrix x in $\mathrm{GL}(n, \Lambda)$ belongs to a net subgroup $G(\sigma)$ if and only if both x and x^{-1} are simultaneously contained in the multiplicative system $e + M(\sigma)$. The question naturally arises of when it might happen that for all nets σ over a ring Λ every invertible matrix in $e + M(\sigma)$ is also in $G(\sigma)$. The answer is given by the following theorem.

THEOREM 1. *In order that for all nets σ over a ring Λ*

$$G(\sigma) = G \cap (e + M(\sigma)), \tag{2}$$

it is necessary and sufficient that, for every natural number n and every ideal $\mathfrak{a}$, the matrix ring $M(n, \Lambda/\mathfrak{a})$ be Dedekind finite, i.e. every element in $M(n, \Lambda/\mathfrak{a})$ that has a one-sided inverse must have a two-sided inverse.

Since this theorem will not be used in the rest of our article, we omit its proof. We merely observe that (2) has been proved in [3] for rings that are finitely generated as modules over their centers. In particular, (2) holds for every commutative ring.

We introduce a partial ordering on the set of all nets σ over Λ of fixed degree n. Namely for given nets σ and τ we write $\sigma \leqslant \tau$ if $\sigma_{ij} \subset \tau_{ij}$ for all i and j. The conditions $\sigma \leqslant \tau$ and $G(\sigma) \leqslant G(\tau)$ are clearly equivalent. All nets of a given degree n (and all net subgroups) constitute a full lattice relative to this partial order.

DEFINITION. Let $\sigma = (\sigma_{ij})$ be any net of ideals in Λ, of degree n. The subgroup of $\mathrm{GL}(n, \Lambda)$ generated by all elementary transvections contained in $G(\sigma)$, i.e. by all matrices $t_{ij}(\xi)$, $\xi \in \sigma_{ij}$ $(i \neq j)$, is called the *elementary net subgroup corresponding to the net* σ, and is denoted by $E(\sigma)$.

If σ_{ij} is the zero ideal whenever i and j are distinct, then $E(\sigma)$ is the trivial subgroup. For the identity net, the corresponding elementary net subgroup coincides with $E(n, \Lambda)$.

NOTATION. For a given D-net σ we let $N(\sigma)$ stand for the normalizer $N_G(G(\sigma))$ of the net subroup $G(\sigma)$ in the full linear group $G = \mathrm{GL}(n, \Lambda)$.

Let ν be an equivalence relation on the index set $I = \{1, \dots, n\}$. If i and j are equivalent with respect to ν, we write $i \sim {}^{\nu}j$, or simply $i \sim j$ when it is clear which equivalence relation we are talking about. Let $I_1, \dots, I_m$ be the equivalence classes defined on I by ν. We will denote by $h(\nu)$ the smallest of the orders $|I_1|, \dots, |I_m|$ of these classes.

We associate to the equivalence relation ν on I a D-net $[\nu]$, defined by specifying that $[\nu]_{ij}$ is the identity ideal Λ whenever $i \sim j$; otherwise it is the zero ideal. The corresponding D-net subgroup $G([\nu])$ will also be denoted by $D(\nu)$. If each of the equivalence classes $I_1, \dots, I_m$ consists of a sequence of successive integers, then $D(\nu)$ is the group of block-diagonal invertible matrices, the diagonal blocks being of degree $|I_1|, \dots, |I_m|$ respectively. It is clear that, for an arbitrary equivalence relation ν, the group $D(\nu)$ can be transformed, through conjugation by a permutation matrix, into an ordinary group of block-diagonal matrices. On the basis of this fact, we are justified, even in the general case, in calling the subgroup $D(\nu) \leqslant \mathrm{GL}(n, \Lambda)$ the "block-diagonal subgroup" of the given type ν.

The elementary net subgroup $E([\nu])$ corresponding to a D-net $[\nu]$ will also be denoted by $E(\nu)$, and will be called an *elementary block-diagonal group of type* ν. In other words, $E(\nu)$ is the subgroup of G generated by all elementary transvections $t_{ij}(\xi)$, where $i \sim {}^{\nu}j$ and $\xi \in \Lambda$.

Let σ be an arbitrary D-net of degree n. We define an equivalence relation ν_σ on I by considering a pair of indices i, j to be equivalent relative to ν_σ if and only if $\sigma_{ij} = \sigma_{ji} = \Lambda$. Since $D(\nu_\sigma) \leqslant G(\sigma)$ and $E(\nu_\sigma) \leqslant E(\sigma)$, it clearly follows that $[\nu_\sigma] \leqslant \sigma$. It is also obvious that if for some equivalence relation ν the subgroup $D(\nu)$ is contained in $G(\sigma)$, then $[\nu] \leqslant [\nu_\sigma]$, seeing that $D(\nu_\sigma)$ is the largest block-diagonal group contained in $G(\sigma)$. Similarly, $E(\nu_\sigma)$ is the largest elementary block-diagonal group contained in $E(\sigma)$. Given a D-net σ, we put $h(\sigma) = h(\nu_\sigma)$.

PROPOSITION 1. *Let σ be an arbitrary D-net, and ν_σ the corresponding equivalence relation on I. If $i \sim r$ and $j \sim s$ (relative to ν_σ), then $\sigma_{ij} = \sigma_{rs}$.*

PROOF. We have

$$\sigma_{ij} = \sigma_{ij}\Lambda = \sigma_{ij}\sigma_{js} \subset \sigma_{is} = \Lambda\sigma_{is} = \sigma_{ri}\sigma_{is} \subset \sigma_{rs}.$$

By symmetry, the opposite inclusion holds as well.

§3. A characterization of the matrices in the normalizer

The matrices a in G belonging to the net subgroup $G(\sigma)$ are characterized by rather simple conditions: $a_{ij} \in \sigma_{ij}$ and $a'_{ij} \in \sigma_{ij}$ for all i and j; and, as a rule, the condition on the entries of the inverse of the matrix can be dropped (Theorem 1). It is important to also have a description of the matrices that belong to the normalizer $N(\sigma)$.

Let σ be an arbitrary net of ideals in Λ, of degree n, and let a be a matrix in G. If for all collections of indices i, j, r, s we have

$$a_{ir}\sigma_{rs}a'_{sj} \subset \sigma_{ij} \qquad (i, j, r, s \in I), \tag{1}$$

then the matrix a obviously normalizes the subgroup $G(\sigma)$, i.e. $a \in N(\sigma)$. The question arises to what extent conditions (1) are sufficient.

Let $a \in N(\sigma)$, and let $\xi \in \sigma_{rs}$ with $r \neq s$. We have the following expressions for the entries of the matrix $b = at_{rs}(\xi)a^{-1}$:

$$b_{ij} = \delta_{ij} + a_{ir}\xi a'_{sj}.$$

It follows from this that when $r \neq s$ the conditions (1) are also necessary for the matrices a in $N(\sigma)$. The inclusions

$$a_{ir}\,\sigma_{rr}a'_{rj} \subset \sigma_{ij} \qquad (i, j, r \in I)$$

have to be considered separately. We limit ourselves here to considering two special cases.

PROPOSITION 2. *Let R be a commutative ring with identity, such that all elements of the form $\varepsilon - 1$, $\varepsilon \in R^*$, together generate the identity ideal, and let σ be an arbitrary D-net of ideals in R. A matrix a in G is contained in $N(\sigma)$ if and only if all the inclusions* (1) *are valid.*

To prove this, one has to consider the matrices $ad_r(\varepsilon)a^{-1}$ (cf. [3], §6).

REMARK. In the case of a noncommutative ring Λ the statement of Proposition 1 would hold for an arbitrary D-net σ, provided that the following two conditions are satisfied: 1) there exists an invertible element ε in Λ such that $\varepsilon - 1$ is also invertible, and 2) every element in Λ can be represented as a sum of invertibles.

The following proposition will be used in §6 to prove Theorem 2.

PROPOSITION 3. *Let Λ be an arbitrary ring and σ a D-net of ideals in Λ satisfying the condition $h(\sigma) \geqslant 2$. A matrix a in G belongs to the normalizer $N(\sigma)$ if and only if for all i and j and all r and s the inclusions* (1) *are valid. In the given situation, all the conditions* (1) *are also equivalent to the inclusions*

$$a_{ir}\sigma_{rs}a'_{sj} \subset \sigma_{ij} \qquad (r \neq s) \tag{3}$$

for all pairs of distinct r and s.

PROOF. We have to verify that, under the hypotheses of Proposition 3, all the inclusions (2) follow from the inclusions (3). We fix some $r \in I$. Since $h(\sigma) \geqslant 2$, we can find a $p \in I$ such that $\sigma_{rp} = \sigma_{pr} = \Lambda$. We have

$$a_{ir}\sigma_{rr}a'_{rj} = a_{ir}\Lambda a'_{rj} = a_{ir}\sigma_{rp} \cdot 1 \cdot \sigma_{pr}a'_{rj} = a_{ir}\sigma_{rp}\Big(\sum_q a'_{pq}a_{qp}\Big)\sigma_{pr}a'_{rj}$$

$$\subset \sum_q \big(a_{ir}\sigma_{rp}a'_{pq}\big)\big(a_{qp}\sigma_{pr}a'_{rj}\big) \subset \sum_q \sigma_{iq}\sigma_{qj} \subset \sigma_{ij},$$

which is what we had to prove.

§4. The associated net

In this section, Λ will be an arbitrary ring and ν a fixed equivalence relation on $i = \{1, \dots, n\}$ such that $h(\nu) \geqslant 2$.

We will consider a subgroup H of the full linear group $\mathrm{GL}(n, \Lambda)$ containing the group $E(\nu)$ of elementary block-diagonal matrices of type ν. To each such subgroup H we will associate some well-defined D-net of ideals σ.

So, let $E(\nu) \leqslant H \leqslant G$. Given an ordered pair of distinct indices i and j, we let σ_{ij} denote the set of all elements ξ in Λ for which $t_{ij}(\xi) \in H$. Clearly, σ_{ij} is an additive group. If $i \sim {}^{\nu}j$, then it follows from the definition of $E(\nu)$ that all the transvections $t_{ij}(\xi)$, $\xi \in \Lambda$, are contained in $E(\nu)$, and hence in H. So in this case $\sigma_{ij} = \Lambda$. Now suppose that the indices i and j are not equivalent, and let $\xi \in \sigma_{ij}$ and $\alpha \in \Lambda$. We choose r such that $r \sim j$ but $r \neq j$. This is possible, by virtue of the condition $h(\nu) \geqslant 2$. The following formulas apply:

$$\begin{gathered} [t_{ij}(\xi), t_{jr}(\alpha)] = t_{ir}(\xi\alpha), \\ [[t_{ij}(\xi), t_{jr}(\alpha)], t_{rj}(1)] = t_{ij}(\xi\alpha). \end{gathered} \tag{1}$$

Since all transvections $t_{ij}(\xi)$, $t_{jr}(\alpha)$ and $t_{rj}(1)$ are contained in H, it follows that $t_{ij}(\xi\alpha)$ is likewise in H, which means that $\xi\alpha \in \sigma_{ij}$. With this we have proved that σ_{ij} is a right ideal in Λ. One verifies in a similar manner that σ_{ij} is a left ideal.

When, in addition, we put $\sigma_{ii} = \Lambda$ for all $i \in I$, we obtain a square array $\sigma = (\sigma_{ij})$ of two-sided ideals of degree n in Λ.

We now make an arbitrary choice of ξ from σ_{ij} and α from σ_{jr}, so that $t_{ij}(\xi) \in H$ and $t_{jr}(\alpha) \in H$. According to (1), the element $t_{ir}(\xi\alpha)$ is also contained in H, i.e. $\xi\alpha \in \sigma_{ir}$. This proves the inclusion $\sigma_{ij}\sigma_{jr} \subset \sigma_{ir}$ for pairwise distinct indices. However, if some of the indices are equal, this inclusion is trivially satisfied. Thus, the ideals σ_{ij} of our array σ satisfy the conditions (1) of §2, which means that σ is a D-net of ideals of degree n in Λ.

DEFINITION. Let $h(\nu) \geqslant 2$ and $E(\nu) \leqslant H \leqslant G$. The D-net σ we have just constructed is called the *D-net associated with the subgroup H*.

Since $[\nu] \leqslant \sigma$, we have $h(\nu) \leqslant h(\sigma)$.

§5. The extraction of transvections

Let ν be a fixed equivalence relation on I such that $h(\nu) \geqslant 2$, and let H be a subgroup of G containing $E(\nu)$. Our aim is to prove the existence of certain transvection in H, as a result of the presence of matrices of a special type in H. We recall that if σ is a D-net associated with H, then the conditions $t_{ij}(\xi) \in H$ and $\xi \in \sigma_{ij}$ are equivalent (for $i \neq j$).

LEMMA 1. *If the pth row of a matrix a in H coincides with the pth row of the identity matrix, i.e. $a_{pj} = \delta_{pj}$ for all j, then*

$$a_{iq} \in \sigma_{iq} \tag{1}$$

for all i and all $q \sim {}^{\nu}p$, $q \neq p$.

PROOF. We fix an index q ($q \sim p$, $q \neq p$). If $i \sim q$, then $\sigma_{iq} = \Lambda$ and the inclusion (1) is obvious. Suppose that i and q are not equivalent. We choose r such that $r \sim i$ and $r \neq i$ (this is possible, since $h(\nu) \geqslant 2$). Let us consider the matrices

$$b = at_{qp}(1)a^{-1}, \qquad c = [b, t_{ri}(-1)].$$

Since $t_{qp}(1) \in H$ and $t_{ri}(-1) \in H$, it follows that $c \in H$. After doing some calculations, we see that

$$b = e + \sum_{s=1}^{n} a_{sq}e_{sp},$$

and, furthermore, $c = t_{rp}(a_{iq})$. Therefore, $a_{iq} \in \sigma_{rp}$. But, according to Proposition 1 (§2), $\sigma_{rp} = \sigma_{iq}$, whence $a_{iq} \in \sigma_{iq}$, and Lemma 1 is proved.

LEMMA 2. *If the qth column of a matrix a in H coincides with the qth column of the identity matrix, i.e.* $a_{iq} = \delta_{iq}$ *for all i, then* $a_{pj} \in \sigma_{pj}$ *for all j and all* $p \sim {}^{\nu}q$, $p \neq q$.

This lemma is proved in the same way as Lemma 1. One has to take

$$b = a^{-1}t_{qp}(1)a, \qquad c = \left[t_{jr}(-1), b\right]$$

(where $r \sim j$, $r \neq j$ and j is not equivalent to p).

LEMMA 3. *Let* $h(\nu) \geqslant 3$. *If the qth column of a matrix a in H coincides with the qth column of the identity matrix, i.e.* $a_{iq} = \delta_{iq}$, *then* $a_{ip} \in \sigma_{ip}$ *for all i and all* $p \sim {}^{\nu}q$.

PROOF. We consider the matrix

$$u = \prod_r t_{qr}\left(a'_{qr}\right) \cdot a,$$

where the product is taken over all indices $r \sim {}^{\nu}q$, $r \neq q$. Clearly, u (which is contained in H) only differs from a in the qth row, and its inverse u^{-1} only differs from a^{-1} in the rth rows, with $r \sim q$, $r \neq q$ and $u'_{qr} = 0$ for all $r \sim {}^{\nu}q$, $r \neq q$. We fix $p \sim q$, $p \neq q$, and consider the matrix

$$v = ut_{pq}(1)u^{-1} = e + \sum_{i,j} u_{ip}u'_{qj}e_{ij},$$

which is likewise contained in H. If $r \sim q$ but $r \neq q$, then $v_{ir} = \delta_{ir}$ for all i (since $u'_{qr} = 0$). Thus, we see that all the rth columns ($r \sim q$, $r \neq q$) of the matrix v coincide with the corresponding columns of the identity matrix, and we can therefore apply Lemma 2 to v. According to this lemma, $v_{pj} \in \sigma_{pj}$ for all j and all $p \sim r$, $p \neq r$. We now apply the premise $h(\nu) \geqslant 3$, which implies that r can have only two possible values. Hence

$$v_{pj} \in \sigma_{pj} \tag{2}$$

for all $p \sim q$ and all j. Furthermore, for $i \neq q$ we have $v_{iq} = u_{ip} = a_{ip}$. The matrix

$$w = v \prod_{j \neq p} t_{pj}(-v_{pj})$$

(here j runs through all indices except p) is contained in H by virtue of (2), and its entries satisfy

$$\begin{aligned} w_{pj} &= \delta_{pj} \qquad (1 \leqslant j \leqslant n), \\ w_{ir} &= v_{ir} \qquad (r \sim p,\ r \neq p). \end{aligned} \tag{3}$$

We now apply Lemma 1 to w. By (3), we have

$$w_{ir} \in \sigma_{ir} \qquad (i \neq p,\ r \sim p,\ r \neq p).$$

For r we may take q, and we then get

$$a_{ip} = u_{ip} = v_{iq} = w_{iq} \in \sigma_{iq} = \sigma_{ip}.$$

This proves Lemma 3.

§6. Inclusion in the normalizer

We now arrive at one of the main results of the paper.

THEOREM 2. *Let R be any commutative ring with identity, $G = \mathrm{GL}(n, R)$ the full linear group over R of degree n, ν some equivalence relation on the set $I = \{1, \dots, n\}$ such that $h(\nu) \geqslant 3$, and $E(\nu)$ the elementary block-diagonal group of type ν. Furthermore, let H be a subgroup of G, with $E(\nu) \leqslant H \leqslant G$. There then exists a unique D-net $\sigma \geqslant [\nu]$ such that $E(\sigma) \leqslant H \leqslant N(\sigma)$.*

To be precise, the net σ for which the last inclusion holds is the D-net associated with the subgroup H.

PROOF. Let a be an arbitrary matrix in H. According to Proposition 3, the proof of Theorem 2 requires that we verify all the inclusions (3) of §3. Let r and s be two distinct indices in I, and ξ an arbitrary element of σ_{rs}. We will consider the matrix

$$b = at_{rs}(\xi)a^{-1}. \tag{1}$$

Since $b_{ij} = \sigma_{ij} + a_{ir}\xi a'_{sj}$, conditions (3) of §3 are equivalent to the inclusion $b_{ij} \in \sigma_{ij}$. Thus, Theorem 2 will have been proved if we can establish that a matrix b of form (1) is contained in $G(\sigma)$.

We will fix three pairwise distinct but equivalent indices p, q, h, and consider the matrix

$$c = bt_{hp}(a_{qr})t_{hq}(-a_{pr})b^{-1}.$$

Since $p \sim q \sim h$, c is contained in H. By using explicit expressions for the entries of the matrices b and b^{-1}, and remembering that R is commutative, we find that for every i

$$c_{ip} = \delta_{ip} + b_{ih}q_{qr}, \quad c_{iq} = \delta_{iq} - b_{ih}a_{pr}, \quad c_{ij} = \delta_{ij} \quad \text{for } j \neq p,\ n \neq q.$$

In particular, $c_{ih} = \delta_{ih}$ for all i. Applying Lemma 3 to the matrix c, we find that $c_{ip} \in \sigma_{ip}$ (and $c_{iq} \in \sigma_{iq}$). Thus

$$b_{ih}a_{pr} \in \sigma_{ih} \qquad (p \sim h,\ p \neq h).$$

We make up yet another matrix

$$u = bt_{hp}\left(-b'_{qp}\right)t_{hq}\left(b'_{pp}\right)b^{-1}.$$

Calculations show that for all i

$$u_{ip} = \delta_{ip}, \qquad u_{iq} = b_{ih}\left(1 - a_{pr}\xi a'_{sp} - a_{qr}\xi a'_{sq}\right).$$

We apply Lemma 3 to the matrix u, to conclude that $u_{iq} \in \sigma_{iq} = \sigma_{ih}$, and, since $b_{ih}a_{pr}$ and $b_{ih}a_{qr}$ are contained in σ_{ih}, we have $b_{ih} \in \sigma_{ih}$ for all i. Here the index h is again arbitrary; hence $b \in G(\sigma)$.

We now show that for every subgroup H containing $E(\nu)$ the D-net σ associated with H is uniquely determined by the inclusions $E(\sigma) \leqslant H \leqslant N(\sigma)$ (even for $h(\nu) \geqslant 2$). Indeed, let $a = t_{ij}(\xi) \in N(\sigma)$. By Proposition 3, we have an inclusion $a_{ij}Ra'_{jj} \subset \sigma_{ij}$, and, since in this case $a_{ij} = \xi$ and $a'_{jj} = 1$, it follows that $\xi \in \sigma_{ij}$. Thus, for $h(\sigma) \geqslant 2$, every elementary transvection contained in $N(\sigma)$ is contained in $E(\sigma)$, which proves the uniqueness of σ.

This completes the proof of Theorem 2.

The following example shows that it is impossible to weaken the condition $h(\nu) \geqslant 3$ to $h(\nu) \geqslant 2$. Namely, let $n = 2l$, and let ν_0 be an equivalence relation on I that partitions I into l two-element equivalence classes $\{2k-1, 2k\}$. The subgroup $E(\nu_0)$ is contained in the symplectic group $\Gamma = \mathrm{Sp}(l, R)$, with the Gram matrix

$$\operatorname{diag}\left(\begin{Vmatrix} 0 & 1 \\ -1 & 0 \end{Vmatrix}, \dots, \begin{Vmatrix} 0 & 1 \\ -1 & 0 \end{Vmatrix}\right).$$

The only elementary transvections contained in Γ are those that belong to $E(\nu_0)$. Hence, the net of ideals associated with Γ coincides with $[\nu_0]$. However, Γ is not contained in $N_G(E(\nu_0))$.

Theorem 2 remains valid if among the equivalence classes $I_1, \dots, I_m$ determined by the equivalence relation ν on I we can find one consisting of either one or two elements (the rest containing at least three elements, with $m \geqslant 2$).

It is interesting to note that for noncommutative division rings the analog of Theorem 2 is true even when $h(\nu) = 2$. This has to do with the nonexistence of the simplectic group in the noncommutative case.

To conclude this section, we observe that Theorem 2 was established in [8] for Dedekind rings, and was carried over in [6] to arbitrary commutative rings (subject to $h(\nu) \geqslant 4$). In the noncommutative case, an analogous theorem applies to rings with a finiteness condition on the stable rank, provided that $h(\nu)$ is sufficiently large [8], [7]. The analog of Theorem 2 for semilocal (not necessarily commutative) rings was established in [11]. The case of a division ring was considered in [12].

§7. Decomposition in terms of transvections

Our next step will be to prove that, subject to the conditions of Theorem 2 with $h(\sigma) \geqslant 3$, the subgroup $E(\sigma)$ is a normal in $N(\sigma)$. In the case of the identity net σ (when all the σ_{ij} are identity ideals), this last statement means that $E(n, R)$ is normal in $\mathrm{GL}(n, R)$. This is Suslin's result [17]. Our proof that $E(\sigma)$ is normal in $N(\sigma)$ basically follows from the first part of §1 of [17].

In this preparatory section we will show that, for a commutative ring, the matrix $b = at_{rs}(\xi)a^{-1}$ decomposes as a product of transvections and matrices that differ from the identity matrix in only one diagonal block of degree 2. All the factors will be explicitly indicated, thus allowing us to use the decomposition later on.

Let R be an arbitrary commutative ring. A row $(\alpha_1, \dots, \alpha_n)$ of elements of R is called *unimodular* if the ideal they generate coincides with R, i.e.

$$\alpha_1\beta_1 + \cdots + \alpha_n\beta_n = 1 \tag{1}$$

for some $\beta_1, \dots, \beta_n$ in R. We will consider the homogeneous linear equation

$$\alpha_1\xi_1 + \cdots + \alpha_n\xi_n = 0 \tag{2}$$

with a unimodular row of coefficients. The following "two term" solutions of this equation are obvious:

$$\alpha_j e_i - \alpha_i e_j, \qquad 1 \leqslant i < j \leqslant n, \tag{3}$$

where $e_i = (0, \dots, 1, \dots, 0)$ is the row having 1 in its ith position, and zeros everywhere else.

LEMMA 4 (SUSLIN'S LEMMA). *Every solution of* (2) *is a linear combination of the two-term solutions* (3), *i.e.*

$$(\xi_1, \dots, \xi_m) = \sum_{i<j} \gamma_{ij}(\alpha_j e_i - \alpha_i e_j), \qquad \gamma_{ij} \in R. \tag{4}$$

PROOF. We need only put

$$\gamma_{ij} = \beta_j \xi_i - \beta_i \xi_j, \tag{5}$$

where $\beta_1, \dots, \beta_n$ are taken from (1).

In [18], the analog of Lemma 4 was introduced for the case of a ring Λ that is finitely generated as a module over its center.

We will now fix a pair of distinct indices r and s, an element $\xi \in R$ and a matrix $a \in \mathrm{GL}(n, R)$. To make our writing a bit more concise, we introduce the notation

$$\alpha_i = a_{ir}, \quad \beta_j = a'_{rj}, \quad \xi_j = \xi a'_{sj}. \tag{6}$$

Then

$$b = a t_{rs}(\xi) a^{-1} = e + \sum_{i,j} \alpha_i \xi_j e_{ij}. \tag{7}$$

Since $\sum_i \alpha_i \beta_i = 1$, the row $(\alpha_1, \dots, \alpha_n)$ must be unimodular, and $\sum_i \alpha_i \xi_i = 0$. We put

$$c(i, j) = \sum_{h=1}^{n} \alpha_h (\alpha_j e_{hi} - \alpha_i e_{hj}).$$

A simple computation shows that the matrices $c(i, j)$ have the following property:

LEMMA 5. *For all indices i, j, i' and j', the product $c(i, j) \subset (i', j')$ is equal to the zero matrix.*

LEMMA 6. *The following product decomposition applies to the matrix* (7):

$$e + \sum_{i,j} \alpha_i \xi_j e_{ij} = \prod_{1 \leqslant i < j \leqslant n} (e + \gamma_{ij} c(i, j)), \tag{8}$$

where the γ_{ij} are defined by (5).

PROOF. According to (4),

$$\sum_j \xi_j e_{hj} = \sum_{i<j} \gamma_{ij} (\alpha_j e_{hi} - \alpha_i e_{hj}),$$

and so

$$b = e + \sum_{i<j} \gamma_{ij} c(i, j),$$

which, by Lemma 5, coincides with the product (8).

We consider the separate factors of (8), with fixed i and j $(i < j)$:

$$u(i, j) = e + \gamma_{ij} \sum_{h=1}^{n} \alpha_h (\alpha_j e_{hi} - \alpha_i e_{hj}).$$

The matrix $u(i, j)$ decomposes as a product

$$u(i, j) = v(i, j) w(i, j), \tag{9}$$

where

$$v(i, j) = e + \gamma_{ij} \left(\alpha_i \alpha_j e_{ii} - \alpha_i^2 e_{ij} + \alpha_j^2 e_{ji} - \alpha_j \alpha_i e_{jj} \right) \tag{10}$$

is a matrix that only differs by one diagonal block of second degree from the identity matrix, and

$$w(i, j) = \prod_{h \neq i,\, h \neq j} t_{hi}(\gamma_{ij}\alpha_h\alpha_j) \cdot t_{hj}(\gamma_{ij}\alpha_h\alpha_i) \tag{11}$$

is a product of transvections.

We now formulate the result we have obtained:

PROPOSITION 4. *Let R be a commutative ring, $r \neq s$, $\xi \in R$, and $a \in \mathrm{GL}(n, R)$. Then*

$$at_{rs}(\xi)a^{-1} = \prod_{1 \leqslant i < j \leqslant n} v(i, j)w(i, j),$$

where $v(i, j)$ and $w(i, j)$ (adopting the notation (6)) *are defined by* (10) *and* (11) *respectively.*

We now show that each of the matrices $v(i, j)$ can be represented in the form of a commutator (subject to a suitable hypothesis regarding the size of n). In order to do so, we may of course limit ourselves to considering the case $i = 1$, $j = 2$.

LEMMA 7. *Let $n = 3$. Then $v(1, 2) = [A, B]$, where*

$$A = \begin{Vmatrix} 1 & 0 & \alpha_1 \\ 0 & 1 & \alpha_2 \\ 0 & 0 & 1 \end{Vmatrix}, \qquad B = \begin{Vmatrix} 1 & 0 & 0 \\ 0 & 1 & 0 \\ \gamma_{12}\alpha_2 & -\gamma_{12}\alpha_1 & 1 \end{Vmatrix}.$$

LEMMA 8. *Let $n \geqslant 4$. Then $v(1, 2) = [X, Y]$, where*

$$X = \begin{Vmatrix} 1 & 0 & \alpha_1\beta_2 & \alpha_1\xi_2 \\ 0 & 1 & \alpha_2\beta_2 & \alpha_2\xi_2 \\ 0 & 0 & 1 & 0 \\ 0 & 0 & 0 & 1 \end{Vmatrix}, \qquad Y = \begin{Vmatrix} 1 & 0 & 0 & 0 \\ 0 & 1 & 0 & 0 \\ \alpha_2\xi_1 & -\alpha_1\xi_1 & 1 & 0 \\ -\alpha_2\beta_1 & \alpha_1\beta_1 & 0 & 1 \end{Vmatrix}.$$

Both lemmas can be verified by a simple calculation.

We will use Lemmas 7 and 8 when $n \geqslant 3$ and $n \geqslant 4$ respectively. If x is a matrix of degree k, we will denote its image under the standard embedding $x \to \mathrm{diag}(x, e_{n-k})$ by the same letter x.

REMARK. If $\mathfrak{a}$ is an ideal in R, we let $E(n, R, \mathfrak{a})$ denote the normal subgroup of $E(n, r)$ generated by all transvections of the form $t_{ij}(\xi)$, $\xi \in \mathfrak{a}$, $i \neq j$. It was shown in [17] that $E(n, R, \mathfrak{a})$ is a normal subgroup of $\mathrm{GL}(n, R)$, provided only that $n \geqslant 3$. Our presentation in this section is also practically a repetition of the proof of this fact from [17]. Indeed, if $\xi \in \mathfrak{a}$, then we see from (5) and (6) that all the γ_{ij} are contained in $\mathfrak{a}$, which means that all the $w(i, j)$ belong to $E(n, R, \mathfrak{a})$. Furthermore, $E(n, R, \mathfrak{a})$ contains the matrix B (of Lemma 7) and $v(1, 2) = ABA^{-1} \cdot B^{-1}$, and hence all $v(i, j)$.

§8. The normalizer of an elementary net subgroup

We have completed all the necessary preparations for the proof that $E(\sigma)$ is normal in $N(\sigma)$ (for $h(\sigma) \geqslant 3$).

THEOREM 3. *Let R be any commutative ring and σ a D-net of ideals of degree n in R. Furthermore, let $E(\sigma)$ be the elementary net subgroup in $G = \mathrm{GL}(n, R)$ corresponding to the D-net σ, and let $N(\sigma)$ be the normalizer in G of the net subgroup $G(\sigma)$. If $h(\sigma) \geqslant 3$, then $E(\sigma)$ is normal in $N(\sigma)$.*

PROOF. We have to show that for every matrix a in $N(\sigma)$ and every element $\xi \in \sigma_{rs}$ $(r \neq s)$ the matrix $b = at_{rs}(\xi)a^{-1}$ is representable as a product of transvections in $E(\sigma)$. In view of Proposition 4, the proof of this fact only requires that we verify that a matrix $v(i, j)$ of the form (10) of §7 can be represented as a product of transvections in $E(\sigma)$, and that each transvection in the product (11) of §7 is contained in $E(\sigma)$.

We will start out by verifying this fact. For every h we have

$$\alpha_h \xi_{ij} = a_{hr}(\xi a'_{rj})a'_{si} - a_{hr}(\xi a'_{sj})a'_{ri} = a_{hr}(\xi a'_{si})a'_{rj} - a_{hr}(\xi a'_{ri})a'_{sj}.$$

From Proposition 3 and these last equations, we deduce that $\sigma_h \gamma_{ij} \in \sigma_{hi}$ and $\alpha_h \gamma_{ij} \in \sigma_{hj}$. This proves that all transvections in $w(i, j)$ are contained in $E(\sigma)$.

We now take up the matrices $v(i, j)$. As we pointed out at the end of §7, we need only analyze the case of the matrices $v(1,2)$. First we suppose that the indices 1 and 2 are equivalent (with respect to the D-net σ; cf. §2). Since $h(\sigma) \geqslant 3$, we can find a third index equivalent to 1 and 2. Suppose that this index is 3, so that $\sigma_{13} = \sigma_{23} = \sigma_{31} = \sigma_{32} = R$. Since both matrices A and B of Lemma 7 are contained in $E(\sigma)$, it follows that $v(1,2)$ and, therefore, $v(i, j)$, for an equivalent pair i, j is a product of transvections in $E(\sigma)$.

Let us now suppose that the indices 1 and 2 are not equivalent. Since $h(\sigma) \geqslant 3$, we can find two other indices that are equivalent to 2. We may assume that $2 \sim 3 \sim 4$, so that $\sigma_{23} = \sigma_{32} = \sigma_{24} = \sigma_{42} = R$. According to Proposition 1, we also have $\sigma_{12} = \sigma_{13} = \sigma_{14}$ and $\sigma_{21} = \sigma_{31} = \sigma_{41}$. By Proposition 3, the following inclusions are valid:

$$\alpha_1 \beta_2 = a_{1r}a'_{r2} \in \sigma_{12} = \sigma_{13}, \qquad \alpha_1 \xi_2 = a_{1r}\xi a'_{s2} \in \sigma_{12} = \sigma_{14},$$
$$\alpha_2 \xi_1 = a_{2r}\xi a'_{s1} \in \sigma_{21} = \sigma_{31}, \qquad \alpha_2 \beta_1 = a_{2r}a'_{r1} \in \sigma_{21} = \sigma_{41}.$$

Hence, both matrices X and Y of Lemma 8 decompose as products of transvections in $E(\sigma)$, which means that $v(1,2)$ is also a product of transvections in $E(\sigma)$. So, we see that if the indices i and j are not equivalent, the matrix $v(i, j)$ is also representable as a product of transvections in $E(\sigma)$.

This proves Theorem 3.

The hypothesis $h(\sigma) \geqslant 3$ in Theorem 3 is essential (there exist commutative rings R such that $E(2, R)$ is not a normal subgroup of GL(2, R)).

We observe that Theorem 3 remains valid if among the equivalent classes $I_1, \ldots, I_m$ defined by ν_σ there is exactly one singleton class, and the remaining classes contain at least three elements.

§9. On a commutator formula

Let Λ be any ring, and $\mathfrak{a}$ a two-sided ideal in Λ. The canonical epimorphism $\Lambda \to \Lambda/\mathfrak{a}$ induces a homomorphism of full linear groups

$$\mathrm{GL}(n, \Lambda) \to \mathrm{GL}(n, \Lambda/\mathfrak{a}). \tag{1}$$

The kernel of (1) is denoted by $\mathrm{GL}(n, \Lambda, \mathfrak{a})$, and is called a *principal congruence subgroup in* $\mathrm{GL}(n, \Lambda)$ *of level* $\mathfrak{a}$ (cf. [1], Chapter V). Furthermore, the normal subgroup of $E(n, \Lambda)$, generated by the transvections $t_{ij}(\xi)$ for which $\xi \in \mathfrak{a}$ is denoted by $E(n, \Lambda, \mathfrak{a})$, and $\mathrm{GL}'(n, \Lambda, \mathfrak{a})$ denotes the full preimage of the center of $\mathrm{GL}(n, \Lambda/\mathfrak{a})$ under the homomorphism (1). Thus, we have a chain of subgroups

$$E(n, \Lambda, \mathfrak{a}) \leqslant \mathrm{GL}(n, \Lambda, \mathfrak{a}) \leqslant \mathrm{GL}'(n, \Lambda, \mathfrak{a}) \leqslant \mathrm{GL}(n, \Lambda).$$

It is known that the following commutator formula applies, subject to the premise that Λ has finite stable rank and $n \geqslant 3$ is large enough:

$$[E(n,\Lambda), \mathrm{GL}'(n,\Lambda,\mathfrak{a})] = E(n,\Lambda,\mathfrak{a}). \tag{2}$$

In this section we show that the auxiliary apparatus we built up in §7 to prove Theorem 3 also enables us to give a proof of (2) for every commutative ring $\Lambda = R$ and every $n \geqslant 3$ (without additional assumptions).

THEOREM 4. *Let R be an arbitrary commutative ring. If $n \geqslant 3$, then*

$$[E(n,R), \mathrm{GL}'(n,R,\mathfrak{a})] = E(n,R,\mathfrak{a}). \tag{3}$$

PROOF. We observe, first of all, that, by virtue of the formula $t_{ij}(\xi) = [t_{ir}(1), t_{rj}(\xi)]$ (in which i, j and r are pairwise distinct), the right side of the above equation is contained in its left side.

We will prove the inclusion

$$[E(n,R), \mathrm{GL}(n,R,\mathfrak{a})] \leqslant E(n,R,\mathfrak{a}). \tag{4}$$

Let $\xi \in R$ and $a \in \mathrm{GL}(n,R,\mathfrak{a})$. The latter implies that a_{ij} and a'_{ij} are congruent to δ_{ij} modulo $\mathfrak{a}$. We consider the commutator

$$[a, t_{rs}(\xi)] = a t_{rs}(\xi) a^{-1} \cdot t_{rs}(-\xi) \tag{5}$$

and apply Proposition 4 to the element $b = a t_{rs}(\xi) a^{-1}$. Since the elements $\alpha_h \alpha_j = a_{hr} a_{jr}$ and $\alpha_h \alpha_i = a_{hr} a_{ir}$ are contained in $\mathfrak{a}$, it follows that all $w(i,j)$ belong to $E(n,R,\mathfrak{a})$. Furthermore,

$$\gamma_{ij} = \xi\left(a'_{rj} a'_{si} - a'_{ri} a'_{sj}\right).$$

To fix our ideas, we will suppose that $r < s$ (the case $s < r$ is completely analogous). If $i \neq r$ or $j \neq s$ (we recall our assumption $i < j$), then $\gamma_{ij} \in \mathfrak{a}$, and hence, by Lemma 7, $v(i,j) \in E(n,R,\mathfrak{a})$ (seeing that $v(1,2) = ABA^{-1} \cdot B^{-1}$). We still have to consider the exceptional case $i = r$, $j = s$. We will assume that $i = 1$ and $j = 2$. We first put

$$\alpha_1 = \alpha_{11} = 1 - \lambda_1, \qquad \alpha_2 = a_{21} = \lambda_2,$$
$$\gamma_{12}\alpha_2 = \mu_1, \qquad -\gamma_{12}\alpha_1 = \xi + \mu_2,$$

where λ_1, λ_2, μ_1 and μ_2 are elements of $\mathfrak{a}$, and then apply Lemma 7. We put

$$\begin{aligned} v(1,2) &= [t_{13}(1)t_{13}(\lambda_1)t_{23}(\lambda_2),\ \ t_{31}(\mu_1)t_{32}(\mu_2)t_{32}(\xi)] \\ &= f[t_{13}(1), t_{32}(\xi)] = f t_{12}(\xi), \end{aligned}$$

where $f \in E(n,R,\mathfrak{a})$. Taking (5) and Proposition 4 into account, we obtain

$$[a, t_{12}(\xi)] = \prod_{\substack{i<j \\ (i,j)\neq(1,2)}} v(i,j)w(i,j) \cdot v(1,2)w(1,2) \cdot t_{12}(-\xi) = Fv(1,2)t_{12}(-\xi),$$

where $F \in E(n,R,\mathfrak{a})$, and, finally, $[a, t_{ij}(\xi)] \in E(n,R,\mathfrak{a})$. This completes the proof of (4).

In order to obtain (3) from (4), we have to apply the group-theoretic Lemma 5.1 of [1], Chapter V. This requires that we bear in mind that $E(n,R,\mathfrak{a})$ is a normal subgroup of $\mathrm{GL}(n,R)$ (cf. the remark at the end of §7), and that we use the equation

$$[E(n,R), E(n,R)] = E(n,R), \tag{6}$$

which is true for every $n \geqslant 3$.

This ends the proof of Theorem 4.

REMARK. Formula (6) is a special case of the following general result.

LEMMA 9. *Let Λ be any ring and σ a D-net of ideals in Λ. If $h(\sigma) \geqslant 3$, then*

$$[E(\sigma), E(\sigma)] = E(\sigma).$$

PROOF. Let $t_{ij}(\xi) \in E(\sigma)$, $i \neq j$. Choose r such that $r \neq i$, $r \neq j$ and $r \sim j$. Then $\xi \in \sigma_{ij} = \sigma_{ir}$,

$$t_{ij}(\xi) = \left[t_{ir}(\xi), t_{rj}(1)\right] \in \left[E(\sigma), E(\sigma)\right],$$

and Lemma 9 is proved.

By virtue of Lemma 9, we can express Theorem 3 in the form of the following commutator formula:

$$[E(\sigma), N(\sigma)] = E(\sigma)$$

(for commutative R and $h(\sigma) \geqslant 3$). By combining Lemma 9 with the last formula, we conclude that in Theorem 2 the equation $[E(\sigma), H] = E(\sigma)$ holds for D-nets σ.

§10. Normal subgroups

We apply the commutator formula (3) of §9 to the problem of the normal structure of $\mathrm{GL}(n, R)$, $n \geqslant 3$. This application is expressed in the following theorem:

THEOREM 5. *Let R be a commutative ring with identity, and $G = \mathrm{GL}(n, R)$ the full linear group over R of degree $n \geqslant 3$. If a subgroup $H \leqslant G$ is normalized by the group $E(n, R)$ generated by all the elementary transvections in G, then*

$$E(n, R, \mathfrak{a}) \leqslant H \leqslant \mathrm{GL}'(n, R, \mathfrak{a}) \tag{1}$$

for some uniquely determined ideal $\mathfrak{a}$ in R.

Wilson proved this theorem in [20] for $n \geqslant 4$. It was obtained, for $n \geqslant 3$, by Golubchik in [14] (no detailed proof has been published). Bass obtained Theorem 5 for the noncommutative case (cf. [1], Chapter V, §§4–5), subject to n being larger than the stable rank of the ring. For the sake of completeness, we give a detailed proof of Theorem 5.

LEMMA 10. *If a subgroup H is not contained in the center of $\mathrm{GL}(n, R)$, and is normalized by $E(n, R)$, then it contains some nontrivial transvection (i.e. an elementary matrix $t_{rs}(\alpha)$, $\alpha \in R$, $\alpha \neq 0$).*

PROOF. Let $a = (a_{ij})$ be an arbitrary noncentral element in H. Starting from this element, we have to show that there is a nontrivial transvection in H. As a first step, we establish the existence in H of a noncentral matrix of the form

$$\begin{Vmatrix} \lambda & w \\ 0 & c \end{Vmatrix}, \tag{2}$$

where $\lambda \in R^*$, w is a row of length $n-1$ and $c \in \mathrm{GL}(n-1, R)$.

For a fixed index r, $1 \leqslant r \leqslant n$, we put

$$u = t_{12}\left(-a'_{3r}\right)t_{13}\left(a'_{2r}\right) \in E(n, R),$$

and we consider the commutator

$$b = \left[a, u\right] = a \cdot ua^{-1}u^{-1}.$$

Since H is normalized by $E(n, R)$, $b \in H$. We have

$$b = \left(e + \sum_{i,j} a_{i1}\Delta_{rj}e_{ij}\right)u^{-1},$$

where $\Delta_{rj} = a'_{2r}a'_{3j} - a'_{3r}a'_{2j}$. Clearly, $\Delta_{rj} = -\Delta_{jr}$ and $\Delta_{rr} = 0$. The matrix $b = b(r)$ depends on r. We first consider the matrix b when $r = 1$. Since $\Delta_{11} = 0$, we have $b_{i1} = \delta_{i1}$ in this case, for all $i = 1, \ldots, n$, i.e. the matrix $b(1)$ has the form (2). If it is noncentral, we will have attained our intermediate goal (of finding in H a noncentral matrix of the form (2)).

If, however, $b(1)$ lies in the center of G, then $b(1) = e$, and hence

$$a_{i1}\Delta_{1j} = 0 \tag{3}$$

for all $i = 2, \ldots, n$ and all $j = 1, \ldots, n$. We then look at the matrix $b(r)$ for $r \geqslant 2$. By (3), for $i \geqslant 2$ we have

$$a_{i1}\Delta_{r1} = -a_{i1}\Delta_{1r} = 0,$$

and this means that $b(r)$ is of the form (2). If for any r the matrix $b(r)$ is not in the center of G, then it qualifies as the desired noncentral matrix in H of the form (2).

It remains to consider the case when all the $b(r)$ are central, i.e. scalar matrices. In this case

$$a_{i1}\Delta_{rj} = 0 \tag{4}$$

for all $i \geqslant 2$ and all j distinct from i. For $j = i$, we have $a_{i1}\Delta_{ri} = -a_{i1}\Delta_{ir}$, and this is zero both for $r = i$ ($\Delta_{ii} = 0$) and for $r \neq i$, by (4). Thus, the equations (4) are valid for all $i \geqslant 2$ and all values of r and j. Now, all the minors Δ_{rj} of degree 2 in a^{-1}, taken from the second and third rows, generate the identity ideal; hence it follows from (4) that $a_{i1} = 0$ for all $i \geqslant 2$. We therefore see that in the case under consideration the matrix a itself is of the form (2).

So, H contains a noncentral matrix of the form (2). We will show, on the basis of this, that there is a matrix in H of the form (2) with $c = \lambda e$ and $w \neq 0$, i.e.

$$\begin{Vmatrix} \lambda & w \\ 0 & \lambda e \end{Vmatrix}, \quad w \neq 0. \tag{5}$$

If $c = \lambda e$ in the matrix (2), then $w \neq 0$ (since the matrix is noncentral). If, however, $c \neq \lambda e$, then, taking x to be a row of length $n - 1$, we consider the commutator

$$\left[\begin{Vmatrix} 1 & x \\ 0 & e \end{Vmatrix}, \begin{Vmatrix} \lambda & w \\ 0 & c \end{Vmatrix}\right] = \begin{Vmatrix} 1 & xc^{-1}(c - \lambda e) \\ 0 & e \end{Vmatrix},$$

which is obviously in H. In our situation $c - \lambda e \neq 0$; hence x can be chosen so that the row $xc^{-1}(c - \lambda e)$ would be nonzero. Thus, in both cases there is a matrix of form (5) in H.

Let $w = (\alpha_2, \ldots, \alpha_n)$. To fix our ideas, we suppose that $\alpha_2 \neq 0$. We take up the matrix (5), which we denote by f, and form the commutator with $t_{23}(1)$, to obtain

$$[f, t_{23}(1)] = t_{13}(\lambda^{-1}\alpha_2) \in H,$$

which proves Lemma 10.

Proof of Theorem 5. Let H be a subgroup of $\mathrm{GL}(n, R)$, normalized by $E(n, R)$. We let $\mathfrak{a}$ denote the set of all elements $\xi \in R$ such that $t_{rs}(\xi) \in H$ for some indices r and s ($r \neq s$). From (1) of §4 it follows that if $t_{rs}(\xi) \in H$, then $t_{rj}(\xi) \in H$ and $t_{is}(\xi) \in H$. Hence, for each ξ in $\mathfrak{a}$, the transvections $t_{ij}(\xi)$ are in H for all $i \neq j$.

Clearly, $\mathfrak{a}$ is a subgroup of the additive group of the ring R. We again call upon (1) of §4 to establish that $\mathfrak{a}$ is an ideal of R. If H is a central subgroup, then $\mathfrak{a}$ is the zero ideal.

We now go over to the quotient ring $R/\mathfrak{a}$ and we let H' denote the image of H under the natural homomorphism (1) of §9 (for $\Lambda = R$). It is clear that H' is normalized by $E(n, R/\mathfrak{a})$. We have to show that H' is in the center of $\mathrm{GL}(n, R/\mathfrak{a})$. Suppose the contrary. Then, by Lemma 10, H' contains a nontrivial transvection $t_{rs}(\bar{\gamma})$, where $\bar{\gamma}$ is a nonzero element of $R/\mathfrak{a}$ represented by $\gamma \in R$ (γ is not in $\mathfrak{a}$). We have

$$t_{rs}(\gamma) = xa, \qquad x \in H,\ a \in \mathrm{GL}(n, R, \mathfrak{a}).$$

Furthermore, if $u = t_{ir}(1)$, where the index i is distinct from r and s, then

$$t_{is}(\gamma) = [t_{ir}(1), t_{rs}(\gamma)] = [u, xa] = [u, x] \cdot x[u, a]x^{-1}.$$

But $[u, x] \in H$ (since $E(n, R)$ normalizes H) and $[u, a] \in E(n, R, \mathfrak{a}) \leqslant H$, according to Theorem 4 (one only has to refer to the inclusion (4) of §9). Therefore, $t_{is}(\gamma) \in H$, i.e. $\gamma \in \mathfrak{a}$, and we have arrived at a contradiction. Thus, H' is a central subgroup of $\mathrm{GL}(n, R/\mathfrak{a})$, i.e. $H \leqslant \mathrm{GL}'(n, R, \mathfrak{a})$.

The uniqueness of the ideal $\mathfrak{a}$ in (1) follows from the formula $[E(n, R), H] = E(n, R, \mathfrak{a})$, which is obtained by combining (3) of §9 with the formula $[E(n, R), E(n, R, \mathfrak{a})] = E(n, R, \mathfrak{a})$.

This completes the proof of Theorem 5.

§11. Polynormality

In order to place Theorem 2 in a broader context, we proceed to introduce some definitions and some of the simpler associated propositions from [4] and [10].

Let G be a group and D a subgroup. Every subgroup H, for which $D \leqslant H \leqslant G$ will be called *intermediate*. We denote the lattice of all intermediate subgroups by $L(D, G)$. Let $\{G_\alpha\}$ be some family of intermediate subgroups. For each subgroup G_α, we let N_α stand for its normalizer in G. If the lattice $L(D, G)$ decomposes as a disjoint union of lattices $L(G_\alpha, N_\alpha)$, i.e.

$$L(D, G) = \bigcup_\alpha L(G_\alpha, N_\alpha),$$

then the family $\{G_\alpha\}$, together with the normalizers N_α, is called a *fan* for the subgroup D. The subgroups G_α are called *basic subgroups of the fan* $\{G_\alpha, N_\alpha\}$. The fan $\{G_\alpha, N_\alpha\}$ can therefore be characterized by the property that for each intermediate subgroup H there exists a unique index α such that $G_\alpha \leqslant H \leqslant N_\alpha$. By virtue of the homomorphism theorem for the fan subgroup D, the subgroups of the lattice $L(D, G)$ are in bijective correspondence with all of the subgroups of all of the quotients N_α/G_α.

Fans do not always exist. A fan subgroup (for which a fan exists) can have more than one fan. If a subgroup is a fan subgroup, and its intermediate subgroups satisfy the minimal condition, then there exists only one fan for this subgroup. This is true in particular for finite groups.

Let A be a subset of G. We will let D^A denote the subgroup generated by the conjugates $D^a = aDa^{-1}$ of D by all elements a of A. An intermediate subgroup F, $D \leqslant F \leqslant G$, is called *D-full* if $D^F = F$, i.e. if the normal closure of D in F coincides with F.

A subgroup D of a group G is called *polynormal* in G if for every $x \in G$ the subgroup $D^{\langle x\rangle}$ is D-full (here $\langle x\rangle$ is the cyclic subgroup generated by the element x). The role of the polynormal subgroups is determined by the following proposition (cf. [10]).

PROPOSITION 5. *A subgroup D of G is polynormal if and only if it is a fan subgroup, and all of the basic subgroups of its fan are D-full.*

For polynormal subgroups, the fan is unique. If $\{G_\alpha, N_\alpha\}$ is a fan for a polynormal subgroup $D < G$, then N_α coincides with the subnormalizer of G_α in G.

A subgroup D of a group G is called *paranormal* if for every $x \in G$ the subgroup $\langle D, D^x\rangle$ is D-full. Every paranormal subgroup is polynormal. The author thinks that the converse is false. However, there is no known example of a polynormal subgroup that is not paranormal. Every pronormal subgroup (cf. [4]) is paranormal.

LEMMA 11. *Let Λ be any ring and ν an equivalence relation on the index set $I = \{1, \dots, n\}$. If $h(\nu) \geqslant 2$ and $\sigma \geqslant [\nu\}$, then $E(\sigma)$ is an $E(\nu)$-full subgroup of* $\mathrm{GL}(n, \Lambda)$.

PROOF. We have to show that every transvection from $E(\sigma)$ is representable as a product of matrices that are conjugates in $E(\sigma)$ of matrices in $E(\nu)$. Let $\xi \in \sigma_{ij}$. If $i \sim {}^\nu j$, then $t_{ij}(\xi) \in E(\nu)$. If, on the other hand, i and j are not equivalent, we choose r such that $j \sim r$ but $j \neq r$. Then $\sigma_{ij} = \sigma_{ir}$ (Proposition 1), and hence $t_{ir}(\xi) \in E(\sigma)$. We have

$$t_{ij}(\xi) = t_{ir}(\xi)t_{rj}(1)t_{ir}(-\xi) \cdot t_{rj}(-1),$$

and Lemma 11 is proved.

LEMMA 12. *Let R be commutative, $h(\nu) \geqslant 2$ and $\sigma \geqslant [\nu]$. If the inclusion $aE(\nu)a^{-1} \leqslant N(\sigma)$ applies to some matrix a in* $\mathrm{GL}(n, R)$, *then*

$$a_{ir}a'_{sj} \in \sigma_{ij}$$

for all i and j and all pairs of equivalent r and s (in particular for $r = s$).

PROOF. Let $r \sim {}^\nu s$, $r \neq s$, and suppose that $b = at_{rs}(1)a^{-1}$, so that $b_{ij} = \delta_{ij} + a_{ir}a'_{sj}$ and $b'_{ij} = \delta_{ij} - a_{ir}a'_{sj}$. Since $b \in N(\sigma)$, it follows from Proposition 3 that

$$b_{ij}b'_{hh} = a_{ir}a'_{sj}\left(1 - a_{hr}a'_{sh}\right) \in \sigma_{ih} = \sigma_{ij},$$
$$b_{ih}b'_{hj} = -a_{ir}a'_{sh}a_{hr}a'_{sj} \in \sigma_{ij},$$

where $i \sim j$, $h \neq j$, i is not equivalent to j and, hence, $h \neq i$. From this we get

$$a_{ir}a'_{sj} = b_{ij}b'_{hh} - b_{ih}b'_{hj} \in \sigma_{ij},$$

and Lemma 12 is proved for $r \neq s$. Suppose now that $p \sim r$ but $p \neq r$. Then

$$a_{ir}a'_{rj} = \sum_{q=1}^{n} a_{ir}a'_{pq}a_{qp}a'_{rj} \in \sum_q \sigma_{iq}\sigma_{qj} \subset \sigma_{ij}.$$

REMARK. The proof of Lemma 12 yields the following result (taking, instead of $t_{rs}(1)$, the elements $t_{rs}(\xi)$, $\xi \in \sigma_{rs}$, for every $r \neq s$): for a commutative ring R and $h(\sigma) \geqslant 2$, the inclusion $aE(\sigma)a^{-1} \leqslant N(\sigma)$, $a \in G$, implies that $a \in N(\sigma)$. Hence, for a D-net σ, satisfying the condition $h(\sigma) \geqslant 3$, the elementary net subgroup $E(\sigma)$ is weakly normal in G (cf. [19]; a subgroup F of a group G is called *weakly normal* if $F^a \leqslant N_G(F)$ implies $F^a = F$).

THEOREM 6. *Under the hypotheses of Theorem* 2, *i.e.* R *is commutative and* $h(\nu) \geqslant 3$, *the elementary block-diagonal subgroup* $E(\nu)$ *is paranormal in* $G = \mathrm{GL}(n, R)$.

PROOF. By Theorem 2, there corresponds to each intermediate subgroup H a unique D-net σ such that $E(\sigma) \leqslant H \leqslant N(\sigma)$. This means that $E(\nu)$ is a fan subgroup in G and the $E(\sigma)$ are (for all $\sigma \geqslant [\nu]$) basic subgroups of its fan. According to Lemma 11, all these $E(\sigma)$ are $E(\nu)$-full, which, by Proposition 5, implies that the subgroup $E(\nu)$ is polynormal.

To prove that $E(\nu)$ is paranormal, we have to verify that if $a \in G$ and $aE(\nu)a^{-1} \leqslant N(\sigma)$, then $aE(\nu)a^{-1} \leqslant E(\sigma)$. Let $r \sim {}^{\nu}s$. We have to show that for every $\xi \in R$ the matrix $b = at_{rs}(\xi)a^{-1}$ is contained in $E(\sigma)$. We turn to the proof of Theorem 3. For fixed r and s ($r \neq s$) the inclusion $b \in E(\sigma)$ was introduced there on the basis of the conditions $a_{ir}\sigma_{rs}a'_{sj} \subset \sigma_{ij}$ (for all i and j), which were guaranteed by the premise $a \in N(\sigma)$. In our case, at present, the matrix a does not necessarily belong to the normalizer $N(\sigma)$. Nevertheless, all the required conditions are fulfilled here as well, by virtue of Lemma 12. We now repeat word for word the proof of Theorem 3.

REMARK. Under the hypotheses of Theorem 2, the group $N(\sigma)$ coincides, for every net $\sigma \geqslant [\nu]$, with the subnormalizer of the subgroup $E(\sigma)$ in $\mathrm{GL}(n, R)$ (compare [9]). This follows from the polynormality of the subgroup $E(\nu)$.

§12. Conjectures

It is natural to ask whether our results carry over to noncommutative rings. The information gathered in this area allows us to conjecture that all the theorems in this paper can be extended in some form to noncommutative rings, and Theorem 1 suggests to us the natural limits to this extension. The authors believe very strongly that the following conjectures are true.

CONJECTURE 1. A necessary and sufficient condition that every elementary block-diagonal subgroup $E(\nu)$, with $h(\nu) \geqslant 3$, of the full linear group $G = \mathrm{GL}(n, R)$ over a ring Λ be paranormal in G, and all elementary net subgroups $E(\sigma)$, $\sigma \geqslant [\nu]$, exhaust the collection of all $E(\nu)$-full intermediate subgroups, is that for all n and all ideals $\mathfrak{a}$ in Λ the matrix ring $M(n, \Lambda/\mathfrak{a})$ be Dedekind finite. In these circumstances, the normalizer in G of the subgroup $E(\sigma)$ coincides with $N(\sigma)$.

If the following formulas are valid for every two-sided ideal $\mathfrak{a}$ in R:

$$[E(n, \Lambda, \mathfrak{a}), \mathrm{GL}(n, \Lambda)] = E(n, \Lambda, \mathfrak{a}),$$

$$[E(n, \Lambda), \mathrm{GL}'(n, \Lambda, \mathfrak{a})] = E(n, \Lambda, \mathfrak{a}),$$

then we will say that *the standard commutator formulas apply in* $\mathrm{GL}(n, \Lambda)$.

Furthermore, we will say that the *standard description of normal subgroups in* $G = \mathrm{GL}(n, \Lambda)$ *applies to* Λ if for every subgroup $H \leqslant G$ normalized by $E(n, \Lambda)$ there exists a unique two-sided ideal $\mathfrak{a}$ of Λ such that

$$E(n, \Lambda, \mathfrak{a}) \leqslant H \leqslant \mathrm{GL}'(n, \Lambda, \mathfrak{a}).$$

CONJECTURE 2. The standard commutator formulas and the standard description of normal subgroups apply in the group $\mathrm{GL}(n, \Lambda)$ for all $n \geqslant 3$ if and only if all the rings $M(n, \Lambda/\mathfrak{a})$ are Dedekind finite.

Another topical problem in the theory of linear groups is the problem of describing the automorphisms of the full linear group over a ring.

CONJECTURE 3. If all the rings $M(n, \Lambda/\mathfrak{a})$ are Dedekind finite for every $n \geqslant 4$, then the automorphism group of $\mathrm{GL}(n, \Lambda)$ is generated by the standard automorphisms. The same is even true for $n = 3$, if one supposes in addition that $2 \in \Lambda^*$. Conversely, if for every $n \geqslant 4$ the description of the automorphisms of $\mathrm{GL}(n, \Lambda)$ is standard, then all the rings $M(n, \Lambda/\mathfrak{a})$ are Dedekind finite.

BIBLIOGRAPHY

1. Hyman Bass, *Algebraic K-theory*, Benjamin, 1968.

2. Z. I. Borevich, *On parabolic subgroups in linear groups over a semilocal ring*, Vestnik Leningrad. Univ. **1976**, no. 13 (Ser. Mat. Mekh. Astr. vyp 3), 16–24; English transl. in Vestnik Leningrad Univ. Math. **9** (1981).

3. ______, *Description of the subgroups of the full linear group that contain the group of diagonal matrices*, Zap. Nauchn. Sem. Leningrad. Otdel Mat. Inst. Steklov. (LOMI) **64** (1976), 12–29; English transl. in J. Soviet Math. **17** (1981), no. 2.

4. ______, *On the distribution of subgroups*, Zap. Nauchn. Sem. Leningrad. Otdel. Mat. Inst. Steklov. (LOMI) **94** (1979), 5–12; English transl. in J. Soviet Math. **19** (1982), no. 1.

5. Z. I. Borevich and N. A. Vavilov, *Subgroups of the full linear group over a semilocal ring that contain the group of diagonal matrices*, Trudy Mat. Inst. Steklov. **148** (1978), 43–57; English transl. in Proc. Steklov Inst. Math. **1980**, no. 4 (148).

6. ______, *On subgroups of the full linear group over a commutative ring*, Sixteenth All-Union Algebra Conf., Abstracts of Reports, Leningrad. Otdel Mat. Inst. Steklov., Leningrad, 1981, pp. 23–24. (Russian)

7. ______, *The distribution of subgroups containing the group of block-diagonal matrices in the full linear group over a ring*, Izv. Vyssh. Uchebn. Zaved. Mat. **1982**, no. 11 (246), 12–16; English transl. in Soviet Math. (Iz. VUZ) **26** (1982).

8. Z. I. Borevich, N. A. Vavilov and V. Narkevich [W. Narkiewicz], *Subgroups of the full linear group over a Dedekind ring*, Zap. Nauchn. Sem. Leningrad. Otdel. Mat. Inst. Steklov. (LOMI) **94** (1979), 13–20; English transl. in J. Soviet Math. **19** (1982), no. 1.

9. Z. I. Borevich and L. Yu. Kolotilina, *The subnormalizer of net subgroups in the full linear group over a ring*, Zap. Nauchn. Sem. Leningrad. Otdel. Mat. Inst. Steklov. (LOMI) **116** (1982), 14–19; English transl. in J. Soviet Math. **26** (1984), no. 3.

10. Z. I. Borevich and O. N. Matsedonskaya, *On the lattice of subgroups*, Zap. Nauchn. Sem. Leningrad. Otdel. Mat. Inst. Steklov. (LOMI) **103** (1980), 13–19; English transl. in J. Soviet Math. **24** (1983), no. 4.

11. N. A. Vavilov, *Description of the subgroups of the full linear group over a semilocal ring that contain the group of diagonal matrices*, Zap. Nauchn. Sem. Leningrad. Otdel. Mat. Inst. Steklov. (LOMI) **86** (1979), 30–33; English transl. in J. Soviet Math. **17** (1981), no. 4.

12. ______, *On the subgroups of the full linear group over a semilocal ring that contain the group of diagonal matrices*, Vestnik Leningrad. Univ. **1981**, no. 1 (Ser. Mat. Mekh. Astr. vyp. 1), 10–15; English transl. in Vestnik Leningrad Univ. Math. **14** (1982).

13. N. A. Vavilov and L. Yu. Kolotilina, *The normal structure of the full linear group over a semilocal ring*, Zap. Nauchn. Sem. Leningrad. Otdel. Mat. Inst. Steklov. (LOMI) **94** (1979), 37–39; English transl. in J. Soviet Math. **19** (1982), no. 1.

14. I. Z. Golubchik, *The full linear group over an associative ring*, Uspekhi Mat. Nauk **28** (1973), no. 3 (171), 179–180. (Russian)

15. V. A. Koĭbaev, *On subgroups of the full linear group that contain the group of elementary block-diagonal matrices*, Vestnik Leningrad. Univ. **1982**, no. 13 (Ser. Mat. Mekh. Astr. vyp. 3), 33–40; English transl. in Vestnik Leningrad Univ. Math. **15** (1983).
16. N. S. Romanovskiĭ, *The subgroups of the special and general linear groups over a ring*, Mat. Zametki **9** (1971), 699–708; English transl. in Math. Notes **9** (1971).
17. A. A. Suslin, *On the structure of the special linear group over polynomial rings*, Izv. Akad. Nauk SSSR Ser. Mat. **41** (1977), 235–252; English transl. in Math. USSR Izv. **11** (1977).
18. M. S. Tulenbaev, *The Schur multiplier of the group of elementary matrices of finite order*, Zap. Nauchn. Sem. Leningrad. Otdel. Mat. Inst. Steklov. (LOMI) **86** (1979), 162–169; English transl. in J. Soviet Math. **17** (1981), no. 4.
19. Karl Hans Müller, *Schwachnormale Untergruppen: Eine gemeinsame Verallgemeinerung der normalen und normalisatorgleichen Untergruppen*, Rend. Sem. Mat. Univ. Padova **36** (1966), 129–157.
20. J. S. Wilson, *The normal and subnormal structure of general linear groups*, Proc. Cambridge Philos. Soc. **71** (1972), 163–177.

Translated by D. GILDENHUYS

Труды Матем. Инст. Стеклов.
165(1984)

Proceedings of the Steklov
Institute of Mathematics
1985, Issue 3

ON EVEN UNIMODULAR EXTREMAL LATTICES

UDC 511.342

B. B. VENKOV

ABSTRACT. This paper is devoted to spherical t-block designs, which are analogous to the classical combinatorial concept of t-block designs. A finite subset of the unit sphere $X \subset S^n$ is called a spherical t-block design if for any orthogonal transformation T and any homogeneous polynomial f of degree $\leqslant t$

$$\sum_{x \in X} f(Tx) = \sum_{x \in X} f(x).$$

The existence of spherical t-block designs with $t \leqslant 11$ is known. This paper contains many new examples of spherical 11-block designs. The main result is the establishment of a connection between spherical block designs and even, unimodular, euclidean lattices. This connection is based on an analogy between the theory of such lattices and the theory of even, self-dual codes on the field of two elements. Formerly coding theory was used to construct nontrivial classical block designs. The author constructs the spherical analogue of this approach.

Bibliography: 9 titles.

§1. Introduction

One of the classical concepts of combinatorics is the concept of a t-block design. A finite set X of v elements together with selected subsets (blocks) is called a t-(v, d, λ)-*block design* if any t distinct elements of X lie in the same number λ of blocks and if all the blocks consist of the same number d of elements. Such t-block designs arise naturally from t-transitive permutation groups and are the geometries through which these groups are usually most simply described. For example, there exists a unique 5-(24, 8, 1)-block design, and its automorphism group is the largest Mathieu group M_{24}.

A finite subset X of the unit sphere $\Omega \subset \mathbf{R}^d$ is called a *spherical t-block design* if for any orthogonal transformation $T \in O(d)$ and for any homogeneous polygon f on $\mathbf{R}^d$ of degree $\leqslant t$ we have

$$\sum_{\xi \in X} f(T\xi) = \sum_{\xi \in X} f(\xi).$$

This concept, recently introduced by Delsarte, Goethals and Seidel [3], turned out to be very useful, and spherical t-block designs are already widely elucidated in the literature (see, for example, the survey [5]). The analogy with the classical definition is explained in [3]. Spherical 2-block designs were studied earlier by Schläfli. They are the orthogonal projections on $\mathbf{R}^d$ of n mutually orthogonal unit vectors (with a

1980 *Mathematics Subject Classification*. Primary 05B05, 05B35; Secondary 94B25.

corresponding normalization). A large number of distinct conditions on $X \subset \Omega$, equivalent to the requirement that a t-block design be spherical, are mentioned in [3] and [5]. For us the following equivalence is important:

PROPOSITION 1 [3]. *A finite subset $X \subset \Omega$ is a spherical t-block design if and only if*

$$\sum_{\xi \in X} f(\xi) = 0 \tag{1}$$

for all homogeneous harmonic polynomials f of degree $\leqslant t$.

The larger t is, the more limiting are the conditions that a t-block design be spherical, and it is not known whether there exist spherical 12-block designs. This is entirely analogous to the fact that it is unknown whether there exist nontrivial classical t-block designs with $t > 5$. Spherical 11-block designs exist (see [3] and [5]), and many other examples are constructed below.

Many nontrivial examples of classical t-block designs were constructed by Assmus and Mattson [2] with the help of coding theory. Thus, for example, they showed the following. Let $V \subset F_2^n$ be an even, self-dual code on the field F_2 of two elements, suppose that $n \equiv 0 \pmod{24}$ and suppose that the code V is extremal—that is, its minimal distance attains the greatest value compatible with the MacWilliams equations. Then the code words of fixed weight are the blocks for a 5-block design. For coding theory terminology, see [7]. For example, the above-mentioned 5-(24, 8, 1)-block design consists of the words of weight 8 in an extremal 24-dimensional code—the Golay code.

As is well known, there are deep analogies between even, self-dual codes on F_2 and integral, unimodular, even, euclidean lattices. In particular, the MacWilliams equations are analogous to the functional equation for a θ-row of a lattice. Therefore a natural question, clearly formulated by Sloane in [9], arises: does there exist a spherical analogue of the theorem of Assmus and Mattson? One of our tasks is to show that the answer to this question is positive. This is done in Theorem 1. The proof is extraordinarily short, but analytical, since it is based on the theory of modular forms. This connection with modular forms enables us to indicate one additional property of the spherical t-block designs constructed in this work. The analogue of this additional condition for classical t-block designs together with a refinement of the theorem of Assmus and Mattson for codes will be discussed in another place. A spherical t-block design with this additional property will be called a $t\frac{1}{2}$-block design.

DEFINITION. A spherical t-block design which is not a spherical $(t + 1)$-block design is called a *$t\frac{1}{2}$-block design* if equation (1) is satisfied also for all homogeneous harmonic polynomials of degree $t + 3$.

It turns out that spherical block designs constructed from extremal lattices of dimension divisible by 24 are $11\frac{1}{2}$-block designs—that is, for them equation (1) is satisfied also for harmonic polynomials of degree 14 (but not of degree 12).

Theorems 2 and 3 are corollaries of Theorem 1.

In the last section certain further properties of extremal lattices of dimension 48 are shown.

Theorem 1 (and, therefore, Theorems 2 and 3 too) may be generalized also to extremal lattices of dimensions not divisible by 24. But as in the theorem of Assmus and Mattson, the resulting spherical t-block designs have t less than 11, and therefore these block designs are less interesting and the corresponding results will not be discussed.

§2. Extremal lattices and associated spherical block designs

Let $\Lambda \subset R^{24k}$ be an even, unimodular, euclidean lattice in R^{24k}. The lattice Λ is called *extremal* if for $\lambda \in \Lambda$, $\lambda \neq 0$, we always have $(\lambda, \lambda) > 2k$. (Do not confuse this definition with another definition—extreme lattices—which involves arbitrary lattices.) A ϑ-row of a lattice Λ

$$\vartheta_\Lambda(q) = \sum_{n=0}^{\infty} a_n q^n, \qquad q = e^{2\pi i \tau},$$

where $a_n = \sum_{\lambda \in \Lambda, (\lambda,\lambda)=2n} 1 = |\Lambda_{2n}|$, is a modular form of weight $12k$ with respect to the full modular group PSL(2, $\mathbf{Z}$). The space of all such forms has dimension $k + 1$. If the lattice Λ is extremal, then we have $k + 1$ linear relations

$$a_0 = 1, \qquad a_1 = a_2 = \cdots = a_k = 0 \tag{2}$$

on the coefficients a_i. This shows that the ϑ-row is uniquely defined for an extremal lattice. For more details on extremal lattices are Sloane [9] and the references listed there.

THEOREM 1. *Let $\Lambda \in \mathbf{R}^{24k}$ be an extremal lattice, $n \geqslant k + 1$ and $\Lambda_{2n} = \{\lambda \in \Lambda | (x, x) = 2n\}$. Then $X = 1/\sqrt{2n}\,\Lambda_{2n}$ is a spherical $11\frac{1}{2}$-block design.*

PROOF. Take a homogeneous harmonic polynomial F on $\mathbf{R}^{24k}$ of degree l, where $0 < l < 16$, $l \neq 12$. One can assume l to be even, since for odd l equation (1) is always trivially satisfied because of the symmetry of X. Consider the following formal power series:

$$\theta_F(q) = \sum_{n=0}^{\infty} b_n q^n = \sum_{n=0}^{\infty} \Big(\sum_{\lambda \in \Lambda_{2n}} F(\lambda) \Big) q^n.$$

It is well known (see, for example, Hecke [4] or Ogg [8]) that $\theta_F(q)$ is a Fourier decomposition for the modular form of weight $12k + l$ with respect to the full modular group. Clearly, under our assumptions about the lattice Λ we have

$$b_0 = b_1 = \cdots = b_k = 0. \tag{3}$$

In the ring of modular forms relative to PSL(2, $\mathbf{Z}$) the forms which satisfy (3) make up the principal ideal generated by Δ^{k+1}, where Δ is the unique normalized parabolic modular form of weight 12. In particular,

$$\theta_F(q) = \Delta^{k+1} g,$$

where g is some modular form. Comparing weights, we obtain $12k + l = 12(k + 1) +$ weight g; that is, $l = 12 +$ weight g. Taking into consideration the fact that $0 < l < 16$, $l \neq 12$, and that there does not exist a nontrivial modular form of negative weight or of weight 2, we find that $g = 0$—that is, $\theta_F(q) = 0$ and, consequently,

$$\sum_{\lambda \in \Lambda_{2n}} F(\lambda) = 0$$

for any n.

It remains only to normalize the set Λ_{2n} so that Λ_{2n} lies on the unit sphere. The theorem is proved.

THEOREM 2. *Let $\Lambda \subset \mathbf{R}^{24k}$ be an extremal lattice, and suppose that $\alpha \in \mathbf{R}^{24k}$. Then*

$$\sum_{n=0}^{\infty}\Bigl(\sum_{\xi\in\Lambda_{2n}}(\xi,\alpha)^{\nu}\Bigr)q^n = \frac{1\cdot 3\cdots(\nu-1)}{24k(24k+2)\cdots(24k+\nu-2)}2^{\nu/2}(\alpha,\alpha)^{\nu/2}\theta^{\nu/2}\vartheta_{\Lambda}$$

for $\nu = 2, 4, 6, 8, 10$. Here ϑ is a ϑ-row of a lattice Λ, and θ is the differential operator qd/dq. Furthermore,

$$\sum_{n}\Bigl(\sum_{\xi\in\Lambda_{2n}}(\xi,\alpha)^{14}\Bigr)q^n = a(\alpha,\alpha)P\sum_{n}\Bigl(\sum_{\xi\in\Lambda_{2n}}(\xi,\alpha)^{12}\Bigr)q^n$$
$$+b(\alpha,\alpha)^7\theta^7\vartheta_{\Lambda} + c(\alpha,\alpha)^7 p\theta^6\vartheta_{\Lambda}.$$

Here, a, b and c are certain constants depending only on k, and p is the Eisenstein series of weight 2 for PSL(2, **Z**) *(which is not a modular form).*

PROOF. This follows with the help of formal computation from the previous theorem. For this one must express the polynomial functions $\xi \to (\xi,\alpha)^{\nu}$ with the help of the Gegenbauer polynomial in terms of harmonic polynomials and $(\xi,\xi)^{\mu}$ and then use Theorem 1. We omit the details.

Theorems 1 and 2 have many corollaries. We note only one of these, related to the action of the automorphism group of a lattice on the vectors of fixed length. Let $\Lambda \subset \mathbf{R}^m = W$ be a lattice and $G = \operatorname{Aut}\Lambda$ its automorphism group. Since the lattice is euclidean, G is a finite group. Now G acts on each of the finite sets Λ_n. We denote by $[\Lambda_n] = W_n$ the corresponding transitivity module. We denote by W_n^+ (respectively W_n^-) the invariant subspace in W_n on which $-1 \in G$ acts like $+1$ (respectively, like -1).

On the other hand, G acts orthogonally on the underlying space W, and one may consider the harmonic degrees $\operatorname{Harm}^i(W)$ of this representation.

THEOREM 3. *Let $\Lambda \subset \mathbf{R}^{24k}$ be an extremal lattice and G its automorphism group. Then each of its nontrivial modules W_n^+ (that is, for $n \geqslant 2(k+1)$) contains*

$$\operatorname{Harm}^0(W) \oplus \operatorname{Harm}^2(W) \oplus \operatorname{Harm}^4(W),$$

and each of the modules W_n^- contains

$$\operatorname{Harm}^1(W) \oplus \operatorname{Harm}^3(W) \oplus \operatorname{Harm}^5(W).$$

PROOF. The representation space W_n can be identified with the space of all complex functions on Λ_n, which has a selected basis indexed with points from Λ_n and consisting of functions with support in that point. We order somehow the elements of Λ_n and we identify the operators in Λ_n with the matrices in this ordered basis. We consider the following matrices:

$$(V_l)_{(x,y)} = (x,y)^l.$$

Clearly they commute with the operators from G—that is, $V_l \in \operatorname{Hom}_G(v,v)$. Furthermore, using Theorem 2 we may compute the products $v_l \cdot v_m$ for $l + m \leqslant 10$, and they will be linear combinations of the matrices V_r with $r \leqslant \max(l,m)$. Then we may select, one at a time, linear combinations l_i of V_l which are mutually orthogonal and idempotent. Below we state the corresponding formulas. For simplicity we restrict the discussion to the case of the shortest vectors $n = 2(k+1)$. The general case is entirely analogous. Let $w = |\Lambda_{2(k+1)}|$. Consider the following matrices:

$$e_0 = \frac{1}{w} V_0, \qquad e_1 = \frac{m}{2(k+1)w} V_1,$$

$$e_2 = \frac{m(m+2)}{2^3(k+1)^2 w} V_2 - \frac{1}{2}(m+2)e_0, \quad e_3 = \frac{m(m+2)(m+4)}{2^4\cdot 3(k+1)^3 w} V_3 - \frac{m+4}{2} e_1,$$

$$e_4 = \frac{m(m+2)(m+4)(m+6)}{2^7\cdot 3(k+1)^4 w} V_4 - \frac{1}{2}(m+6)e_2 - \frac{1}{2^3}(m+4)(m+6)e_0,$$

$$e_5 = \frac{m(m+2)(m+4)(m+6)(m+8)}{2^8\cdot 3\cdot 5(k+1)^5 w} V_5 - \frac{1}{2}(m+8)e_3 - \frac{(m+6)(m+8)}{2^3} e_1.$$

Here $m = 24k = \dim W$.

Then $e_i^2 = e_i$ and $e_i \cdot e_j = 0$ for $0 \leqslant i,\ j \leqslant 5$ and $i \neq j$. Furthermore, it can be shown that $\operatorname{Im} e_i$ as a $C[G]$-module is isomorphic to $\operatorname{Harm}^i(v)$. Clearly $\operatorname{Im} e_{2i} \subset W^+_{2(k+1)}$, and $\operatorname{Im} e_{2i+1} \subset W^-_{2(k+1)}$.

The theorem is proved.

COROLLARY. *Let $\Lambda \subset \mathbf{R}^{24}$ be the Leech lattice and let W_4^- be the odd part of the transitivity O-module acting on the shortest vectors of the Leech lattice. Then*

$$W_4^- = W \oplus \operatorname{Harm}^3(W) \oplus \operatorname{Harm}^5(W).$$

PROOF. This follows from Theorem 3 and a calculation of dimensions. For the even part of W_4^+ the corresponding equation is already not satisfied. Aside from $\operatorname{Harm}^0(W) \oplus \operatorname{Harm}^2(W) \oplus \operatorname{Harm}^4(W)$, W_4^+ has one more irreducible representation of dimension $2\cdot 3^3\cdot 5\cdot 13\cdot 23$.

§3. Extremal lattices of dimension 48

Since we are considering only dimensions divisible by 24 (Theorems 1, 2 and 3 hold only for these), 48 is the next case after 24 for which the extremal lattice is the Leech lattice. Already for this case the classification of extremal lattices is not known. Only two specific examples of extremal 48-dimensional lattices, constructed by Leech and Sloane [6], are known. Applying Theorem 1 to them, we may obtain many new examples of $11\frac{1}{2}$-block designs. However we can obtain further information about extremal lattices of dimension 48 which is unknown for greater dimensions. In this section we present several facts without proof.

Let Λ be an arbitrary extremal lattice in $\mathbf{R}^{48}$.

First, it may be shown that Λ, as a lattice, is generated by its shortest vectors. Therefore, instead of Λ it suffices to study the finite configuration of shortest vectors Λ_6. This is a spherical $11\frac{1}{2}$-block design consisting of $2^{10}\cdot 3^2\cdot 5^2\cdot 17\cdot 19$ vectors lying in the sphere $(\alpha,\alpha) = 6$. The inner product of any two $\alpha, \beta \in \Lambda_6$, $\alpha \neq \pm\beta$, can take only the values $0, \pm 1, \pm 2, \pm 3$. We fix $\alpha \in \Lambda_6$ and denote by $n_i(\alpha)$ the number of $\beta \in \Lambda_6$ such that $(\alpha, \beta) = i$. It turns out that the numbers $n_i(\alpha) = n_i$ do not depend on α and are equal to

$$n_1 = n_{-1} = 2^4\cdot 3^6\cdot 23\cdot 47, \qquad n_2 = n_{-2} = 3^6\cdot 7^2\cdot 47,$$

$$n_3 = n_{-3} = 2^4\cdot 7^2\cdot 47, \qquad n_0 = 2^4\cdot 5\cdot 297337.$$

If we fix a pair of elements $\alpha, \beta \in \Lambda_6$ and denote by $n_{i,j}^{\{\alpha,\beta\}}$ the number of $\gamma \in \Lambda_6$ such that $(\gamma, \alpha) = i$ and $(\gamma, \beta) = j$, it turns out that in the majority of cases the numbers $n_{i,j}^{\{\alpha,\beta\}}$ do not depend on the specific choice of the pair α, β, but only on their inner product. This will be the case if $(\alpha, \beta) \neq 0$. If $(\alpha, \beta) = 0$, then in general the $n_{i,j}^{\{\alpha,\beta\}}$ depend on the pair $\{\alpha, \beta\}$, but they can all be expressed by a universal formula in terms of one of the coefficients, for example $n_{3,3}^{\{\alpha,\beta\}}$. The situation is analogous to the one which was examined in [1] for 32-dimensional lattices.

Finally, from the configuration Λ_6 and a fixed $\alpha \in \Lambda_6$ one can construct a strongly regular graph with parameters $(2^3 \cdot 7^2 \cdot 47, 2^3 \cdot 3^2 \cdot 23, 2 \cdot 139)$. This graph is constructed as follows. Let X be the set of those $\beta \in \Lambda_6$ for which $(\beta, \alpha) = 3$, and let σ: $X \to X$ be the involution which acts on X so that

$$\sigma\beta = \alpha - \beta.$$

For the vertex-set of our graph we take the quotient space $X/\{\sigma\}$—that is, we identify β and $\sigma\beta$. We obtain a set consisting of $2^3 \cdot 7^2 \cdot 47$ elements. Further, if β, $\gamma \in X$ and $\gamma \neq \beta$, $\sigma\beta$, then (β, γ) may take only the values 0, 1, 2, 3. We define a graph structure on X by considering β and γ to be joined by an edge if $(\beta, \gamma) = 0$ or 3. It is easy to check that this structure can be extended to $X/\{\sigma\}$, and that the graph so obtained is strongly regular. It always has the parameter

$$(2^3 \cdot 7^2 \cdot 47, 2^3 \cdot 3^2 \cdot 23, 2 \cdot 139).$$

Bibliography

1. B. B. Venkov, *Even modular Euclidean lattices of dimension* 32, Zap. Nauchn. Sem. Leningrad. Otdel. Mat. Inst. Steklov. (LOMI) **116** (1982), 44–55; English transl. in J. Soviet Math. **26** (1984), no. 3.

2. E. F. Assmus, Jr., and H. F. Mattson, Jr., *New 5-designs*, J. Combinatorial Theory **6** (1969), 122–151.

3. P. Delsarte, J. M. Goethals and J. J. Seidel, *Spherical codes and designs*, Geometriae Dedicata **6** (1977), 363–388.

4. E. Hecke, *Analytische Arithmetik der positiven quadratischen Formen*, Danske Visl. Selsk. Math.-Fys. Medd. **17** (1940), no. 12, reprinted in his *Mathematische Werke*, Vandenhoeck & Ruprecht, Göttingen, 1959.

5. J. M. Goethals and J. J. Seidel, *Spherical designs*, Relations between Combinatorics and Other Parts of Mathematics, Proc. Sympos. Pure Math., vol. 34, Amer. Math. Soc., Procidence, R. I., 1979, pp. 255–272.

6. John Leech and N. J. A. Sloane, *Sphere packings and error-correcting codes*, Canad. J. Math. **23** (1971), 718–745.

7. F. J. MacWilliams, and N. J. A. Sloane, *The theory of error-correcting codes*. I, II, North-Holland, 1977.

8. Andrew Ogg, *Modular forms and Dirichlet series*, Benjamin, 1969.

9. N. J. A. Sloane, *Self-dual codes and lattices*, Relations between Combinatorics and Other Parts of Mathematics, Proc. Sympos. Pure Math., vol. 34, Amer. Math. Soc., Providence, R. I., 1979, pp. 273–308.

Translated by T. R. WALSH

Труды Матем. Инст. Стеклов.
165(1984)

Proceedings of the Steklov
Institute of Mathematics
1985, Issue 3

POISSON BRACKETS AND COMPLEX TORI

UDC 531.314 + 514.12

A. P. VESELOV AND S. P. NOVIKOV

ABSTRACT. This paper is devoted to a theory, introduced by the authors, of a broach class of completely integrable Hamiltonian systems and Poisson brackets on the space of finite-zone potentials or a fibering of hyperelliptic Jacobians. Analysis shows that the Poisson brackets corresponding to known examples of integrable systems on this phase space belong to the class studied. The geometry of the "action-angle" variables and the coordination condition for the structures introduced with KdV theory is investigated; examples are considered.

Bibliography: 30 titles.

Introduction

As is known, at the end of the sixties Gardner, Greene, Miura, and Lax discovered the "method of the inverse scattering problem" for solving the Korteweg-de Vries equation (KdV) in the class of rapidly decreasing functions in x.

A history of this discovery and further development of the method can be found in [1]. The problem of finding solutions periodic in x led to the discovery of the so-called method of finite-zone integration created in 1974–1975 (see [2]–[6] and the survey [7] or [1], Chapter II). The method of finite-zone integration is based on a synthesis of the theory of classical, completely integrable Hamiltonian systems, the spectral theory of linear operators with periodic and quasiperiodic coefficients (further developed in connection with KdV even for the classical Hill or Schrödinger operator), and, finally, the classical algebraic geometry of Riemann surfaces and Abelian varieties, including the theory of multidimensional theta functions. Further development of the method and a number of new applications of it can be found in the surveys [8]–[11].

Already in the case of the ordinary KdV equation the so-called finite-zone solutions (including the formal complex analogue of them) form a finite-dimensional manifold of dimension $3g + 1$ in the space of functions of x for each integer $g \geqq 1$. The collection of "higher KdV" on this manifold in one realization is a commutative pencil of Hamiltonian systems in the space $\mathbf{C}^{2g}$ depending rationally on $g + 1$ parameters and having polynomial Hamiltonians. Another realization of this same pencil of phase spaces identifies it with the affine part of the $(3g + 1)$-dimensional space of a fibering of hyperelliptic Jacobians; the base of this fibering is the moduli

---

1980 *Mathematics Subject Classification*. Primary 58F05; Secondary 35Q20, 14K30.

space of hyperelliptic curves (more precisely, the symmetrized set of roots λ_0, $\lambda_1, \ldots, \lambda_{2g}$ of the polynomial $R_{2g+1}(\lambda) = \prod_j(\lambda - \lambda_j)$ for the realization of the curve Γ in the form $y^2 = R_{2g+1}(\lambda)$; the fiber is the Jacobi variety $J(\Gamma)$ of the curve Γ, i.e., an Abelian torus of complex dimension g). The "higher KdV" in the second realization are rectilinear windings of the tori $J(\Gamma)$. As I. R. Shafarevich pointed out after a talk by S. P. Novikov in 1974 before the algebraic geometry seminar at Moscow State University, the fact of the unirationality of the space of moduli of hyperelliptic Jacobians $J(\Gamma)$ which follows from a comparison of these relations was previously not known in algebraic geometry (see the theorem in [12]). Of course, this fact could have been derived also from other systems of classical mechanics and geometry known already in the nineteenth century to be integrable in theta functions (for example, the case of Kovalevskaya for genus $g = 2$, as indicated in [7], or the geodesic flow on ellipsoids of all dimension in elliptic coordinates discovered by Jacobi for the integration of geodesics [13]). Earlier, however, before the appearance of the method of finite-zone integration in KdV theory, algebraic geometry did not exploit the connection with the theory of classical integrable Hamiltonian systems (although classical mechanicists, on the other hand, who discovered familiar integrable cases, used theta functions, especially of genus $g = 2$). Further more profound applications of the method of finite-zone integration of nonlinear systems in the theory of Abelian varieties (in particular, in the well-known Schottky problem) can be found in the survey [9]. A number of new applications of the theory of vector bundles over algebraic curves Γ and projective space $\mathbf{C}P^3$ to the integration of equations of mathematical physics can be found in the surveys [10] and [11].

This paper is devoted to further development of the theory of Hamiltonian systems completely integrable in the Liouville sense on the phase space already discussed above of finite-zone potentials (solutions of KdV) or of hyperelliptic Jacobians $J(\Gamma)$; this theory was begun in [14]. The rich collection of classical and new integrable systems, connected, as analysis shows, with precisely this phase space, leads to a multitude of distinct Poisson brackets on it which possess an interesting algebro-geometric structure that is, in a particular sense, common to all of them. It is interesting that many Poisson brackets lead to pencils of complex tori that are not even Abelian varieties (see the end of §1).

§1. Phase spaces (complex). Analytic Poisson brackets. Complex tori

We prescribe some algebraic family (variety) M^n whose points are some collection of parameters determining algebraic curves which are not necessarily hyperelliptic, $\Gamma \in M^n$. The same curve up to equivalence may be encountered several times; n is the complex dimension. For any integer $k \geqslant 1$ there is the natural fibering

$$N^{n+k} \overset{S^k\Gamma}{\to} M^n, \tag{1}$$

where the fiber over a point $\Gamma \in M^n$ is the symmetric degree of the curve Γ: $(P_1, \ldots, P_k) \sim (P_{i_1}, \ldots, P_{i_k})$. For the special case $k = g$ (the genus of the curve) the fiber is birationally equivalent to the Jacobi variety $S^k\Gamma \cong J(\Gamma)$.

There are two most important examples in the theory of integrable systems.

EXAMPLE 1. $\Gamma \in M^{2g+1}$ is given in the form

$$y^2 = \prod_{j=0}^{2g} (\lambda - \lambda_j) = R(\lambda). \tag{2}$$

Here Γ is determined by the symmetrized collection $(\lambda_0, \ldots, \lambda_{2g})$. In some problems the covering $\hat{M}_g \to M^{2g+1}$ determined by the unsymmetrized collection $(\lambda_0, \ldots, \lambda_{2g}) \in \hat{M}_g$ arises.

EXAMPLE 2. $\Gamma \in M^{2g+2}$ is given in the form

$$y^2 = \prod_{j=0}^{2g+1} (\lambda - \lambda_j) = R(\lambda), \tag{3}$$

where the polynomial $R(\lambda)$ has degree $2g + 2$,

$$(\lambda_0, \ldots, \lambda_{2g+1}) \sim (\lambda_{i_0}, \ldots, \lambda_{i_{2g+1}}) \in M^{2g+2}.$$

For convenience we shall henceforth always assume that the family of algebraic curves M^n is given to us in the form of m-sheeted coverings of the λ-plane, $\Gamma \xrightarrow{\lambda} \mathbf{C}^1$, so that a "general" point P has the function $\lambda(P)$ as local coordinate. A symmetrized collection $(P_1, \ldots, P_k) \in S^k\Gamma$ has a standard collection of local coordinates at a "general" point $\lambda(P_1) = \gamma_1, \ldots, \lambda(P_k) = \gamma_k$, where $P_j = (\gamma_j, \varepsilon_j)$ and ε_j is the index of the sheet, which assumes m values. In the most important hyperelliptic case of Examples 1 and 2 we have $m = 2$ and $\varepsilon_j = \pm\sqrt{1}$. Sometimes we shall somewhat carelessly denote the point P_j itself by γ_j if in the given context this causes no confusion. Thus, points of the space N^{n+k} have the form

$$N^{n+k} = \{(\Gamma; (\gamma_1, \varepsilon_1), \ldots, (\gamma_k, \varepsilon_k)), (\gamma_i, \varepsilon_i) \in \Gamma, \Gamma \in M^n\}. \tag{4}$$

The analytic Poisson brackets of importance for us are determined as follows:

a) There is given a subsheaf of rings A in the sheaf of germs of meromorphic functions on M^n depending only on a point of the base $\Gamma \in M^n$. Subrings of the form A_U for open domains U play the role of the "annihilator" of the Poisson bracket which is actually concentrated on subvarieties $N_A \subset N^{n+k}$ where $f = \text{const}$ for all $f \in A$: $N_A \to M_A \subset M^n$.

b) There is given a meromorphic differential 1-form $Q(\Gamma)$ on the Riemann surface Γ or its covering $\hat{\Gamma} \to \Gamma$ depending on $\hat{\Gamma}$ as a parameter. In local notation we have

$$Q(\Gamma) = Q(\Gamma, \lambda)\, d\lambda.$$

It is required that the derivatives of $Q(\Gamma)$ along all directions of the base tangent to the manifolds M_A be globally defined meromorphic differential forms on the algebraic curve Γ itself (rather than on the covering $\hat{\Gamma}$)

c) In all the most important examples it has been found that either the form Q is meromorphic on Γ from the very beginning or it is meromorphic on the regular covering $\hat{\Gamma}$ with Abelian monodromy group $\hat{\Gamma} \to \Gamma$, where the image $\pi_1(\hat{\Gamma}) \to \pi_1(\Gamma) \to H_1(\Gamma, Z)$ is generated by a collection of cycles $a_1, \ldots, a_k$ with zero pairwise intersection indices $a_j \circ a_s = 0$.

REMARK. We have $k = g$ (the genus) for the KdV and sine-Gordon equations and the Toda lattice, and $k = g + 1$ for the NS equation (the nonlinear Schrödinger equation) (see [18]). For general matrix systems of order $l + 1$ we have $k = g + l$.

DEFINITION. If the closed 2-form

$$\Omega_Q = \sum_{i=1}^{k} dQ(\Gamma, \gamma_i) \wedge d_{\gamma_i} \tag{5}$$

is nondegenerate at a "general" point, i.e. in a Zariski-open region of the manifold N_A where the pair (A, Q) possesses properties a), b), and c), then we say that an *analytic Poisson bracket with annihilator* A is given on an open region of the manifold N^{n+k}. In this case the dimension of the manifold N_A must be equal to $2k$, $\dim M_A = k \leqslant n$.

By definition, the Poisson bracket (5) is given by the properties

$$\begin{aligned} &1) \quad \{\gamma_i, \gamma_j\} = 0 \qquad (i, j = 1, \dots, k), \\ &2) \quad \{Q(\gamma_i), Q(\gamma_j)\} = 0 \qquad (i, j = 1, \dots, k), \\ &3) \quad \{Q(\gamma_j), \gamma_i\} = \delta_{ji} \qquad (i, j = 1, \dots, k), \\ &4) \quad \{f, \gamma_j\} = \{f, Q(\gamma_k)\} = 0, \qquad f \in A, \end{aligned} \tag{6}$$

where the annihilator A contains only functions of the form $f(\Gamma)$, $\Gamma \in M^n$. The following straightforward result holds.

PROPOSITION 1. *Any analytic Poisson bracket possesses the following property: any two functions g, h depending only on a point of the base $\Gamma \in M^n$ (i.e., only on the Riemann surface) are in involution*:

$$\{g(\Gamma), h(\Gamma)\} = 0. \tag{7}$$

PROOF. Let $\lambda_1, \dots, \lambda_k$ be coordinates on the manifold M_A; then it follows easily that the form Ω_Q contains in its expansion only terms of the form $d\lambda_i \wedge d\gamma_j$, i.e., the matrix of the form has block form:

$$\omega = \begin{Vmatrix} 0 & A \\ -A^t & 0 \end{Vmatrix}, \qquad A_{ij} = \frac{\partial Q}{\partial \lambda_i}(\gamma_j).$$

The Poisson bracket is determined by the inverse matrix which has the same block structure, whence the involutivity of λ_i and (7) follow. The proposition is proved.

REMARK 1. Below it will be very important (see the end of this section) that the Poisson bracket is given only on an open region in the manifold N^{n+k}. It will be clear that the "true" phase space of the "action-angle" variables, which we shall call the complex phase space of Liouville variables, if it has a natural compactification may lead to fiberings different from (1) where the fiber is bimeromorphically (birationally) nonisomorphic to the variety $S^k\Gamma$.

REMARK 2. The rest of the theory is not altered in an essential way if γ_j is replaced by $\tilde{\gamma}_j = \gamma_j + f_j(\Gamma)$ in the definition of the bracket. Although this is apparently precisely the case in some concrete systems, we shall not investigate this case.

We further consider pairs (A, Q) such that all derivatives of Q along tangential directions $\tau_1, \dots, \tau_k$ to the manifolds M_A at a "general" point $\Gamma \in M_A$ constitute a collection of meromorphic 1-forms $\nabla_{\tau_i} Q$ on Γ with the following properties.

1. The forms $\nabla_{\tau_i} Q$ can be represented in the form

$$\nabla_{\tau_i} Q = \omega_i + \tilde{\omega}_i + \sum_{t=1}^{l} \tilde{\tilde{\omega}}_{it},$$

where the forms ω_i are holomorphic on Γ, and the forms $\tilde{\omega}_i$ are meromorphic with zero residues at all poles; the forms $\tilde{\tilde{\omega}}_{it}$ have a pair of poles of first order (P_t', P_t'') at which the residues differ only in sign.

2. If $k \geqslant g$, then it is required that the forms ω_i generate the one-dimensional cohomology group $H^{1,0}(\Gamma) = C^g$ and $\omega_i = 0$, $i > g$.

For the distinguished canonical basis of cycles

$$(a_1, \dots, a_g, b_1, \dots, b_g), \quad a_i \circ a_j = b_i \circ b_j = 0, \quad a_i \circ b_j = \delta_{ij}$$

we require the normalization

$$\oint_{a_j} \omega_i = \delta_{ij} \quad (i \leqslant g), \qquad \oint_{a_j} \tilde{\omega}_q = \oint_{a_j} \tilde{\tilde{\omega}}_{qt} = 0.$$

The case where the pairs of poles of first order (P_t', P_t'') do not intersect pairwise or with the other poles of higher orders we shall call the case of "general position".

The residues of the form Q whose gradients on M_A are the residues of the forms $\tilde{\tilde{\omega}}_{it}$ at the poles of first order P_t' depend only on Γ. We denote them by $\varphi_t(\Gamma)$. The transcendence degree of the collection of functions $\varphi_t(\Gamma)$ on a "general" manifold M_A (i.e., the functional dimension modulo the annihilator A) we denote by κ.

According to Proposition 1, all Hamiltonians of the form $H(\Gamma)$ commute with one another. The standard Liouville procedure for introducing formal "angle" variables (complex for the time being) which have linear time dependence leads to the following assertion for Hamiltonians of the form $H(\Gamma)$.

PROPOSITION 2. *If a Poisson bracket of general position is given by a pair* (A, Q) *satisfying conditions* 1 *and* 2 (*see above*), *then the following complex variables at a point* $(\Gamma = \Gamma_0, (\gamma_1, \varepsilon_1), \dots, (\gamma_k, \varepsilon_k))$ *of general position are independent and have linear dynamics in time*:

$$\Psi_j = \sum_{i=1}^{k} \int_{P_0}^{(\gamma_i, \varepsilon_i)} \nabla_{\tau_j} Q. \tag{8}$$

The variables Ψ_j are defined up to the lattice in the space $\mathbf{C}^k$ generated by the $2g + l$ vectors $e_1, \dots, e_g, e_1', \dots, e_g', \eta_i, \dots, \eta_k$ of the form

$$e_q^i = \oint_{a_q} \nabla_{\tau_i} Q = \delta_q^i, \quad e_q^{\prime i} = \oint_{b_q} \nabla_{\tau_i} Q = b_q^i, \quad \eta_s^i = \oint_{\delta_s} \nabla_{\tau_i} Q. \tag{9}$$

The cycles δ_s are small circular contours enclosing the poles of first order P_s', $s = 1, \dots, l$. Among these vectors at a point of general position $(\Gamma_0, \gamma_1, \dots, \gamma_k)$ there are precisely $2g + \kappa$ which are linearly independent over $\mathbf{Z}$. The vectors τ_i are tangent to the surface of the constant annihilator $M_A \subset M^n$ in the base of the fibering (1) and generate the tangent space to M_A. The following inequality holds:

$$2g + \kappa \leqslant 2k, \qquad \kappa \leqslant l.$$

The proof of the proposition follows immediately from the classical Liouville procedure (see, for example, [15], §50).

REMARK 1. It makes sense to discuss the special choice of the vectors τ_i corresponding to differentiation of Q along the so-called action variables canonically conjugate to the "angles" on the tori varying from 0 to 2π only for the real theory in the case where the level surfaces are compact, i.e., represent tori T^k (see §2).

REMARK 2. The compact complex torus T^{2k} can arise for the variables Ψ in the case where $2g + \kappa = 2k$. The most important case is $\kappa = 0$ $(k = g)$. An Abelian torus is by no means always obtained (on the contrary, very rarely). We point out that the present situation can arise for systems whose original phase space was an algebraic variety, while the integrals were polynomials or rational functions. Therefore, the level surfaces in the complex sense are algebraic varieties. The replacement (8) may possess ambiguity and singularities. Therefore, the complex space of Liouville variables, generally speaking, is bimeromorphically nonequivalent to the original phase space. Their real parts may nonetheless be analytically diffeomorphic. If the torus T^{2k} is non-Abelian, then there is no theory of theta functions connected with it, and there are no good formulas for solutions in terms of meromorphic special functions on the torus where the argument varies linearly with time. It is interesting that only tori of the form $J(\Gamma)$ arise from the method of the inverse problem.

We point out that in correspondence with the preceding results we construct a theory of integration only for Hamiltonians of the form $H(\Gamma)$. We shall first consider a system given only on an open region of points of general position (the Poisson bracket may have singularities on a special submanifold). Thus, for example, on the fibers $S^k\Gamma$ there are distinguished only subregions $V \subset S^k$ on which the replacement (8) is one-to-one.

Conclusion. The family of algebraic curves M^n with 1-form $Q(\Gamma)$ and annihilator A makes it possible, using (8), to construct a completely different complex fibering distinct from (1), where the fiber is the factor $J_Q(\Gamma) = \mathbf{C}^k/D$ and D is the lattice of (9).

It is possible that the fibering (10) can be constructed nonsingularly only over "general" level surfaces of the annihilator A:

$$N_{(Q,A)} \overset{J_Q(\Gamma)}{\rightarrow} M_A. \tag{10}$$

We shall consider the important case $\kappa = 0$ or $k = g$ of general position when the variables (8) are actually defined on the compact complex torus $J_Q(\Gamma)$. The following result is a summary of the foregoing considerations.

THEOREM 1. *The usual Abel transformation $S^g\Gamma \to J(\Gamma)$ linearizes the dynamics of all Hamiltonians of the form $H(\Gamma)$ for Poisson brackets defined by pairs (A, Q) of general position possessing properties* 1 *and* 2 (*see above*) *if and only if the derivatives $\nabla_{\tau_i} Q$ in all directions tangent to the "general" manifold of the constant annihilator M_A provide a basis of the holomorphic* 1*-forms of the Riemann surface* Γ.

The proof follows immediately from Propositions 1 and 2 by comparing (8) with the Abel transformation.

§2. The real structure. "Action-angle" variables

We shall assume that the family of algebraic curves M^n is given as a manifold with an antiholomorphic involution σ: $M^n \to M^n$, $\sigma^2 = 1$, extended to an anti-involution of the fibering with fiber Γ: $N^{n+1} \to M^n$. The fixed manifold M_A^σ thus consists only of curves Γ with an antiholomorphic involution σ_Γ: $\Gamma \to \Gamma$, $\sigma_\Gamma^2 = 1$. The form $Q = Q(\Gamma, \lambda)\, d\lambda$ and the annihilator A here must also be coordinated in the natural way with the involutions σ and σ_Γ of the fibering $M^{n+1} \to M^n$ with fiber Γ:

$$\begin{aligned} &\text{a)} \quad \sigma_\Gamma^* Q = \overline{Q} \quad \text{on } N^{n+1}, \\ &\text{b)} \quad \sigma^* A = \overline{A} \quad \text{on } M^n. \end{aligned} \tag{11}$$

The simplest class of real structures, which we shall call "elementary" for the Hamiltonian systems of interest to us, can be described as follows. We consider M- or $(M-1)$-curves Γ where the number of fixed ovals of the anti-involution σ_τ is equal either to $g+1$ (M-curves) or g (($M-1$)-curves). For M-curves $k=g$ or $k=g+1$ is possible. For $(M-1)$-curves only $k=g$ is possible (we assume that $k \geqq g$; see §1; the case $k<g$ is also interesting, but we do not consider it for the time being). The Hamiltonian $H(\Gamma)$ must be real, i.e., $H(\sigma\Gamma)=\overline{H}(\Gamma)$.

LEMMA. *If the Poisson bracket* (A, Q) *is coordinated with the elementary real structure, i.e., it possesses property* (11), *then the real collections* $\gamma_1, \dots, \gamma_k$ *are invariant under the dynamics generated by* $H(\Gamma)$, *where the* γ_i *all lie on pairwise distinct, fixed ovals of the anti-involution*:

$$\gamma_j \in a_j, \quad a_i \cap a_j = \Phi \qquad \left(i = i_1, \dots, i_g \text{ or } i = 1, 2, \dots, g+1\right). \tag{12}$$

For M-curves and $k=g$ the possible collections $\gamma_i \in a_i$ form the $g+1$ connected component which is isomorphic to the real torus T^g. For M-curves and $k=g+1$ or $(M-1)$-curves and $k=g$ there is only one connected component—the real torus T^{g+1} or T^g.

The lemma is proved by an obvious verification. We have the real structure with the tori (12) in the case $k=g$ for KdV and the case $k=g+1$ for NS_+ (the nonlinear Schrödinger equation with attraction).

Elementary real structures of more general type, by definition, represent a transformation $S^k\sigma_\Gamma$: $S^k\Gamma \to S^k\Gamma$ generated by the original anti-involution σ_Γ on the curve; the "real" subset (in the complex case the torus T^k) may be distinguished among the fixed points $S^k\sigma_\Gamma(x)=x$ in some manner distinct from (12).

Nonelementary real structures for $k=g$ arise in the important example of the very familiar sine-Gordon equation and for $k=g+1$ for the NS_- equation (the nonlinear Schrödinger equation with repulsion). For all systems admitting general Lax matrix representations of order $l+1$ we have $k=g+l$ (see [7], Chapter III, and [16]); if the realness conditions are such that the so-called monodromy matrix after a period (in the periodic problem) for the corresponding linear Lax L-operator belongs to a compact Lie group (similar to the group SU(2) for the sine-Gordon and NS equations), then the real structure is nonelementary.

Suppose the Poisson bracket is given by a pair (A, Q) and there is given an anti-involution

$$\sigma\colon M^n \to M^n, \qquad \sigma_\Gamma\colon N^{n+1} \to N^{n+1},$$

while the Hamiltonian is real: $H(\sigma x)=\overline{H}(x)$, $x \in M^n$, X a Riemann surface. According to §1, we have the fibering (10)

$$N_{(A,Q)} \xrightarrow{J_Q\Gamma} M_A$$

over a "general" level surface of the annihilator A.

DEFINITION. A *nonelementary real structure* is an arbitrary anti-involution τ: $N_{(A,Q)} \to N_{(A,Q)}$ commuting with the fibering (10), i.e., with the usual involution on the family of Riemann surfaces: σ, σ_Γ. The realness conditions for A and Q remain the same (11). On fibers $J_Q(\Gamma)$ the transformation τ must be a superposition of translation and an isomorphism of the real commutative groups $J_Q(\Gamma)$.

Real submanifolds in phase space are distinguished proceeding from conditions (13) or (14):

$$\sigma(\Gamma) = \Gamma, \qquad \tau(\eta) = \eta; \tag{13}$$

$$\sigma(\Gamma) = \Gamma, \qquad \tau(\eta) = -\eta + \eta_0^{(\alpha)}. \tag{14}$$

The case (14), which is encountered for the sine-Gordon and NS_- equations, is especially interesting. For a given curve the constant vector $\eta_0^{(\alpha)}$ may assume a finite number of values $\alpha = 1, \ldots, m$. Calculation of these quantities for the sine-Gordon equation was first carried out in [17]; the problem of explicitly distinguishing real solutions in the language of theta-function formulas on $J(\Gamma)$ was thus solved.

However, the involution τ is badly described in the variables $\gamma_1, \ldots, \gamma_k$; apparently, in the cases of NS_- and sine-Gordon it is not possible to explicitly and effectively distinguish the collections $\gamma_1, \ldots, \gamma_k$ for real solutions.

In the joint work of Dubrovin and Novikov the homological image in

$$H_1(\Gamma \setminus (P_1 \cup P_2), \mathbf{Z})$$

was determined and computed corresponding to "real" collections $\gamma_1, \ldots, \gamma_k$ for these two equations, where P_1 and P_2 are the infinitely distant points for NS_- (the curve Γ has the form $y^2 = R_{2g+2}(\lambda)$); for the sine-Gordon equation Γ is given in the form $y^2 = \prod_{j=0}^{2g}(\lambda - \lambda_j)$, $\lambda_0 = 0$; $P_1 = 0$, $P_2 = \infty$. The homological image is a collection of cycles $a_1, \ldots, a_k$, where $a_j \circ a_i = 0$, "corresponding" to real tori T^k.

The next assertion, which shows the importance of this collection of elements in $H_1(\Gamma, \mathbf{Z})$, follows from the proof of the Liouville theorem (see [14]).

PROPOSITION 3. *For analytic Poisson brackets satisfying elementary and nonelementary realness conditions the action variables J_j canonically conjugate to the angular coordinates on the tori T^k varying from 0 to 2π are given by*

$$J_j = \frac{1}{2\pi} \oint_{a_j} Q(\Gamma, \lambda)\, d\lambda. \tag{15}$$

PROOF. Formula (15) represents, by definition, the quantity $J_j = \oint_{\tilde{a}_j} p\, dq$, where the $\tilde{a}_j$ are the basis 1-cycles on the tori T^k. For the class of Poisson brackets we are studying the variables J_j acquire an important interpretation not only as integrals over the elements $\tilde{a}_j$ from the group $H_1(T^k, \mathbf{Z})$ according to Liouville, but also as integrals over the elements a_j of the group $H_1(\Gamma \setminus P, \mathbf{Z})$ where P is the set of poles of the form Q or the forms $\nabla_{\tau_j} Q$ for its derivatives along M_A; this proves the proposition.

LEMMA. *Compact real tori T^k are possible only when*

$$k \leqslant 2g + \kappa, \tag{16}$$

where κ is the number of essential residues of the poles of the form Q modulo the annihilator A.

PROOF. As indicated in §1, the number $2g + \kappa$ coincides for "general" brackets given by the pair (A, Q) with the rank of the lattice D in the space $\mathbf{C}^k$ of the variables (8). Since the real anti-involution τ is an automorphism of the real group $J_Q(\Gamma)$ and the dimension of the torus over $\mathbf{R}$ is equal to k, inequality (16) means simply that the dimension of the torus cannot be greater than the rank of the lattice. The lemma is proved.

We shall call $\delta = 2k - 2g - \kappa$ (the corank of the lattice) the number of variables of "phase type", if the real parts are compact tori.

The varieties $J_Q(\Gamma)$ are fibered in the form

$$J_Q(\Gamma) \overset{\mathbf{C}^s/\mathbf{Z}^\delta}{\rightarrow} \tilde{J}_Q^{k-s}(\Gamma), \qquad s \geqslant \delta,$$

where $\tilde{J}_Q(\Gamma)$ is a compact complex torus. Under the conditions of the lemma we have $s = \delta$.

§3. Poisson brackets coordinated with KdV theory. The most important examples

DEFINITION. A Poisson bracket on the phase space (4), where Γ are hyperelliptic curves of the form $y^2 = \prod_{j=0}^{2g}(\lambda - \lambda_j) = R(\lambda)$ is said to be *coordinated with KdV theory* if all the higher KdV are Hamiltonian in this bracket. Here $k = g$.

In [14] a number of examples were given of brackets coordinated with KdV theory (see below). For such brackets $J_Q(\Gamma) = J(\Gamma)$ always, i.e., the usual Abel transformation defined by means of the collection of basis holomorphic differential forms linearizes the dynamics of virtue of Hamiltonians of the form $H(\Gamma)$.

THEOREM 2. a) *An analytic Poisson bracket is coordinated with KdV theory if and only if all the forms $\nabla_\tau Q$—the derivatives in directions τ tangent to a level surface of functions from the annihilator A—are holomorphic and generate the group $H^{1,0}(\Gamma)$.*

b) *If the bracket is coordinated with KdV, then the coefficients of the expansion*

$$Q(\Gamma,\lambda) = \sum_{k=-N}^{\infty} \left(\frac{z}{2}\right)^k q_k(\Gamma), \qquad z = \lambda^{-1/2},$$

are such that $q_{2l+3}(\Gamma) = h_l(\Gamma)$ are Hamiltonians of the higher KdV with index $l \geqslant 0$, while the remaining coefficients q_k belong to the annihilator A.

REMARK. Part b) was established by the authors in [14].

PROOF. The usual Abel transformation linearizes the dynamics of all higher KdV; the necessity of the condition is therefore obvious. Suppose now that there is given a bracket (A, Q) such that $\nabla_\tau Q = \omega_\tau$ are holomorphic differentials generating the entire space $H^{1,0}(\Gamma)$. Since $z = \lambda^{-1/2}$, from the holomorphicity of the derivatives $\nabla_\tau Q$ along tangent directions to M_A it follows that the expansion for large λ has the form

$$Q(\Gamma, \lambda) = \sum_{k=0}^{\infty} \left(\frac{z}{2}\right)^{2k+3} H_k(\Gamma) + \text{Ann},$$

where Ann is a series with coefficients in the annihilator. We shall prove that $H_k(\Gamma)$ for $k = 0, 1, \dots, g-1$ are the Hamiltonians of precisely the first g higher KdV. We consider the standard set of holomorphic forms $\omega_1, \dots, \omega_g$ normalized by the conditions

$$\oint_{a_j} \omega_s = 2\pi i \delta_{js},$$

where a_j, b_j is the basis of cycles on Γ such that

$$a_i \circ a_j = b_i \circ b_j = 0, \qquad a_i \circ b_j = \delta_{ij}.$$

We define two collections of variables φ_i $(i = 1, \ldots, g)$ and ψ_j $(j = 0, \ldots, g-1)$:

$$\varphi_s = \sum_{j=1}^{g} \int_{P_0}^{(\gamma_j, \varepsilon_j)} \omega_s, \quad \psi_s = \sum_{j=1}^{g} \int_{P_0}^{(\gamma_j, \varepsilon_j)} \alpha_s, \qquad \alpha_s = \left.\frac{\partial Q}{\partial H_s}\right|_A .$$

The forms α_s are holomorphic and have at $\lambda = \infty$ the expansion $(z = \lambda^{-1/2})$

$$\alpha_s = \left[2\left(\frac{z}{2}\right)^{2s} + O(z^{2g})\right] dz \qquad \left(s = 0, \ldots, g^{-1}\right).$$

The transition matrix $\omega_s = A_{sj}\alpha_j$ is formed from the coefficients of the expansion of the forms ω_s, namely

$$\omega_s = \sum_{j=0}^{\infty} c_{sj} z^{2j}\, dz,$$

at the point $\lambda = \infty$, so that

$$A_{sj} = 2^{1-2j} c_{sj}.$$

For Abelian differentials of second kind Ω_j with zero a-periods and having a single pole at $\lambda = \infty$ with asymptotics $\Omega_j = -d(z^{2j-1}) + O(1)$ the following equality holds (see [19], §10–3):

$$\oint_{b_s} \Omega_j = c_{sj} = 2^{2j-1} A_{sj}.$$

From this it follows that the columns of the matrix A_{ij} coincide with the vectors which on the Jacobians $J(\Gamma)$ give the direction of the flows of the higher KdV with the corresponding indices $j = 0, 1, \ldots, g-1$. On the other hand, these columns, by definition, are the direction vectors of Hamiltonian flows with Hamiltonians $H_0, \ldots, H_{g-1}$. Thus, for the first g flows coordination has been proved. For the remaining flows it is automatically satisfied, since they can be expressed in terms of the first g on our phase space.

The proof of the theorem is complete.

We now list the most important examples.

EXAMPLE 1. *The standard bracket* [20], [21]. *From the results of* [22] *we derive*

$$Q = 2ip(\lambda)\, d\lambda, \qquad A = \left\{T_1, \ldots, T_g, \bar{u} = \lim_{T \to \infty} \frac{1}{T} \int_0^T u\, dx\right\},$$

where the T_i are the periods of the quasiperiodic potential $u(x)$, $p(\lambda)$ is the quasimomentum, and $dp(\lambda)$ is a differential of second kind with a single pole at $\lambda = \infty$,

$$dp(\lambda) = d(z^{-1}) + O(1), \qquad z = \lambda^{-1/2},$$

$$\oint_{a_j} dp(\lambda) = 0, \qquad j = 1, \ldots, g.$$

EXAMPLE 2. *The pencil of local brackets for all higher KdV* (see [23]), where the Hamiltonian systems have the form

$$\dot{u} = \left(al + b\frac{\partial}{\partial x}\right)\frac{\delta H}{\delta u}, \qquad l = \frac{1}{2}\frac{\partial^3}{\partial x^3} + u\frac{\partial}{\partial x} + \frac{\partial}{\partial x}u.$$

Here we have

$$Q = 2ip(\lambda)(a\lambda + b)^{-1}\, d\lambda.$$

If I_k are the Kruskal integrals, then

$$l\frac{\delta I_k}{\delta u} = \frac{\partial}{\partial x}\frac{\delta I_{k+1}}{\delta u}.$$

There is a relation among the integrals I_k—the stationary higher KdV equation which the potential $u(x)$ satisfies

$$\sum_{k=0}^{g+1} c_k \frac{\delta I_{g-k}}{\delta u} = 0, \qquad c_0 = 1.$$

The quantity $J = \Sigma_0^g c_k I_{g-k-1}$ belongs to the annihilator of the bracket for $b = 0$, since

$$l\frac{\delta J}{\delta u} = \frac{\partial}{\partial x}\frac{\delta}{\delta u}\left(\sum_{k=0}^{g} c_k I_{g-k}\right) = \frac{\partial}{\partial x}\text{const} = 0.$$

We thus obtain (for $b = 0$)

$$A = (T_1, \dots, T_g, J).$$

EXAMPLE 3. *The Hamiltonian formalism of the stationary problem for higher KdV generated by the variational calculus.* Coordination with KdV theory is proved in [24]; the Hamiltonians of the higher KdV are computed in [25]. Here

$$Q = \sqrt{-R(\lambda)}\, d\lambda, \qquad R(\lambda) = \prod_{i=0}^{2g} (\lambda - \lambda_i).$$

This result can be derived from [26] and [27]. As is known, the annihilator is generated by the first $g + 1$ symmetric polynomials in λ'_i.

EXAMPLE 4. *The Hamiltonian structure generated by the so-called latent isomorphism of Moser-Trubowitz* (see [28] and [13]) between the KdV dynamics on the space of finite-zone potentials and the classical Neumann systems describing the dynamics on the sphere S^g in $\mathbf{R}^{2g+1}$,

$$S^g = \left\{x \colon |x|^2 = 1, x \in \mathbf{R}^{g+1}\right\},$$

under the action of the quadratic potential

$$U(x) = \frac{1}{2}\sum_{i=0}^{g} a_i x_i^2.$$

The coefficients a_i, representing the set of end points of the zones λ_i, here play the role of the annihilator. Using [28] and [13], we obtain (see also [27])

$$Q = \sqrt{-R(\lambda)}\prod_{j=0}^{g} (\lambda - \lambda_{2j})^{-1}\, d\lambda, \qquad A = \{\lambda_0, \lambda_2, \dots, \lambda_{2g}\}.$$

EXAMPLE 5. *The integrable case of Goryachev and Chaplygin in the dynamics of a rigid body with a fixed point* [30]. Here

$$Q(\Gamma, \lambda)\arcsin\frac{1}{\mu}\left(\frac{\lambda^2}{2} - \frac{1}{2}H - \frac{2G}{\lambda}\right),$$

where H is the energy of the gyroscope, G is the Goryachev-Chaplygin integral, and μ is a parameter of the problem. The curve Γ is given by the equation

$$y^2 = 4\mu^2\lambda^2 - (\lambda^3 - H\lambda - 4G)^2.$$

EXAMPLE 6. In the well-known Kovalevskaya case the action variables were not computed previously. Let I_1, I_2 and I_3 be, respectively, the energy, area, and Kovalevskaya integrals, and let $\gamma_i = s_i$ be the Kovalevskaya variables (see [30]). Calculations lead to the result

$$Q(\Gamma, \lambda) = \frac{1}{2\sqrt{-\lambda}} \ln\left[\sqrt{-\lambda}\left((\lambda - I_1)^2 - I_3^2\right) + \frac{\nu^2}{2\sqrt{-\lambda}}(\lambda - 2I_2^2) + \sqrt{-R(\lambda)}\right],$$
$$R(\lambda) = \left(\lambda\left((\lambda - I_1)^2 + \nu^2 - I_3^2\right) - 2\nu^2 I_2^2\right)\left((\lambda -{}_1)^2 - I_3^2\right),$$

while the curve Γ is given by the equation $y^2 = R(\lambda)$. Integrating $Q = Q(\Gamma, \lambda)\,d\lambda$ over the "real" cycles a_j on which the γ_i lie, we obtain the action variables J_j. In correspondence with the result of Kovalevskaya the Abel transformation linearizes the dynamics (we note that the replacement of the time and Hamiltonian structure proposed by G. V. Kolosov (see [29]) leads for this system to non-Abelian tori).

The integrable Steklov case for a solid body in a fluid is also of major interest.

BIBLIOGRAPHY

1. S. P. Novikov (editor), *Theory of solitons*, "Nauka", Moscow, 1980; English transl., Plenum Press, New York, 1984.

2. S. P. Novikov, *A periodic problem for the Korteweg-de Vries equation*. I, Funktsional. Anal. i Prilozhen. **8** (1974), no. 3, 54–66; English transl. in Functional Anal. Appl. **8** (1974).

3. B. A. Dubrovin and S. P. Novikov, *Periodic and conditionally periodic analogs of the many-soliton solutions of the Korteweg-de Vries equation*, Zh. Èksper. Teoret. Fiz. **67** (1974), 2131–2144; English transl. in Soviet Phys. JETP **40** (1974).

4. A. R. It·s and V. B. Matveev, *Schrödinger operators with finite-zero spectrum and N-soliton solutions of the Korteweg-de Vries equation*, Teoret. i Mat. Fiz. **23** (1975), 51–68; English transl. in Theoret. Math. Phys. **23** (1975).

5. Peter D. Lax, *Periodic solutions of the KdV equation*, Comm. Pure Appl. Math. **28** (1975), 141–188.

6. H. P. McKean and P. van Moerbeke, *The spectrum of Hill's equation*, Invent. Math. **30** (1975), 217–274.

7. V. A Dubrovin, V. B. Matveev and S. P. Novikov, *Nonlinear equations of Korteweg-de Vries type, finite-zone linear operators, and Abelian varieties*, Uspekhi Mat. Nauk **31** (1976), no. 1 (187), 55–136; English transl. in Russian Math. Surveys **31** (1976).

8. I. M. Krichever, *Methods of algebraic geometry in the theory of nonlinear equations*, Uspekhi Mat. Nauk. **32** (1977), no. 6 (198), 183–208; English transl. in Russian Math. Surveys **32** (1977).

9. B. A. Dubrovin, *Theta functions and nonlinear equations*, Uspekhi Mat. Nauk **36** (1981), no. 2 (218), 11–80; English transl. in Russian Math. Surveys **36** (1981).

10. I. M. Krichever and S. P. Novikov, *Holomorphic bundles over algebraic curves and nonlinear equations*, Uspekhi Mat. Nauk **35** (1980), no. 6 (216), 47–68; English transl. in Russian Math. Surveys **35** (1980).

11. V. G. Drinfel'd et al., *Methods of algebraic geometry in contemporary mathematical physics*, Soviet Sci. Rev. Sect C: Math. Phys. Rev., vol. 1, Harwood Academic Publ., Chur, 1980, pp. 1–54.

12. B. A. Dubrovin and S. P. Novikov, *A periodic problem for the Korteweg-de Vries and Sturm-Liouville equations. Their connection with algebraic geometry*, Dokl. Akad. Nauk SSSR **219** (1974), 531–534; English transl. in Soviet Math. Dokl. **15** (1974).

13. J. Moser, *Various aspects of integrable Hamiltonian systems*, Dynamical Systems (C.I.M.E. Summer School, Bressanone, 1978), Progress in Math., vol. 8, Birkhäuser, 1980, pp. 233–289; also published in Dynamical Systems (Bressanone, 1978), Liguori, Naples, 1980, pp. 137–195.

14. A. P. Veselov and S. P. Novikov, *On Poisson brackets compatible with algebraic geometry and Korteweg-de Vries dynamics of the set of finite-zone potential*, Dokl. Akad. Nauk SSSR **266** (1982), 533–537; English transl. in Soviet Math. Dokl. **26** (1982).

15. V. I. Arnol'd, *Mathematical methods in classical mechanics*, "Nauka", Moscow, 1974; English transl., Springer-Verlag, 1978.

16. I. M. Krichever, *Commutative rings of ordinary linear differential operators*, Funktsional. Anal. i Prilozhen. **12** (1978), no. 3, 20–31; English transl. in Functional Anal. Appl. **12** (1978).

17. B. A Dubrovin and S. M. Natanzon, *Real two-zone solutions of the sine-Gordon equation*, Funktsional. Anal. i Prilozhen **16** (1982), no. 1, 27–43; English transl. in Functional Anal. Appl. **16** (1982).

18. B. A. Dubrovin and S. P. Novikov, *Algebro-geometric Poisson brackets for real finite-zone solutions of the sine-Gordon equation and the nonlinear Schrödinger equation*, Dokl. Akad. Nauk SSSR **267** (1982), 1295–1300; English transl. in Soviet Math. Dokl. **26** (1982).

19. Georege Springer, *Introduction to Riemann surfaces*, Addison-Wesley, 1957.

20. Clifford S. Gardner, *Korteweg-de Vries equation and generalizations*, IV, J. Mathematical Phys. **12** (1971), 1548–1551.

21. V. E. Zakharov and L. D. Faddeev, *The Korteweg-de Vries equation is a fully integrable Hamiltonian system*, Funktsional. Anal. i Prilozhen. **5** (1971), no. 4, 18–27; English transl. in Functional Anal. Appl. **5** (1971).

22. H. Flaschka and D. W. McLaughlan, *Canonically conjugate variables for the Korteweg-de Vries equation and the Toda lattice with periodic boundary conditions*, Progr. Theoret. Phys. **55** (1976), 438–456.

23. Franco Magri, *A simple model of the integrable Hamiltonian equation*, J. Mathematical Phys. **19** (1978), 1156–1162.

24. O. I. Bogoyavlenskiĭ and S. P. Novikov, *The connection between the Hamiltonian formalisms of stationary and nonstationary problems*, Funktsional. Anal. i Prilozhen. **10** (1976), no. 1, 9–13; English transl. in Functional Anal. Appl. **10** (1976).

25. O. I. Bogoyavlenskiĭ, *The integrals of higher stationary KdV equations and the eigenvalues of the Hill operator*, Funktsional. Anal. i Prilozhen. **10** (1976), no. 2, 9–12; English transl. in Functional Anal. Appl. **10** (1976).

26. S. I. Al'ber [Alber], *Investigation of equations of Korteweg-de Vries type by the method of recurrence relations*, J. London Math. Soc. (2) **19** (1979), 467–480. (Russian; English summary).

27. ______, *On stationary problems for equations of Korteweg-de Vries type*, Comm. Pure Appl. Math. **34** (1981), 259—272.

28.* A. P. Veselov, *Finite-zone potentials and integrable systems on the sphere with a quadratic potential*, Funktsional. Anal. i Prilozhen. **14** (1980), no. 1, 48–50; English transl. in Functional Anal. Appl. **14** (1980).

29. Yu. A. Arkhangel'skiĭ, *Analytic dynamics of rigid body*, "Nauka", Moscow, 1977. (Russian)

30. V. V. Kozlov, *Methods of qualitative analysis in the dynamics of a rigid body*, Moskov. Gos. Univ., Moscow, 1980. (Russian)

Translated by J. R. SCHULENBERGER

*Editor's note. There is no item [28] in the Russian original. The context suggests that the restoration given here is what was intended.

Труды Матем. Инст. Стеклов.
165(1984)

Proceedings of the Steklov
Institute of Mathematics
1985, Issue 3

G-DIMENSION AND GENERALIZED PERFECT IDEALS

UDC 512.717

E. S. GOLOD

ABSTRACT. A finitely generated module K over a Noetherian local ring A is called *suitable* if $\operatorname{End}_A(K) = A$ and $\operatorname{Ext}^i_A(A, K) = 0$ for $i \geqslant 1$. Relative to such a module one can define G_K-dimension, G_K-dim M, for finitely generated A-modules M (see H.-B. Foxby, Math. Scand. **31** (1973), 261–284). An ideal I of the ring is called *G_K-perfect* if $G_K\text{-}\dim A/I = \operatorname{grade} I$. The fundamental result: Let K be a suitable A-module, I a G_K-perfect ideal not in A, $\operatorname{grade} I = s$, $K' = \operatorname{Ext}^s_A(A/I, K)$, and M an A/I-module; then $G_K\text{-}\dim_A M = G_{K'}\text{-}\dim_{A/I} M + s$.

Bibliography: 5 titles.

We consider a Noetherian local ring A with maximal ideal m and residue class field A/m. All modules under discussion will be assumed to be finitely generated. We choose and fix some (nonzero) A-module K. For any A-module M let M^* (or, when necessary, M^*_K) denote $\operatorname{Hom}_A(M, K)$. We call a module M *K-reflexive* if the canonical homomorphism $M \to M^{**}$ is bijective. The given module K determines a class of A-modules $G_K(A)$: $P \in G_K(A)$ if the module P is K-reflexive and $\operatorname{Ext}^i_A(P, K) = \operatorname{Ext}^i_A(P^*, K) = 0$ for $i \geqslant 1$. If $A \in G_K(A)$, then the module K will be called *suitable*. Relative to a suitable module K, it is possible, by analogy with the definition of G-dimension [1], [4], when the ring A itself serves in the capacity of K, to define G_K-dimensionality: G_K-dim M for an A-module M is the minimum of the lengths of G_K-resolvents of the module M, i.e. of exact sequences

$$0 \to P_n \to P_{n-1} \to \cdots \to P_0 \to M \to 0,$$

where $P_i \in G_K(A)$, $i = 0, 1, \ldots, n$; if in fact M does not have any finite G_K-resolvents, then we consider that $G_K\text{-}\dim M = \infty$. Such a generalization of G-dimension (even in somewhat more general situations) was first considered by Foxby [2]. In particular, he proved that every A-module has finite G_K-dimension if and only if K is a canonical module, i.e. is a Gorenstein module of rank 1 [2], [5] (and in the more general situation considered by him, a Gorenstein module of arbitrary rank). In the present article the fundamental stress is on the generalization of the idea of a perfect ideal connected with G_K-dimension. Recall that an A-module M is called *perfect* if $\operatorname{pd} M = \operatorname{grade} M$, where $\operatorname{pd} M$ denotes the projective dimension of M, and $\operatorname{grade} M$ is the maximal length of an A-regular sequence contained in $\operatorname{Ann} M$; $\operatorname{grade} M$ is characterized also as $\min\{i\colon \operatorname{Ext}^i_A(M, A) \neq 0\}$. If I is an ideal in A, then

1980 *Mathematics Subject Classification*. Primary 13C15; Secondary 13C13.

in place of grade A/I we usually write grade I. An A-module will be called *G_K-perfect* if G_K-dim M = grade M. Such modules were introduced under the name *quasi-perfect* by Foxby [3]. An ideal I of the ring A is called *G_K-perfect* if A/I is a G_K-perfect A-module. Our basic goal is to establish a connection between the G_K-dimension over the ring A and $G_{K'}$-dimension over the factor ring A/I of the ring A by a G_K-perfect ideal I, relative to an appropriate type of a chosen suitable A/I-module K'. In this, even if over the ring A we are considering the usual G-dimension, i.e. $K = A$, the A/I-module K' is not bound to be isomorphic to A/I.

We enumerate basic facts, necessary for the sequel, which touch upon G_K-dimension and G_K-perfect modules.

1. If the module M is K-reflexive, then $\operatorname{Supp} M = \operatorname{Supp} M^* \subseteq \operatorname{Supp} K$ and $\operatorname{Ass} M = \operatorname{Ass} M^* = \operatorname{Ass} K \cap \operatorname{Supp} M$. If K is a suitable A-module and $M \in G_K(A)$, then $\operatorname{Supp} K = \operatorname{Spec} A$, $\operatorname{Ass} K = \operatorname{Ass} A$, $\operatorname{depth} A = \operatorname{depth} K = \operatorname{depth} M$ and every A-regular sequence is also K-regular and M-regular.

2. Let $0 \to P' \to P \to P'' \to 0$ be an exact sequence of A-modules. If $P', P'' \in G_K(A)$ or $P, P'' \in G_K(A)$, then the third module is also an element of $G_K(A)$. If $P', P \in G_K(A)$ and $\operatorname{Ext}^1_A(P'', K) = 0$ then $P'' \in G_K(A)$.

3. Let K and M be A-modules, and let $f \in A$ be at the same time an A-, K- and M-regular element. Then $M \in G_K(A)$ if and only if $M/fM \in G_{K/fK}(A/fA)$, and K is a suitable A-module if and only if K/fK is a suitable A/fA-module.

In assertions 4–10 we assume further that K is a suitable A-module.

4. Let M be an A-module. If G_K-dim $M < \infty$, then

$$G_K\text{-dim}\, M = \max\{i\colon \operatorname{Ext}^i_A(M, K) \neq 0\}.$$

5. If G_K-dim $M \leqslant n$ and $0 \to P_n \to P_{n-1} \to \cdots \to P_0 \to M \to 0$ is an exact sequence in which $P_i \in G_K(A)$ for $i = 0, \ldots, n-1$, then $P_n \in G_K(A)$.

6. If $0 \to M' \to M \to M'' \to 0$ is an exact sequence of A-modules and two of them have finite G_K-dimension, so does the third.

7. Let $f \in A$ be at the same time an A- and M-regular element. Then $G_K\text{-dim}_A M = G_{K/fK}\text{-dim}_{A/fA} M/fM$.

8. If G_K-dim $M < \infty$, then G_K-dim M + depth M = depth A.

9. If A is a Cohen-Macauley ring, then the A-module M is G_K-perfect if and only if G_K-dim $M < \infty$ and M is a Cohen-Macauley module.

10. Let M be a G_K-perfect A-module and grade $M = s$; then $\operatorname{Ext}^s_A(M, K)$ is a G_K-perfect A-module, grade $\operatorname{Ext}^s_A(M, K) = s$, $\operatorname{Ext}^s_A(\operatorname{Ext}^s_A(M, K), K) \simeq M$ and $\operatorname{Ann} \operatorname{Ext}^s_A(M, K) = \operatorname{Ann} M$. In particular, if I is a G_K-perfect ideal, then $\operatorname{Ann} \operatorname{Ext}^s_A(A/I, K) = I$.

DEFINITION. Let K be some A-module. An ideal I of the ring A will be called *weakly G_K-perfect* if $\operatorname{Ext}^i_A(A/I, K) = 0$ for $i \neq \operatorname{grade}_K I$.

We remark that if K is a suitable module, then a weakly G_K-perfect ideal I is G_K-perfect if and only if G_K-dim $A/I < \infty$.

LEMMA. *Let K be some A-module and let I be a weakly G_K-perfect ideal, $s = \operatorname{grade}_K I$. If*

$$0 \to K \to E^0 \to E^1 \to \cdots \to E^i \xrightarrow{d_i} E^{i+1} \to \cdots$$

is a minimal injective resolvent for K, then

$$0 \to \operatorname{Ext}_A^s(A/I, K) \to \operatorname{Hom}_A(A/I, E^s) \to \operatorname{Hom}_A(A/I, E^{s+1}) \to \cdots$$

is itself a minimal injective resolvent for the A/I-module $\operatorname{Ext}_A^s(A/I, K)$.

PROOF. Let $j = \min\{i\colon \operatorname{Hom}(A/I, E^i) \neq 0\}$. Since E^j is a proper extension of the submodule $\operatorname{Ker} d_j$, then $\operatorname{Hom}_A(A/I, E^j) \cap \operatorname{Ker} d_j \neq 0$ ($\operatorname{Hom}_A(A/I, E^j)$ is identified in a canonical way with a submodule of E^j), and therefore $\operatorname{Ext}_A^j(A/I, K) \neq 0$, whence $j = s$, and, moreover, there is an injective homomorphism $\operatorname{Ext}_A^s(A/I, K) \to \operatorname{Hom}_A(A/I, E^i)$. Exactness at the remaining terms of the sequence follows from the fact that $\operatorname{Ext}_A^i(A/I, K) = 0$ for $i > s$. Injectiveness of the A/I-module $\operatorname{Hom}_A(A/I, E^i)$ and minimality of the resulting resolvent is well known.

COROLLARY. *If I is a weakly G_K-perfect ideal of A and $s = \operatorname{grade}_K I$, then in the category of A/I-modules there is an isomorphism of functors*:

$$\operatorname{Ext}_{A/I}^n(M, \operatorname{Ext}_A^s(A/I, K)) \simeq \operatorname{Ext}_A^{n+s}(M, K).$$

In the ensuing, in the situation described in the lemma, the A/I-module $\operatorname{Ext}_A^s(A/I, K)$ will be denoted by K'.

PROPOSITION 1. *Let K be some A-module, and I a weakly G_K-perfect ideal. Then K is a canonical module for A if and only if that K' is a canonical module for A/I.*

PROOF. On the strength of the preceding corollary, we have an isomorphism

$$\operatorname{Ext}_{A/I}^n(k, K') \simeq \operatorname{Ext}_A^{n+s}(k, K).$$

Therefore for K and K' the conditions, that the modules $\operatorname{Ext}_A^i(k, K)$ (respectively $\operatorname{Ext}_{A/I}^i(k, K')$) reduce to zero for all values of i, with one exception, when these modules are isomorphic to k, are equivalent. And this condition, as is known, characterizes a canonical module.

DEFINITION. Let K be a suitable A-module. A G_K-perfect ideal I in A is called *G_K-Gorenstein* if $\operatorname{Ext}_A^s(A/I, K)$ is a cyclic module, where $s = \operatorname{grade} I$. In this case $\operatorname{Ext}_A^s(A/I, K) \simeq$ A/I.

PROPOSITION 2. *Let K be a suitable A-module and let I be an ideal in A. Then any two of the following conditions imply the third*: 1) *I is a G_K-Gorenstein ideal.* 2) *K is a canonical module for A.* 3) *A/I is a Gorenstein ring.*

PROOF. 1), 2) $\Rightarrow$ 3). Since I is a G_K-Gorenstein ideal, $K' \simeq A/I$. By Proposition 1, A/I is a canonical module for A/I, and this means that A/I is a Gorenstein ring.

1), 3) $\Rightarrow$ 2). Since $K' \simeq A/I$ is itself a canonical module for A/I, by Proposition 1, K is a canonical module for A.

2), 3) $\Rightarrow$ 1). As long as there is a canonical module over A, A is a Cohen-Macauley ring. Since A/I is also a Cohen-Macauley ring and $G_K\text{-}\dim_A A/I < \infty$ (in view of the fact that K is a canonical module), then I is a G_K-perfect ideal and, by Proposition 1, K' is a canonical module for A/I. But A/I is a Gorenstein ring, so consequently $K' \simeq A/I$, and this means that I is a G_K-Gorenstein ideal.

PROPOSITION 3. *Let K be a suitable A-module, I a weakly G_K-perfect ideal,* $\operatorname{grade} I = s$, *and let M be some A/I-module. Then the following conditions are equivalent*: 1) $M \in G_{K'}(A/I)$; 2) *M is a G_K-perfect A-module and* $\operatorname{grade}_A M = s$.

PROOF. 2) $\Rightarrow$ 1). We have $M^* = \operatorname{Hom}_{A/I}(M, \operatorname{Ext}^s_A(A/I, K))$. By the corollary to the lemma, $M^* \simeq \operatorname{Ext}^s_A(M, K)$ and $M^{**} \simeq \operatorname{Ext}^s_A(M^*, K) \simeq M$; moreover, $\operatorname{Ext}^i_{A/I}(M, K') \simeq \operatorname{Ext}^{i+s}_A(M, K) = 0$ for $i \geqslant 1$ and analogously for M^*. It remains to show that the canonical homomorphism $M \xrightarrow{g} M^{**}$ is bijective. Since $M \simeq N^*$, where $N = M^*$, we have homomorphisms $N^* \underset{h}{\overset{g}{\leftrightarrows}} N^{***}$, with $hg = \operatorname{Id}_{N^*}$, where h is the homomorphism conjugate to the canonical homomorphism $N \to N^{**}$. Consequently, h is surjective, and since the modules N^* and N^{***} are isomorphic to one another and Noetherian, h is bijective. From this it follows that g is also bijective.

1) $\Rightarrow$ 2). We have $\operatorname{Ext}^i_A(M, K) \simeq \operatorname{Ext}^{i-s}_{A/I}(M, K') = 0$ for $i \neq s$. Therefore it remains to show that $G_K\text{-}\dim M < \infty$. For this we carry out an induction on s. Let $s = 0$. Then $K' = \operatorname{Hom}_A(A/I, K)$, and therefore K-reflexivity and K'-reflexivity for M mean one and the same thing, and since $M \in G_{K'}(A/I)$, then M is K-reflexive and $\operatorname{Ext}^i_A(M, K) \cong \operatorname{Ext}^i_{A/I}(M, K') = 0$ for $i \geqslant 1$ and analogously for M^*, i.e. $M \in G_K(A)$. Now let $s > 0$ and let f be an A-regular element, contained in I; then f is also K-regular and P-regular for all $P \in G_K(A)$. Let $\bar{A} = A/fA$, $\bar{K} = K/fK$ and $\bar{I} = I/fA$. Then $\operatorname{grad}_{\bar{A}} \bar{I} = s - 1$ and $\operatorname{Ext}^i_{\bar{A}}(\bar{A}/\bar{I}, \bar{K}) \simeq \operatorname{Ext}^{i+1}_A(A/I, K) = 0$ for $i \neq s - 1$, i.e. $\bar{I}$ is a weakly $G_{\bar{K}}$-perfect ideal. By the induction assumption, $G_{\bar{K}}\text{-}\dim_{\bar{A}} M < \infty$. Let

$$P^{\cdot}\colon \cdots \to P_n \xrightarrow{d_n} P_{n-1} \to \cdots \to P_2 \xrightarrow{d_2} P_1 \xrightarrow{d_1} P_0 \to M \to 0$$

be an arbitrary $G_K(A)$-resolvent of the A-module M. Since the element f is P_i-regular for all $i = 0, 1, \ldots$, the modules of the homologies of the complex

$$P^{\cdot}/fP^{\cdot}\colon \cdots \to P_2/fP_2 \xrightarrow{\bar{d}_2} P_1/fP_1 \xrightarrow{\bar{d}_1} P_0/fP_0 \to 0$$

are isomorphic to $\operatorname{Tor}^A_i(A/fA, M)$; and since $fM = 0$, then $H_i(P^{\cdot}/fP^{\cdot}) \simeq M$ for $i = 0, 1$ and $H_i(P^{\cdot}/fP^{\cdot}) = 0$ for $i \geqslant 2$. Writing $\operatorname{Ker} \bar{d}_1 = \bar{N}_1$ and $\operatorname{Im} \bar{d}_2 = \bar{N}_2$, we obtain exact sequences

$$0 \to \bar{N}_1 \to P_1/fP_1 \to P_0/fP_0 \to M \to 0$$

and

$$0 \to \bar{N}_2 \to \bar{N}_1 \to M \to 0.$$

The first of these shows that $G_{\bar{K}}\text{-}\dim \bar{N}_1 < \infty$, and the second that $G_{\bar{K}}\text{-}\dim \bar{N}_2 < \infty$. Let $N_2 = \operatorname{Im} d_2$. Since $H_2(P^{\cdot}/fP^{\cdot}) = 0$, the canonical surjective homomorphism $N_2/fN_2 \to \bar{N}_2$ is bijective. Using the N_2-regularity of the element f, we find that $G_K\text{-}\dim N_2 < \infty$. This entails that $G_K\text{-}\dim M < \infty$.

COROLLARY 1. *Let K be a suitable A-module, and I a weakly G_K-perfect ideal. Then K' is a suitable A/I-module if and only if I is a G_K-perfect ideal.*

DEFINITION. A weakly G_K-perfect ideal I, $\operatorname{grade}_K I = s$, is called *weakly G_K-Gorenstein* if $\operatorname{Ext}^s_A(A/I, K)$ is a cyclic module.

COROLLARY 2. *Let K be a suitable A-module. A weakly G_K-Gorenstein ideal I,* $\operatorname{grade} I = s$, *is G_K-Gorenstein if and only if* $\operatorname{Ext}_A^s(A/I, K) \simeq A/I$.

PROPOSITION 4. *Let K be a suitable A-module, I a G_K-Gorenstein ideal,* $\operatorname{grade} I = s$, *$J \supseteq I$, J a G_K-perfect ideal, and* $\operatorname{grade} J = s$. *Let $L = I\colon J$. Then L is a G_K-perfect ideal, $I\colon L = J$ and* $\operatorname{grade} L = s$. *Moreover, J is a G_K-Gorenstein ideal if and only if L/I is a principal ideal in A/I.*

PROOF. Since the ideal I is G_K-Gorenstein, then $K' \simeq A/I$ and, by Proposition 3, $A/J \in G_{A/I}(A/I)$. We have $\operatorname{Hom}_{A/I}(A/J, A/I) = L/I$, and the canonical mapping $A/J \to \operatorname{Hom}_{A/I}(L/I, A/I)$ is bijective. Injectivity of this homomorphism means that $I\colon L = J$, and its surjectivity allows us to draw from the exact sequence

$$0 \to \operatorname{Hom}_{A/I}(A/L, A/I) \to \operatorname{Hom}_{A/I}(A/I, A/I)$$
$$\to \operatorname{Hom}_{A/I}(L/I, A/I) \to \operatorname{Ext}^1_{A/I}(A/L, A/I) \to 0$$

(respectively the exact sequence $0 \to L/I \to A/I \to A/I \to A/L \to 0$) the conclusion that $\operatorname{Ext}^1_{A/I}(A/L, A/I) = 0$. Since L/I, $A/I \in G_{A/I}(A/I)$, then $A/L \in G_{A/I}(A/I)$. By Proposition 3, L is a G_K-perfect ideal and $\operatorname{grade} L = s$. Finally, then, the fact that J is a G_K-Gorenstein ideal is equivalent to the fact that $\operatorname{Ext}_A^s(A/J, K) \simeq \operatorname{Hom}_A(A/J, A/I) = L/I$ is a cyclic module.

PROPOSITION 5. *Let K be a suitable A-module, I a G_K-perfect ideal,* $\operatorname{grade} I = s$, *and M an A/I-module. Then*

$$G_K\text{-}\dim_A M = G_{K'}\text{-}\dim_{A/I} M + s.$$

PROOF. Since $\operatorname{Ext}_A^{n+s}(M, K) \simeq \operatorname{Ext}_A^n(M, K')$, it is sufficient to prove that

$$G_K\text{-}\dim_A M < \infty \Leftrightarrow G_{K'}\text{-}\dim_{A/I} M < \infty.$$

We carry out an induction on that one of the dimensions $G_K\text{-}\dim_A M$ or $G_{K'}\text{-}\dim_{A/I} M$ which is finite. The base of the induction $G_{K'}\text{-}\dim_{A/I} M = 0$ or $G_K\text{-}\dim_A M = s$ (we always have $G_K\text{-}\dim_A M \geqslant \operatorname{grade}_A M \geqslant s$) forms the content of Proposition 3. Let $\dim M$ denote that one of the dimensions of M which is assumed finite; we may take $\dim M = t > \dim A/I$. From the exact sequence $0 \to N \to F \to M \to 0$, where F is a free A/I-module, taking into account that $\dim F < t$, we obtain an exact sequence for the corresponding functor Ext:

$$\cdots \to \operatorname{Ext}^{i-1}(N, \quad) \to \operatorname{Ext}^i(M, \quad) \to 0$$
$$\to \cdots \to 0 \to \operatorname{Ext}^i(N, \quad) \to \operatorname{Ext}^{i+1}(M, \quad) \to 0,$$

from which we find that $\operatorname{Ext}^i(N, \quad) = 0$ for $i \geqslant t$, i.e. $\dim N \leqslant t - 1$. By the inductive assumption, both dimensions for the module N are finite, and consequently they are finite also for M.

BIBLIOGRAPHY

1. Maurice Auslander and Mark Bridger, *Stable module theory*, Mem. Amer. Math. Soc. No. 94 (1969).
2. Hans-Bjørn Foxby, *Gorenstein modules and related modules*, Math. Scand. **31** (1972), 267–284 (1973).
3. ______, *Quasi-perfect modules over Cohen-Macauley rings*, Math. Nachr. **66** (1975), 103–110.
4. *Anneaux de Gorenstein, et torsion en algèbre commutative*, Sém. Algèbre Commutative Dirigé par P. Samuel 1966/67, Secrétariat Math., Paris, 1967.
5. Rodney Y. Sharp, *Gorenstein modules*, Math. Z. **115** (1970), 117–139.

Translated by A. J. BIAGIOLI

Труды Матем. Инст. Стеклов.
165(1984)

Proceedings of the Steklov
Institute of Mathematics
1985, Issue 3

GENERATORS AND RELATIONS IN GROUPS OF BIRATIONAL AUTOMORPHISMS OF TWO CLASSES OF RATIONAL SURFACES

UDC 512.7

V. A. ISKOVSKIKH

ABSTRACT. Let F be a minimal del Pezzo surface of degree 1 or 2 over a perfect field k such that F can be realized as a conic bundle π: $F \to C$ over a curve of genus zero. We denote by $\{F_i | i \in I(F)\}$ the set of all del Pezzo surfaces F_i which are birationally equivalent to F over k and over C. To each F_i one can associate a biregular involution r_i. The author studies the groups of birational k-automorphisms $\operatorname{Bir} F$ for such del Pezzo surfaces F of degree 1 or 2. We prove that if $\deg F = 1$, then $\operatorname{Bir} F$ is almost a free product of $\mathcal{G}$, the group of birational automorphisms carrying the pencil π: $F \to C$ into itself, and all involutions r_i, $i \in I(F)$. In the case $d = 2$, he explicitly describes the complete system of relations in the representation of $\operatorname{Bir} F$ as a quotient of such a free product.

Bibliography: 10 titles.

§1. Introduction and survey of results

1. Let k be an arbitrary field and let F be a smooth projective surface over k. Recall that F is called a rational surface if the surface $\bar{F} = F \otimes \bar{k}$ over the algebraic closure $\bar{k}$ of k is birationally equivalent to the plane $\mathbf{P}^2_{\bar{k}}$. We denote by $\operatorname{Bir} F$ the group of all birational transformations over k of F into itself, i.e. all birational k-automorphisms of F. The study of the group of all birational automorphisms of rational surfaces attracts attention in connection with arithmetic questions of such surfaces (see, for example, [1]), as well as in connection with rationality problems and the description of the groups of birational automorphisms of three-dimensional algebraic varieties over algebraically closed fields (see [2]).

A birational transformation h: $F' \to F$ yields the isomorphism h^*: $\operatorname{Bir} F \to \operatorname{Bir} F'$. Hence we can restrict our attention to $\operatorname{Bir} F$, where F is a surface minimal over $\bar{k}$. The minimal smooth rational surfaces over an arbitrary field are described in [3]. We have the following list:

(i) $\mathbf{P}^2_k$;

(ii) a quadric $Q \subset \mathbf{P}^3_k$ with $\operatorname{Pic} Q \simeq \mathbf{Z}$;

(iii) a surface F such that $\operatorname{Pic} F \simeq \mathbf{Z}$ is generated by the ample anticanonical sheaf $Q_F(-K_F)$ (such a surface is called a del Pezzo surface of the first kind);

(iv) a surface F with $\operatorname{Pic} F \simeq \mathbf{Z} + \mathbf{Z}$ such that there exists a morphism π: $F \to C$ onto a smooth curve C of genus zero whose generic fiber F_η, $\eta \in C$, is also a smooth rational curve (such a surface is called a standard conic bundle; if, in addition, $O_F(-K)$ is ample, then it is called a del Pezzo surface of the second kind).

1980 *Mathematics Subject Classification*. Primary 14E05; Secondary 14J25, 14M20.

2. We give a short survey of known results on the group of birational automorphisms of rational surfaces over a perfect nonclosed field k. For the surface of type (iv), it is proved by the author in [4] that if $(K_F \cdot K_F) \leqslant 0$, then $\operatorname{Bir} F$ can be realized as an extension

$$1 \to \operatorname{Bir} F_\eta \to \operatorname{Bir} F \to G \to 1, \tag{1}$$

where $\operatorname{Bir} F_\eta = \operatorname{Aut} F_\eta$ is the automorphism group of the generic fiber F_η (i.e., canonically over $k(\eta)$) of the structure morphism π: $F \to C$, where $\eta \in C$ is a generic point and G is a (finite) group which can be realized as a subgroup of $\operatorname{Aut} C$. We observe that, for the rest of the surfaces of (i)–(iv), we have $0 \leqslant (K_F \cdot K_F) \leqslant 9$.

The integer $d = (K_F \cdot K_F)$ is called the degree of the surface F. For the rest of the surfaces, it is enough to compute the group $\operatorname{Bir} F$ for the del Pezzo surfaces of the first and second kind only, in view of the fact that for F of type (iv) with $d > 0$, the group $\operatorname{Bir} F$ can be realized in the form (1) if and only if no birational C-map (preserving the degree d) onto a del Pezzo surface of the first kind exists (it can happen only if $d = 1$, 2, and 4) (see [3], [5] and [6]). Moreover, a surface F of (i)–(iv) with $d = 5$ or 7 is always birationally trivial over k (see [7], [9] and [3]); hence $\operatorname{Bir} F \simeq \operatorname{Bir} \mathbf{P}_k^2$. The same is true for F with $d = 6$, 8, or 9, provided $F(k) \neq \varnothing$, where $F(k)$ is the set of rational k-points of F. Thus it remains to consider only a few cases. We now consider these cases in detail.

3. For del Pezzo surfaces of the first kind with $d = 1$, 2, and 3, Manin has investigated the group $\operatorname{Bir} F$ (see [1] and [8]). He proved that if $d = 1$, then $\operatorname{Bir} F$ coincides with the group $\operatorname{Aut} F$ of (biregular) automorphisms, which is, in this case, a subgroup of the Weyl group E_8. If $d = 2$, the group $\operatorname{Bir} F$ can be realized as a semidirect product of the subgroup $\operatorname{Aut} F$ and a normal subgroup $B(F)$, the last being a free product of birational involutions associated with the rational k-points of F (see [8]). For the minimal cubic surfaces F in $\mathbf{P}^3$ (i.e., the del Pezzo surfaces of the first kind and degree 3), the generators and relations in $\operatorname{Bir} F$ were described in [1]. In this case, the generators are the birational involutions t_x and s_x, corresponding to the points $x \in F$ of degree one and the points $y \in F$ of degree two, and the elements of $\operatorname{Aut} F$.

For the del Pezzo surfaces of the first kind with $d = 4$, the group $\operatorname{Bir} F$ was investigated in [5]. If $F(k) = \varnothing$, the group $\operatorname{Bir} F$ looks as in the case $d = 2$ except for the birational involutions which now correspond to the points of degree 2. If $F(k) \neq \varnothing$, then F is birationally equivalent over k to a del Pezzo surface of the second kind with $d = 3$ (cubic surfaces in $\mathbf{P}^3$ with a line defined over k). For such surfaces, the generators of $\operatorname{Bir} F$ were described in [5]. We also observe that the generators of $\operatorname{Bir} \mathbf{P}_k^2$ ($k = \mathbf{Q}$) were described in a classical paper by Kantor [10].

Thus, for the del Pezzo surface F of the first kind and degree d, the group $\operatorname{Bir} F$ remains to be investigated only when $d = 6$, 8, and 9, provided F has no k-points; but if $F(k) = \varnothing$, then $\operatorname{Bir} F \simeq \operatorname{Bir} \mathbf{P}_k^2$ in these cases, and the question about relations in the later group remains open.

4. For the del Pezzo surfaces F of the second kind and degree d, the generators of $\operatorname{Bir} F$ are described in [5] and [6], provided $d = 1$, 2, 3, or 4. In this paper we study the relations between the generators for $d = 1$ and 2. Among the surfaces of this type, only the groups $\operatorname{Bir}(C \times C^1)$ remain to be investigated, where C and C' are curves of genus zero without k-points, and the relations are not described for $d = 3$ and 4.

5. In order to state our results we must introduce some notation. Let F be a del Pezzo surface of the second kind and degree d. Let $\pi\colon F \to C$ be its realization as a standard conic bundle. Let $\{F_i | i \in I(F)\}$ be the set of all standard conic bundles $\pi_i\colon F_i \to C$, birationally equivalent over C to F, that are del Pezzo surfaces (of the second kind). To each surface F_i, $i \in I(F)$, the following groups are associated: the group $\mathcal{G}_i$ of all birational automorphisms of F_i carrying the standard conic bundle $\pi_i\colon F_i \to C$ into itself; it can be realized as an extension

$$1 \to \operatorname{Aut} F_{i\eta} \to \mathcal{G}_i \to B_i \to 1, \tag{2}$$

where $\operatorname{Aut} F_{i\eta}$ is the group of all automorphisms of $F_{i\eta}$, the generic fiber of $\pi_i\colon F_i \to C$ over the residue field $k(\eta) \simeq k(C)$ of the generic point $\eta \in C$, and B_i is a finite group (for $d \leqslant 5$); the group $G_i = \operatorname{Aut} F_i$ of k-automorphisms of F_i and its subgroup $A_i \subset G_i$ carrying the bundle $\pi_i\colon F_i \to C$ into itself; in the cases $d = 1$ and 2, an involution τ_i is distinguished in G_i (the Bertini involution, when $d = 1$, and the Heiser involution when $d = 2$) such that G_i is a semidirect product of groups

$$1 \to A_i \to G_i \to \{1, \tau_i\} \to 1 \tag{3}$$

(for $d \leqslant 5$ the group G_i is always finite [1]); it is clear that $A_i \subset \mathcal{G}_i$.

The involution τ_i acts on A_i so that, for every element $a_i \in A_i$, one can find an element $a_i' \in A_i$ such that

$$\tau_i a_i = a_i' \tau_i, \quad \forall a_i \in A_i,\ i \in I(F). \tag{4}$$

For each $i \in I(F)$, we fix a birational C-isomorphism $h_i\colon F \to F_i$. Let $\mathcal{G}$ denote the group of all birational automorphisms of F carrying the sheaf $\pi\colon F \to C$ into itself. Then h_i yields an isomorphism $h_i^*\colon \mathcal{G}_i \simeq \mathcal{G}$, it determines an embedding $h_i^*\colon A_i \hookrightarrow \mathcal{G}$, $h_i^*(a_i) = h_i^{-1} a_i h_i$, and it lifts the involution τ_i to a birational involution $h_i^*(\tau_i) = h_i^{-1}\tau_i h_i$. Now the relations (4) can be rewritten in the form

$$h_i^*(\tau_i) h_i^*(a_i) = h_i^*(a_i') h_i^*(\tau_i). \tag{5}$$

6. The following result is proved in [5] (see also §2 of the present article).

THEOREM 1. *Let F be a del Pezzo surface of the second kind with $d = (K_F \cdot K_F) = 1$ or 2, and let $\pi\colon F \to C$ be a realization of F as a standard conic bundle. Then with the above notation the group* $\operatorname{Bir} F$ *is generated by the subgroup $\mathcal{G}$ and involutions $h_i^*(\tau_i)$, $\forall i \in I(F)$.*

Let $B(F) = \mathcal{G} *_{i \in I} \tau_i$ denote a free product of $\mathcal{G}$ and all groups of the second order $\{1, h_i^*(\tau_i)\}$, $i \in I(F)$. By Theorem 1, the natural homomorphism $p\colon B(F) \to \operatorname{Bir} F$ is surjective. Hence we get the exact sequence

$$1 \to R(F) \to B(F) \to \operatorname{Bir} F \to 1, \tag{6}$$

where $R(F)$ is the kernel of p. The first main result of the present paper is the following:

THEOREM 2. *Under the assumptions of Theorem* 1 *with $d = 1$, the relations* (5) *or equivalently* (4) *generate a complete system of relations between $\mathcal{G}$ and the involutions τ_i, $i \in I(F)$, i.e. the corresponding words in $B(F)$ generate the normal subgroup $R(F)$. In particular, if $A_i = \{1\}$, $\forall i \in I(F)$, then* $\operatorname{Bir} F = B(F)$ *is a free product of $\mathcal{G}$ and the involutions τ_i, $i \in I(F)$.*

7. To state the corresponding result in the case $d = 2$ we introduce some definitions and notation. For every F_i, $i \in I(F)$, we denote by $F_i^0(k)$ the set of closed points $x \in F_i$ of degree 1 outside the exceptional curves of the surface F_i. A point $x \in F_i^0(k)$ is called a good point of degree 1 (one can similarly give the definition of a good point of degree > 1). We observe that almost all points of $F_i(k)$ are good except, perhaps, for a finite set of points of the intersections of the exceptional curves on $\overline{F}_i$. To each $i \in I(F)$ and a good point $x \in F_i^0(k)$ we associate a birational map e_{ij}: $F_i \to F_j$ over C of F_i onto F_j, $i, j \in I(F)$, which is called an elementary transformation with center x. It is defined as the composition $e_{ij}(x) = \sigma_{x'}\sigma_x^{-1}$ of the blowing up σ_x: $F_i' \to F_i$ of x followed by the blowing down $\sigma_{x'}$: $F_i' \to F_j$ of the proper transform of the fiber $F_{i\pi(x)}$ passing through x.

Let F be a del Pezzo surface of the second kind with $d = 2$. Let F_0 be a surface of $\{F_i | i \in I(F)\}$, and let $x_0 \in F_0$ be a good point of degree 1. We consider the composition of maps

$$F_0 \overset{e_{01}(x_0)}{\to} F_1 \overset{\tau_1}{\to} F_1 \overset{e_{12}(x_1)}{\to} F_2 \overset{\tau_2}{\to} F_2 \overset{e_{23}(x_2)}{\to} F_3 \overset{\tau_3}{\to} F_3 \overset{e_{34}(x_3)}{\to} F_4 \overset{\tau_4}{\to} F_4, \tag{7}$$

where $e_{ss+1}(x_s)$, $s = 0, 1, 2, 3$, is an elementary transformation with center $x_s \in F_s$, $x_s = \tau_s(x_s')$, and x_{s+1}' is the image, under $e_{ss+1}(x_s)$, of the contracted fiber passing through x_s.

PROPOSITION 1. *Let $F_0 \in \{F_i | i \in I(F)\}$ be a surface. Assume $d = 2$ and consider a good point $x_0 \in F_0$ of degree one. Then all the maps given in* (7) *are well defined, $F_4 \simeq F_0$, and with the identifications $F_4 = F_0$ and $\tau_4 = \tau_0$ the composition of maps* (7) *determines an element $a_0 \in A_0$,*

$$a_0 = \tau_0 \cdot e_{34}(x_3) \cdot \tau_3 \cdot e_{23}(x_2) \cdot \tau_2 \cdot e_{12}(x) \cdot \tau_1 \cdot e_{01}(x_0). \tag{8}$$

With the aid of the maps h_i, $i \in I(F)$, one can rewrite (8) as the following relation in the group $\operatorname{Bir} F$:

$$h_0^{-1} \cdot a_0 \cdot h_0 = h_0^{-1} \cdot \tau_0 \cdot h_0 \cdot h_0^{-1} \cdot e_{34}(x_3) \cdot h_3 \cdot h_3^{-1} \cdot \tau_3 \\ \cdots h_1^{-1} \cdot \tau_1 \cdot h_1^{-1} \cdot e_{01}(x_0) \cdot h_0. \tag{9}$$

We can now state our second main result.

THEOREM 3. *Under the assumptions of Theorem* 1 *with $d = 2$, the relations of type* (5) *and* (9) (*equivalently* (4) *and* (8)) *for all $F_0 \in \{F_i | i \in I(F)\}$ and all good closed points $x_0 \in F_0^0(k)$ of degree* 1 *generate a complete system of relations between $\mathcal{G}$ and the involutions τ_i, $i \in I(F)$, i.e. the corresponding words in $B(F)$ generate the normal subgroup $R(F)$. In particular, if $F^0(k) = \varnothing$, then $F_i^0(k) = \varnothing, \forall i \in I(F)$, and the group* $\operatorname{Bir} F$ *looks as in Theorem* 2.

We prove Theorem 1 in §2 and Theorems 2 and 3 in §3. The field k is assumed to be perfect.

§2. The group Bir F for del Pezzo surfaces of the first kind with $d = 1$ and 2. Relations

1. Let π_i: $F_i \to C$, $i \in I(F)$ (see §1.5). We denote by f_i the class of the geometric fiber of the structure morphism π_i: $F_i \to C$, i.e. the positive generator of the group $\pi_i^* \operatorname{Pic} \overline{C} \simeq \mathbf{Z}$ (we recall that $\overline{C} = C \otimes \overline{k}$ and $\operatorname{Pic} \overline{C} = (\operatorname{Pic} \overline{C})^G \simeq \mathbf{Z}$, where $G = \operatorname{Gal}(\overline{k}/k)$; see [4]). We observe that we always have $C \simeq \mathbf{P}_k^1$ for d odd, and for d

even we have two possibilities:

(a) $C(k) \neq \varnothing \Rightarrow C \simeq \mathbf{P}^1_k$.

(b) $C(k) = \varnothing \Rightarrow C$ is a smooth plane conic without k-point.

We observe the following properties of del Pezzo surfaces of the second kind and degree $d = 1$ and 2. Let $\pi\colon F \to C$ be such a surface. Then the following assertions are true:

(i) $\operatorname{Pic} F \simeq \pi^* \operatorname{Pic} C + \mathbf{Z}K_F$, where K_F is the canonical class.

(ii) In case $d = 1$, there exists a biregular involution τ on F (the Bertini involution) acting on $\operatorname{Pic} F$ by the formula (see, for example, [5])

$$\tau(f) = -f - 4K_F, \qquad \tau(K_F) = K_F; \tag{10}$$

we now always have $C \simeq \mathbf{P}^1_k$.

(iii) In case $d = 2$, there exists a biregular involution τ (the Heiser involution) acting on $\operatorname{Pic} F$ by the formula (see [5])

$$\tau(f) = -f - 2K_F, \qquad \tau(K_F) = K_F; \tag{11}$$

we now have both cases: $C \simeq \mathbf{P}^1_k$ and $C(k) = \varnothing$.

(iv) There exist exactly two pencils of rational curves on every del Pezzo surface F of the second kind with $d = 1$ and 2 which realize F as a standard conic bundle: $|f|$ and $|\tau(f)|$.

REMARK 1. The author does not know an example of a standard conic bundle $F \to C$ with $d = 1, 2$, or 4 that has no birational C-maps on a del Pezzo surface (of the second kind) and, as we mentioned in §1.2, the groups $\operatorname{Bir} F$ can be realized in the form (1).

2. We will freely use the technique and even some notation from Manin [1], Chapter V. We mention a few deviations. We denote by $Z^*(F)$ the group $(\operatorname{Pic} \bar{F})^G + Z^0(F)$, where $(\operatorname{Pic} F)^G \subset \operatorname{Pic} \bar{F}$ is the subgroup of $G = \operatorname{Gal}(\bar{k}/k)$-invariant elements and $Z^0(F)$ is a free Abelian group generated by the closed points of the foam space $E(F)$, and not by the geometric k-points as in [1], Chapter V, §3. We consider $(\operatorname{Pic} \bar{F})^G$, not just $\operatorname{Pic} F$, in the definition of $Z^*(F)$ because in case $C(k) = \varnothing$ the class of the geometric fiber f belongs to $(\operatorname{Pic} \bar{F})^G$, and not to $\operatorname{Pic} F$ (see [4]).

Let $u\colon F' \to F$ be a birational map of surfaces which are standard conic bundles. Let f' and f be the classes of the geometric fibers of the corresponding pencils. Then the image of f' in $Z^*(F)$ under the map u has the following form (see [1], Chapter V, §5):

$$z = u(f') = af - bK_F - \sum n_i x_i, \tag{12}$$

where $a, b, n_i \in \mathbf{Z}$, $n_i \geqslant 0\ \forall i$, $b \geqslant 0$, and $x_i \in E(F)$.

With the notation of [1], Chapter V, we have $z \in Z^{+*}(F)$, whence $0 \leqslant (z \cdot f) = 2b \Rightarrow b \geqslant 0$, and if $d = (K_F \cdot K_F)$, then

$$\begin{aligned} (z \cdot z) &= 4ab + b^2 d - \textstyle\sum \deg x_i n_i^2 = 0, \\ -\Omega(z) &= 2a + bd - \textstyle\sum \deg x_i n_i = 2. \end{aligned} \tag{13}$$

LEMMA 1. *With the above notation, the following assertions are true*:

1) *If* $x_j < x_i$ *in the sense of* [1], *Chapter* V, §3 (*i.e., if* x_j *is lying over* x_i), *then* $n_i \geqslant n_j \geqslant 0$.

2) $n_i \leqslant 2b$, $\forall i$.

3) *If* $n_i \leqslant b$, $\forall i$, *and* $b > 0$, *then* $d < 0$.

4) *If* $a < 0$ *and* $d = (K_F \cdot K_F) \geqslant 1$, *then the sheaf* $O_F(-K)$ *is ample, i.e.,* F *is a del Pezzo surface of the second kind.*

5) *If* $x_1, \ldots, x_r$ *are the points with* $n_i \geqslant b > 0$ *and* $a \leqslant 0$, *then* $\Sigma_1 \deg x_i < d$.

PROOF. 1) follows from the fact that $z \in Z^{+*}(F)$ (see [1], Chapter V, 3.11). Now let $F_{\pi(x_i)} = \pi^{-1}(\pi(x_i))$ be the fiber of the pencil $\pi\colon F \to C$ passing through $x_i \in F$. Then $\deg x_i f - x_i \in Z^*_+(F)$ (in the notation of [1], Chapter V, §2). Hence

$$0 \leqslant (\deg x_i f - x_i \cdot z) = 2\deg x_i b - \deg x_i n_i$$

and therefore $n_i \leqslant 2b$, proving 2). 3) follows at once from (13) and the fact that $z \in Z^{+*}(F)$. Indeed, since $n_i \leqslant b$, the first equality of (13) yields

$$4ab + b^2 d \leqslant b \sum \deg x_i n_i.$$

Therefore, by the second equality of (13), $4ab + b^2 d \leqslant b(2a + bd - 2)$ or $a \leqslant -1$, because $b > 0$. We verify 4) using the numeric criterion of ampleness. Let $X \sim pf - qK_F$, $p, q \in \mathbf{Z}$, be an arbitrary irreducible curve on F; then $(X \cdot f) \geqslant 0 \Rightarrow q \geqslant 0$. We assume that $(X \cdot -K_F) \leqslant 0$. Then $(X \cdot -K_F) = 2p + qd \leqslant 0$ and $(X \cdot af - bK_F) = 2aq + (2p + qd)b < 0$, because $a < 0$. However that is impossible, because $z \in Z^{+*}(F)$ ($|af - bK_F|$) has no fixed components. 5) follows at once from the second equality of (13).

The lemma is proved.

LEMMA 2. *With the above notation, let* $u\colon F' \to F$ *be a birational map of surfaces which are standard conic bundles.*

1) *If* F' *and* F *are del Pezzo surfaces of degree* d' *and* $d = 1$ *or* 2 *and* $u(-K_{F'}) = -K_F$ *in the group* $Z^*(F) = (\operatorname{Pic} \tilde{F})^G + Z^0(F)$, *then* u *is an isomorphism.*

2) *If* $u = e(x')$ *and* $F' \to F$ *is an elementary transformation with center in the closed point* $x' \in F'$, *then* $d = d'$ *and the corresponding* $e(x')\colon Z^*(F') \to Z(F)$ *is of the form*

$$\begin{aligned} &e(x')(f') = f,\\ &e(x')(-K_{F'}) = \deg x' f - K_F - 2x,\\ &e(x')(x') = \deg x' f - x,\\ &e(x')(y') = y,\ y' \neq x', \end{aligned} \tag{14}$$

where x *is the image of the blown down fiber under the elementary transformation* $e(x')$.

3) *If* $u(f') = f$, *then there exists an automorphism* $a \in A$ ($A \subset \operatorname{Aut} F$ *is the subgroup of elements carrying the pencil* $|f|$ *into itself*) *such that* au *can be realized as a composition of elementary transformations with centers in closed points.*

PROOF. 1) follows at once from [1], Chapter V, 3.12. 2) easily follows from the behavior of $Z^*(F)$ under monoidal transformations (see [1], Chapter V, §§2 and 3) and the definition of $e(x')$ as a composition of the blowing up of x' and the blowing down of the proper transform of the fiber passing through x'. We prove 3). Let $u(-K_{F'}) = pf - qK_F - \sum m_i x_i$ in $Z^*(F)$. Since $u(f') = f$, we get

$$2 = (f' \cdot -K_{F'}) = (u(f') \cdot u(-K_{F'})) = \left(f \cdot pf - qK_F - \sum m_i x_i\right) = 2q.$$

Hence $q = 1$. Therefore $0 \leqslant m \leqslant 2q = 2$. We have

$$(K_{F'} \cdot K_{F'}) = 4pq + q^2 d - \sum \deg x_i m_i^2 = d, \tag{15}$$
$$-\Omega(-K_{F'}) = 2p + qd - \sum \deg x_i m_i = d.$$

Hence, as before, if $m_i \leqslant q$, $\forall i$, then

$$2qp + (q-1)d \leqslant 0. \tag{16}$$

Therefore, for $q = 1$, conditions (16) and (15) imply the equalities $p = 0$ and $m_i = 0$, $\forall i$; that is, $u(-K_{F'}) = -K_F$. If there exists a point x_i with $m_i > q$, we can assume that $x_i \in F$. In that case, x_i cannot belong to degenerate fibers of the morphism π: $F \to C$ and the degree of x_i on $F_{\pi(x_i)}$, the fiber passing through x_i, over the residue field $k(x_i)$ is equal to 1. Therefore one can define the elementary transformation $e(x_i)$ with center x_i. Repeating this procedure and applying (14), we get a transformation e: $F \to F''$ over C, which is a composition of elementary transformations with centers in closed points, such that $u' = e \cdot u$: $F' \to F''$ has the properties

$$u'(f') = f'', \qquad u'(-K_{F'}) = -K_{F''}.$$

Therefore u' yields an isomorphism $\operatorname{Pic} F' \to \operatorname{Pic} F''$. By Lemma 3.11 in Chapter V of [1], u is an isomorphism. The condition 3) follows at once.

The lemma is proved.

PROOF OF THEOREM 1. This theorem was, in fact, proved in [5]. We reproduce the proof here in order to illustrate the method and for convenient references in the proof of Theorems 2 and 3. Let F be a del Pezzo surface of the second kind with $d = 1$ or 2. Let π: $F \to C$ be its realization as a standard conic bundle and let u: $F \to F$ be a birational transformation. Then $z = u(f) = af - bK_F - \sum n_i x_i$ in the group $Z^*(F) = (\operatorname{Pic} \overline{F})^G + Z^0(F)$, and it satisfies (13). If $b = 0$, then by Lemma 1, 2), we get $n_i = 0$, $\forall i$, $a = 1$, and $u(f) = f$. Therefore $u \in \mathcal{G}$. But if $b > 0$ and $n_i \leqslant b$, $\forall i$, then, by Lemma 1, 3), we get $a < 0$. Since F is a del Pezzo surface of degree 1 or 2, it has an involution τ acting on $Z^*(F)$ by (10) or (11). Applying it to z, we get

$$d = 1;\ z' = \tau(z) = \tau \cdot u(f) = -af - (b + 4a)K_F - \sum n_i \tau(x_i),$$
$$d = 2:\ z' = \tau(z) = \tau \cdot u(f) = -af - (b + 2a)K_F - \sum n_i \tau(x_i).$$

We have $b + 4a < b$ and $b + 2a < b$, because $a < 0$; so we have a descent of the nonnegative integral coefficient at $-K_F$.

If there exists a point x_i with $n_i > b$, then, by Lemma 1, 1), there exists a point, say x_1, with $n_1 > b$ and $x_1 \in F$. It follows from $(f \cdot z) = 2b$ that x_1 cannot belong to a degenerate fiber of the morphism π, and its degree on the fiber $F_{\pi(x_1)}$ over the residue field $k(x_1)$ is equal to 1. Therefore one can define an elementary transformation $e(x_1)$: $F \to F'$ with center x_1. Moreover, π': $F' \to C$ is also a standard conic bundle with $d' = d$ and the same b, and $z' = e(x_1)(z)$. With the aid of a finite number of elementary transformations of type $e(x_1)$ we can get rid of all points with multiplicity $> b$, by passing to a surface $F_1 \in \{F_i | i \in I(F)\}$. Let g_{01}: $F = F_0 \to F_1$ be the composition of all those elementary transformations; then $z_1 = g_{01}(z) = a_1 f_1 - bK_{F_1} - \sum n_{1i} x_{1i}$, where $n_{1i} \leqslant b$, $\forall i$, $a_1 < 0$, and F_1 is a del Pezzo surface of degree d, by Lemma 1, 3) and 4). Hence we can apply an involution τ_1 and make the coefficient at $-K_{F_1}$ smaller. Repeating this procedure, we get $-K_{F_r}$ with coefficient zero after an rth step for some r. Hence $z_r = f_r$ in $Z^*(F_r)$. Then $F \to F_r$, the composition of the maps starting with u, carry f into f_r and, by Lemma 2, 3), it either is an isomorphism or becomes an isomorphism after a birational C-map $F_r \to F$. Thus, we obtain the decomposition of the form

$$u^{-1} = a g_{r0}\tau_r g_{r-1r}\tau_{r-1} \cdots g_{12}\tau_1 g_{01}, \tag{17}$$

where τ_s is an involution on F_s, $s = 1, \dots, r$, $g_{s-1s}\colon F_{s-1} \to F_s$ and $g_{r0}\colon F_r \to F_0 = F$ are birational maps over C, $F_s \in \{F_i | i \in I(F)\}$, and $a \in A$. We observe that we may get $g_{01} = 1$ (i.e., it is an identity map), $g_{r0} = 1$, and $a = 1$. With the aid of the birational transformations h_i, $i \in I(F)$, and the inverse transformations to these birational transformations, one can rewrite (17) in the form

$$u^{-1} = a g_{r0} h_r h_r^{-1} \cdots \tau_2 h_2 h_2^{-1} g_{12} h_1 h_1^{-1} \tau_1 h_1 h_1^{-1} g_{01}. \tag{18}$$

It is clear that the right-hand side of the equality is a composition of maps of $\mathcal{G}$ and the involutions $h_i^*(\tau_i) = h_i^{-1}\tau_i h_i$.

The theorem is proved.

§3. The group $\operatorname{Bir} F$ for del Pezzo surfaces of the second kind with $d = 1$ and 2. Relations

1. Proof of Theorem 2. An element of $B(F)$ can be written as a word without cancellations, consisting of elements of $\mathcal{G}$ and involutions $h_i^*(\tau_i)$, $i \in I(F)$. Set $F = F_0$. With the aid of the maps h_i a word can be written in the form

$$R = g_{r0}\tau_r g_{r-1r}\tau_{r-1} \cdots \tau_2 g_{12}\tau_1 g_{01}, \tag{19}$$

where $g_{s-1s}\colon F_{s-1} \to F_s$, $s = 1, \dots, r$, and $g_{r0}\colon F_r \to F_0$ is a birational map carrying f_{s-1} into f_s and f_r into f_0. They are not always maps over C, but they become such maps after multiplying g_{s-1s} from the left by an element $a_s \in A_s$ or from the right by an element $a_{s-1} \in A_{s-1}$. We observe that some of the g_{s-1s}'s can be isomorphisms. Then with the identifications $F_{s-1} = F_s$ they are elements of $A_{s-1} = A_s$. Moreover, one or both extreme elements g_{01} and g_{r0} can be absent in (19).

We get rid of the elements g_{s-1s} in (19) which are isomorphisms by (4), i.e. by the element $a_{s-1} \in A_{s-1}$, by transferring them from right to left using (4) and by cancelling two adjacent elements τ, because $\tau^2 = 1$. Similarly, with the aid of (4) we can attain that all the g_{s-1s}'s are birational maps over C (which are not isomorphisms) except, perhaps, for the last map g_{r0}, by adjusting g_{s-1s} from the left by an element a_s of A_s (i.e. by considering $a_s^{-1}a_s g_{s-1s}$) and by replacing $\tau_s a_s^{-1}$ by $a_s'\tau$. We shall assume that the word (19) already has such a normal form.

For each birational map $u_{ij}\colon F_i \to F_j$, $i, j \in I(F)$, we define the integral-valued functions $a(u_{ij})$, $b(u_{ij})$, and $n_x(u_{ij})$ from the representation

$$u_{ij}(f_i) = a(u_{ij})f_j - b(u_{ij})K_{F_j} - \sum n_x(u_{ij})x$$

in the group $Z^*(F_j) = (\operatorname{Pic}\overline{F}_j)^G + Z^0(F_j)$.

For each index $s = 1, \dots, r$ we now write the following collections of integers using the word $R \in B(F)$ in the normal form (19):

$$\begin{aligned} a_s &= a(\tau_s \cdot g_{s-1s} \cdots g_{01}), \\ b_s &= b(\tau_s \cdot g_{s-1s} \cdots g_{01}), \\ n_s(x) &= n_x(\tau_s \cdot g_{s-1s} \cdots g_{01}), \\ a'_s &= a(g_{s-1s} \cdot \tau_{s-1} \cdots g_{01}), \\ b'_s &= b(g_{s-1s} \cdot \tau_{s-1} \cdots g_{01}), \\ n'_s(x) &= n_x(g_{s-1s} \cdot r_{s-1} \cdots g_{01}). \end{aligned} \tag{20}$$

By (2), (3), and Lemma 2, 2) and 3), these numbers satisfy the following recurrence relations:

$$\begin{aligned} a_s &= -a'_s, \qquad b_s = b'_s + \frac{4}{d}a'_s, \\ n_s(x) &= n'_s(\tau_s(x)), \qquad b'_{s+1} = b_s. \end{aligned} \tag{21}$$

Moreover, $b_s \geqslant 0$, $2b_s \geqslant n_s(x) \geqslant 0$, and $2b'_s \geqslant n'_s(x) \geqslant 0$.

Now let the word $R \in R(F)$, i.e. it represents the identity map. Then the following initial and final conditions must be satisfied:

$$\begin{aligned} &a'_1 = 1, \qquad b'_1 = 0, \qquad n'_1(x) = 0, \quad \forall x \in E(F_1). \\ &a_1 = -1, \qquad b_1 = 4/d, \qquad n_1(x) = 0, \\ &a_r = 1, \qquad b_r = 0, \qquad n_r(x) = 0, \quad \forall x \in E(F_r). \end{aligned} \tag{22}$$

We assume $R \neq 1$ and derive a contradiction, provided $d = 1$. Since $R \neq 1$, we get $r > 1$ because, for $r = 1$, the initial and final conditions (22) are absurd, or $r = 0$ and $R = g_{01} = 1$. So, there exists an index $s \geqslant 1$ satisfying the conditions

1) $b_s \geqslant b_{s-1} \geqslant \cdots \geqslant b_1$;

2) $b_{s+1} < b_s$.

We shall show that this is impossible when $d = 1$. Since $b_{s+1} = b'_{s+1} + 4a'_{s+1} = b_s + 4a'_{s+1}$ by (21), and $b_{s+1} < b_s$ by 2), we get $a'_{s-1} < 0$. On the other hand, by 1), we get $a'_s \geqslant 0$; that is, $a_s = -a'_s \leqslant 0$. By Lemma 3 (below), $a'_{s+1} = a_s$ and g_{ss+1} is an isomorphism, contradicting the normality of the word R. Therefore $R = 1$, as required.

It remains to prove the following lemma.

LEMMA 3. *Let F and F' be standard conic bundles over C with $d = d' = 1$ or 2. Let $z = af - bK_F - \sum n_i x_i$ and $z' = a'f' - bK_{F'} - \sum n'_i x'_i$ be the classes of the pencils of rational curves on F and F' with $b > 0$ and $a' \leqslant a \leqslant 0$. Assume the existence of a map g: $F \to F'$ birational over C such that $g(z) = z'$.*

1) *If $d = 1$, then g is an isomorphism.*

2) *If $d = 2$, then g is either an isomorphism or an elementary transformation with center in the closed point x of degree 1 outside the degenerate fibers of π: $F \to C$.*

PROOF. We represent g as a composition of elementary transformations: $g = e(x_m) \cdots e(x_1)$ (Lemma 2). We choose such a representation with the smallest value of m. Then, by Lemma 2,

$$\begin{gathered} g(x_i) = \deg x_i f' - x_i', \qquad i = 1, \dots, m, \\ g(-K_F) = a''f' - K_{F'} - \sum_{i=1}^{m} 2x_i', \\ g(f) = f', \\ g(x) = x', \qquad x \neq x_1, \dots, x_m, \end{gathered} \tag{23}$$

where $a'' = \sum_1^m \deg x_i$. We use the fact that elementary transformations with centers on fibers over distinct points of the base commute up to a C-isomorphism and that a sequence of elementary transformations with centers in points belonging to a fiber over a point of the base is uniquely defined (up to a C-isomorphism). Using (23), we obtain

$$\begin{aligned} z' &= a'f' - bK_{F'} - \sum n_i'x_i' = g(z) = g\left(af - bK_F - \sum n_i x_i\right) \\ &= af' - b\left(-a''f' + K_{F'} + \sum_{i=1}^{m} 2x_i'\right) - \sum_{i=1}^{m} n_i\left(\deg x_i f' - x_i'\right) - \sum_{i>m} n_i x_i' \\ &= \left[a + \sum_{i=1}^{m} \deg x_i(b - n_i)\right] - bK_{F'} \\ &\quad - \sum_{i=1}^{m} (2b - n_i)x_i' - \sum_{i>m} n_i x_i'. \end{aligned} \tag{24}$$

Hence $a' = a + \sum_1^m \deg x_i(b - n_i)$, $n_i' = 2b - n_i$, $i = 1, \dots, m$, and $n_i' = n_i$ for $i > m$. Since, by assumption, $a' \leqslant 0$ and $a \leqslant 0$, by Lemma 1, 5) with $d = 1$ we get $n_i' < b$ and $n_i < b$, $\forall i$. Therefore, we have $m = 0$ in (24) provided $d = 1$, i.e., g is an isomorphism, $a = a'$, and $n_i = n_i'$, $\forall i$.

In case $d = 2$, by Lemma 1, 5), we are in a similar situation as when $d = 1$, or there exists only one point, say x_1, with $n_1 \geqslant b$ and $\deg x_1 = 1$ (similarly, x_1' with $n_1' \geqslant b$ and $\deg x_1' = 1$). Therefore $m = 0$ or 1. In the first case g is an isomorphism, and in the second case $g = e(x_1)$ is an elementary transformation with center in the closed point x_1 of degree 1 outside the degenerate fiber, because otherwise the elementary transformation does not exist.

The lemma is proved.

2. Proof of Theorem 3. As in the proof of Theorem 2 in subsection 1, we assume the word R of $R(F)$ is written in the form (19). We also assume that it is already reduced to a normal form by the relations (14). We must show that in case $d = 2$ the relations (7) allow us to reduce the word R, in the normal form (19), to 1. To do this we apply (21) and (22) for the collections of integers (20), as in the proof of Theorem 2. As above, the index s will be determined by 1) and 2) of subsection 1.

First, we observe that g_{ss+1} must be an elementary transformation with center in a closed point $x_s \in F_s$ of degree 1. Indeed, since R is normal, it is not an isomorphism. Moreover, since $a_{s+1}' < 0$ and $a_s \leqslant 0$ by 1) and 2) above, it satisfies the conditions of Lemma 3. Therefore, by Lemma 3, $g_{ss+1} = e(x_s)$. Since F_s does not contain points x_s with $\deg x_s = 1$, we derive a contradiction with the assumption $R \neq 1$, as in Theorem 2.

Let x'_{s+1} be the image of the fiber blown down under the map $e(x_s)$: $F_s \to F_{s+1}$. We put $a'_{s+1} = a$, $b'_{s+1} = b_s = b$, and $n'_{s+1}(x'_{s+1}) = n$. We have $a < 0$ and $b \geqslant 2$. Since $0 \geqslant -a'_s = a_s = a + b$, we get $n \geqslant a + b$.

By the recurrence formulas (21) and formula (14), we obtain

$$\begin{gathered} a_{s+1} = -a, \quad a_s = a + b - n, \quad a'_s = -a - b + n, \\ b_{s+1} = b + 2a, \quad b_s = b, \quad b'_s = 2a + 3b - 2n, \\ n_{s+1}(x_{s+1}) = n, \quad n_s(x_s) = 2b - n, \quad n'_s(x'_s) = 2b - n, \\ x_{s+1} = \tau_{s+1}(x'_{s+1}), \qquad x'_s = \tau_s(x_s). \end{gathered} \tag{25}$$

It follows at once from (25) and the initial conditions (22) that $s \geqslant 2$. We separately consider the cases $n = a + b$ and $n > a + b$.

Case $n = a + b$. We shall show that g_{s-1s} is also the elementary transformation $e(x_{s-1})$: $F_{s-1} \to F_s$ inverse to the elementary transformation $e(x'_s)$ with center $x'_s = \tau_s(x_s)$. We get

$$a'_s = 0, \quad b'_s = b, \quad n'_s(x'_s) = b - a, \quad a_{s-1} = a, \quad b_{s-1} = b.$$

Therefore, the statement $g_{s-1s} = e(x_{s-1})$ follows as in the case of g_{ss+1} of Lemma 3.

The word R can be written in the form

$$R = g_{r0}\tau_r \cdots g_{s+1s+2}\tau_{s+1}e(x_s)\tau_s e(x_{s-1})\tau_{s-1}e(x_{s-2})e^{-1}(x_{s-2})g_{s-2s-1} \cdots g_{01},$$

where $e(x_{s-2})$ is the elementary transformation inverse to $e(x'_{s-1})$: $F_{s-1} \to F'_{s-2}$. Using (8) we replace the part $\tau_{s+1}e(x_s)\tau_s e(x_{s-1})\tau_{s-1}e(x_{s-2})$ of the word R by the part $e(x'_{s-2})\tau'_{s-2}a$ of the word

$$R' = g_{r0}\tau_r \cdots g_{s+1s+2}e(x'_{s-2})\tau'_{s-2}ae(x'_{s-1})g_{s-2s-1} \cdots g_{01},$$

which also belongs to $R(F)$ and contains $r - 2 < r$ elements of type τ. It remains to apply induction on r.

Case $n > a + b$. In this case we can write R in the form

$$R = g_{r0}\tau_r \cdots g_{s+1s+2}\tau_{s+1}e(x_s)\tau_s e(x_{s-1})e^{-1}(x_{s-1})g_{s-1s} \cdots g_{01}.$$

Using (8) we replace the part $\tau_{s+1}e(x_s)\tau_s e(x_{s-1})$ by $e(x'_s)\tau'_s e(x'_{s-1})\tau''_{s-1}a$, where $e(x'_{s-1})$: $F''_{s-1} \to F'_s$ is the elementary transformation with center $x'_{s-1} = \tau''_{s-1}a(x''_{s-1})$, $a \in A(F''_{s-1})$, and $e(x'_s)$: $F'_s \to F_s$ is the elementary transformation with center $x'_s = \tau'_s(x''_s)$, where x''_s is the image of the fiber blown down under the map $e(x'_{s-1})$. We obtain the word

$$R' = g_{r0}\tau_r \cdots g_{s+1s+2}e(x'_s)\tau'_s e(x'_{s-1})\tau''_{s-1}ae(x''_{s-1})g_{s-1s} \cdots g_{01}$$

of $R(F)$ with the same number r of elements of type τ, but with $b'_s = 3b + 2a - 2n = b + 2(a + b - n) < b_s$. Moreover, the remaining b_i remain unchanged. Therefore, by induction on $\Sigma_1^r b_i$ we obtain the required result. So the main statement of Theorem 3 is proved. But if $F_i^0(k) = \varnothing$, then, as we have already mentioned in the proof, all the relations are among the relations (4), and we obtain the same result as in the case $d = 1$.

3. Proof of Proposition 1. If all maps in (6) are defined, then applying (11) and (14) we easily obtain that α,the composition of the maps (7), yields the following map $Z^*(F_0) \to Z^*(F_4)$:

$$\alpha(f_0) = f_4,$$
$$\alpha(-K_{F_0}) = -K_{F_4},$$
$$\alpha(x) = x', \qquad x \in E(F_0),\ x' \in E(F_4).$$

Hence, by Lemma 2, 1), α: $F_0 \to F_4$ is an isomorphism carrying the pencil $|f_0|$ into the pencil $|f_4|$. So, it remains to verify that the composition of maps (6) is well defined. Since $x_0 \in F_0$ is a good point of a del Pezzo surface with $d = 2$, by blowing up this point, σ_{x_0}: $F_0' \to F_0$, we obtain the del Pezzo surface F_0' with $d = 1$ (see [1], Chapter IV, §2). By blowing down the proper transform of the fiber through x_0 on F_0, $\sigma_{x_1'}$: $F_0' \to F_1$, we obtain a del Pezzo surface F_j ([1], Chapter IV, §2). Thus, F_1 is a del Pezzo surface with $d = 2$, and the map τ_1 is defined; here $l_{01}(x_0) = \sigma_{x_1'}\sigma_{x_0}^{-1}$: $F_0 \to F_1$. Since F_0' is a del Pezzo surface obtained by the blow-up $\sigma_{x_1'}$: $F_0' \to F_1$, the point $x_1' \in F_1$ does not belong to the exceptional curves of the first kind. Furthermore, $x_1 = \tau_1(x_1')$ does not belong to the exceptional curves of the first kind of the surface $\overline{F}_1$, because the involution τ_1 carries exceptional curves into exceptional curves. Thus, by iterating the same argument four times, we obtain the required statement, and the proposition is also proved.

Bibliography

1. Yu. I. Manin, *Cubic forms*, "Nauka", Moscow, 1972; English transl., North-Holland, 1973.

2. V. A. Iskovskikh, *Birational automorphisms of three-dimensional algebraic varieties*, Itogi Nauki: Sovremennye Problemy Mat., vol. 12, VINITI, Moscow, 1979, pp. 159–236; English transl. in J. Soviet Math. **13** (1980), no. 6.

3. ______, *Minimal models of rational surfaces over arbitrary fields*, Izv. Akad. Nauk SSSR Ser. Mat. **43** (1979), 19–43; English transl. in Math. USSR Izv. **14** (1980).

4. ______, *Rational surfaces with a pencil of rational curves*, Mat. Sb. **74 (116)** (1967), 608–638; English transl. in Math. USSR Sb. **3** (1967).

5. ______, *Rational surfaces with a pencil of rational curves and with positive square of the canonical class*, Mat. Sb. **83 (125)** (1970), 90–119; English transl. in Math. USSR Sb. **12** (1970).

6. ______, *Birational properties of a surface of degree* 4 *in* $\mathbf{P}^4$, Mat. Sb. **88 (130)** (1972), 31–37; English transl. in Math. USSR Sb. **17** (1972).

7. Ju. I. Manin, *Rational surfaces over perfect fields*. I, Inst. Hautes Études Sci. Publ. Math. No 30 (1966), 55–113; English transl. in Amer. Math. Soc. Transl. (2) **84** (1969).

8. ______, *Rational surfaces over perfect fields*. II, Mat. Sb. **72 (114)** (1967), 161–192; English transl. in Math. USSR Sb. 1 (1967).

9. H. P. F. Swinnerton-Dyer, *Rational points on del Pezzo surfaces of degree* 5, Algebraic Geometry, Oslo, 1970 (Proc. Fifth Nordic Summer School Math.), Noordhoff, 1972, pp. 287–290.

10. S. Kantor, *Rationale Zerlegung der birationalen Transformationen in ihre Primfactoren*, Monatsh. Math. Phys. **10** (1899), 54–74.

Translated by R. TREGER

Труды Матем. Инст. Стеклов.
165(1984)

Proceedings of the Steklov
Institute of Mathematics
1985, Issue 3

INVARIANT LATTICES OF TYPE G_2 AND THEIR AUTOMORPHISM GROUPS

UDC 512.54 + 512.6

A. I. KOSTRIKIN, I. A. KOSTRIKIN AND V. A. UFNAROVSKIĬ

ABSTRACT. Let L be a simple complex Lie algebra of one of the types G_2, E_8, B_{2^m-1} or D_{2^m}. To a multiplicative orthogonal decomposition (MOD) of such an algebra, $L = \oplus_0^h H_k$, where h is the Coxeter number and $[H_i, H_j] = H_k$ (see MR **83c**: 17014), with the group $G = \mathrm{Aut}_{\mathrm{MOD}}(L)$ acting transitively on the set $\{H_k \mid k = 0, 1, \ldots, h\}$ of Cartan subalgebras, we assign a G-invariant positive definite integral lattice $\Lambda = \oplus_0^h \Lambda_k$, with the metric induced by the Killing form on L. We propose to study all the G-invariant sublattices of Λ (they are called lattices of the same type as algebra L). The hypothesis is made that the number of similarity classes of G-invariant nondegenerate lattices of a given type is finite. The main result is the following theorem:

The lattice Λ of type G_2 contains, up to similarity and isometry, just five proper nondegenerate G-invariant sublattices Λ_i, $0 \leqslant i \leqslant 4$. Among these, exactly one lattice, namely Λ_0, does not contain 2Λ but contains 4Λ. Its automorphism group, $\mathrm{Aut}(\Lambda_0)$, is isomorphic to $\mathbf{Z}_2 \times G_2(3)$, where $G_2(3)$ is the simple Dickson-Chevalley group of order 4,245,696.

We used the computer ÈS-1035 for the study of Aut Λ_0. Earlier, P. Smith and J. Thompson constructed using a computer a G-invariant lattice of type E_8 whose automorphism group is the Thompson sporadic simple group, but so far there does not exist a description of all the G-invariant lattices of this type.

Bibliography: 12 titles.

One way of studying finite irreducible linear groups over $\mathbf{Z}$ is lattice theory, whose foundations one can learn from the books [1] and [2]. Interest in lattices and their automorphism groups has greatly increased due to the results of Conway [3], [4], and in recent years under the strong influence of the Monster, F_1 (see [5]), in whose construction the Leech lattice, the affine Lie algebra of type E_8, and the elements of the theory of modular functions of one variable are wonderfully, though not quite explicably, intertwined.

The present paper provides a beginning to the systematic study of lattices of a special type. They arise naturally from orthogonal decompositions of complex simple Lie algebras [6], [7]. In the first place, one should study lattices related to the multiplicative orthogonal decompositions (MOD) of an algebra L of one of the types G_2, E_8, B_{2^m-1} or D_{2^m}:

$$L = \bigoplus_{i=0}^{h} H_i \quad (h \text{ is the Coxeter number}); \qquad [H_i, H_j] = H_k.$$

1980 *Mathematics Subject Classification*. Primary 17B20, 17B25, 17B40, 17E65; Secondary 17D06, 17D08.

The group $G = \mathrm{Aut}_{\mathrm{MOD}}(L)$ acts transitively on the set $\{H_i | i = 0, 1, \dots, h\}$ of Cartan subalgebras. Taking in one of these subalgebras H_0 a lattice Λ^0 is isomorphic (or almost isomorphic) to the lattice of fundamental weights of the algebra L, and "disseminating" it by G, we obtain a G-invariant lattice $\Lambda = \oplus_0^h \Lambda^k$, with the metric induced by the Killing form on L. All the MOD of the algebra L are conjugate (see [7]), so that the construction of Λ has an invariant character.

The lattice Λ is not interesting in itself from the point of view of automorphisms; however, the authors became convinced that among the G-invariant sublattices of Λ (they will all be called lattices of the same type as the starting algebra L) there exists at least one lattice with a rich automorphism group. Without yet making this statement precise, we add to it the important (at least for MOD) conjecture that the number of similarity classes of G-invariant nondegenerate lattices of a given type is finite. Our considerations rely on the well-known integral representation of Thompson's sporadic group of order $2^{15} \cdot 3^{10} \cdot 5^3 \cdot 7^2 \cdot 13 \cdot 19 \cdot 31$ (see [8] and [9]), and on the following theorem, which is the main result of the present paper.

THEOREM. *The lattice Λ of type G_2 contains up to similarity and isometry only five proper nondegenerate G-invariant sublattices Λ_i $(0 \leqslant i \leqslant 4)$. Among these, exactly one lattice, namely Λ_0, does not contain 2Λ, but contains 4Λ. Its automorphism group* $\mathrm{Aut}(\Lambda_0)$ *is isomorphic to $Z_2 \times G_2(3)$, where $G_2(3)$ is the simple Dickson-Chevalley group of order* 4,245,696.

More detailed facts on the lattices Λ_i are contained in §§2 and 3. §1 gives a representation for the group $\mathrm{Aut}_{\mathrm{MOD}}(G_2)$. We remark that Thompson's group corresponds to a G-invariant lattice of type E_8 constructed using a computer, but as far as we know there is no description of all the G-invariant lattices of this type. In our considerably simpler case, a computer was used only at the final stage. The tables given at the end of the paper give the impression of complete clarity. However, in retrospect it would have been desirable to find a more conceptual and convenient interpretation of both the lattice Λ_0 and the lattice realizing Thompson's group.

§1. $\mathrm{Aut}_{\mathrm{MOD}}(G_2)$

The classical Lie algebra L of type G_2 is given by its multiplication table (see Table 1) with respect to the Chevalley basis $\{h_\alpha, h_\beta, X_\gamma\}$, where α and β are simple roots ($\alpha > \beta$), and γ is an arbitrary root. We introduce for convenience the symbols J^d_{abc} with $a, b, c, d \in \{0, 1\}$, putting:

$$J^d_{000} = 0; \quad J^0_{001} = h_\alpha; \quad J^1_{001} = h_\beta;$$

$$J^1_{01c} = X_{2\alpha+3\beta} - (-1)^c X_{-2\alpha-3\beta}$$

and

$$J^d_{abc} = X_{a\alpha+(b+2d)\beta} - (-1)^c X_{-a\alpha-(b+2d)\beta}$$

in the remaining cases. We remark that if we interpret a, b, c, d on the left as elements of Z_2 and on the right as integers, then the spaces

$$H_w = H_{(abc)} = \langle J^d_{abc} | d = 0, 1 \rangle$$

TABLE 1

	$X_{-2\alpha-3\beta}$	$X_{-\alpha-3\beta}$	$X_{-\alpha-2\beta}$	$X_{-\alpha-\beta}$	$X_{-\alpha}$	$X_{-\beta}$	h_α	h_β	X_β	X_α	$X_{\alpha+\beta}$	$X_{\alpha+2\beta}$	$X_{\alpha+3\beta}$	$X_{2\alpha+3\beta}$
$X_{-2\alpha-3\beta}$	0	0	0	0	0	0	$X_{-2\alpha-3\beta}$	0	0	$-X_{-\alpha-3\beta}$	$-X_{-\alpha-2\beta}$	$X_{-\alpha-\beta}$	$X_{-\alpha}$	$-2h_\alpha-h_\beta$
$X_{-\alpha-3\beta}$	0	0	0	0	$-X_{-2\alpha-3\beta}$	0	$-X_{-\alpha-3\beta}$	$3X_{-\alpha-3\beta}$	$-X_{-\alpha-2\beta}$	0	0	$X_{-\beta}$	$-h_\alpha-h_\beta$	X_α
$X_{-\alpha-2\beta}$	0	0	0	$-3X_{-2\alpha-3\beta}$	0	$-3X_{-\alpha-3\beta}$	0	$X_{-\alpha-2\beta}$	$-2X_{-\alpha-\beta}$	0	$2X_{-\beta}$	$-3h_\alpha-$ $-2h_\beta$	X_β	$X_{\alpha+\beta}$
$X_{-\alpha-\beta}$	0	0	$3X_{-2\alpha-3\beta}$	0	0	$-2X_{-\alpha-2\beta}$	$X_{-\alpha-\beta}$	$-X_{-\alpha-\beta}$	$3X_{-\alpha}$	$-X_{-\beta}$	$-3h_\alpha-$ $-h_\beta$	$2X_\beta$	0	$-X_{\alpha+2\beta}$
$X_{-\alpha}$	0	$X_{-2\alpha-3\beta}$	0	0	0	$X_{-\alpha-\beta}$	$2X_{-\alpha}$	$-3X_{-\alpha}$	0	$-h_\alpha$	$-X_\beta$	0	0	$-X_{\alpha+3\beta}$
$X_{-\beta}$	0	0	$3X_{-\alpha-3\beta}$	$2X_{-\alpha-2\beta}$	$-X_{-\alpha-\beta}$	0	$-X_{-\beta}$	$2X_{-\beta}$	$-h_\beta$	0	$3X_\alpha$	$-2X_{\alpha+\beta}$	$-X_{\alpha+2\beta}$	0
h_α	$-X_{-2\alpha-3\beta}$	$X_{-\alpha-3\beta}$	0	$-X_{-\alpha-\beta}$	$-2X_{-\alpha}$	$X_{-\beta}$	0	0	$-X_\beta$	$2X_\alpha$	$X_{\alpha+\beta}$	0	$-X_{\alpha+3\beta}$	$X_{2\alpha+3\beta}$
h_β	0	$-3X_{-\alpha-3\beta}$	$-X_{-\alpha-2\beta}$	$X_{-\alpha-\beta}$	$3X_{-\alpha}$	$-2X_{-\beta}$	0	0	$2X_\beta$	$-3X_\alpha$	$-X_{\alpha+\beta}$	$X_{\alpha+2\beta}$	$3X_{\alpha+3\beta}$	0
X_β	0	$X_{-\alpha-2\beta}$	$2X_{-\alpha-\beta}$	$-3X_{-\alpha}$	0	h_β	X_β	$-2X_\beta$	0	$X_{\alpha+\beta}$	$-2X_{\alpha+2\beta}$	$-3X_{\alpha+3\beta}$	0	0
X_α	$X_{-\alpha-3\beta}$	0	0	$X_{-\beta}$	h_α	0	$-2X_\alpha$	$3X_\alpha$	$-X_{\alpha+\beta}$	0	0	0	$-X_{2\alpha+3\beta}$	0
$X_{\alpha+\beta}$	$X_{-\alpha-2\beta}$	0	$-2X_{-\beta}$	$3h_\alpha+h_\beta$	X_β	$-3X_\alpha$	$-X_{\alpha+\beta}$	$X_{\alpha+\beta}$	$2X_{\alpha+2\beta}$	0	0	$-3X_{2\alpha+3\beta}$	0	0
$X_{\alpha+2\beta}$	$-X_{-\alpha-\beta}$	$-X_{-\beta}$	$3h_\alpha+2h_\beta$	$-2X_\beta$	0	$2X_{\alpha+\beta}$	0	$-X_{\alpha+2\beta}$	$3X_{\alpha+3\beta}$	0	$3X_{2\alpha+3\beta}$	0	0	0
$X_{\alpha+3\beta}$	$-X_{-\alpha}$	$h_\alpha+h_\beta$	$-X_\beta$	0	0	$X_{\alpha+2\beta}$	$X_{\alpha+3\beta}$	$-3X_{\alpha+3\beta}$	0	$X_{2\alpha+3\beta}$	0	0	0	0
$X_{2\alpha+3\beta}$	$2h_\alpha+h_\beta$	$-X_\alpha$	$-X_{\alpha+\beta}$	$X_{\alpha+2\beta}$	$X_{\alpha+3\beta}$	0	$-X_{2\alpha+3\beta}$	0	0	0	0	0	0	0

are, for $w \neq (000)$, Cartan subalgebras of L leading to the MOD:

$$L = \bigoplus_{w \in W \setminus \{0\}} H_w, \qquad [H_w, H_{w'}] = H_{w+w'}.$$

Here $W = \{w = (abc) | a, b, c \in Z_2\}$ is a 3-dimensional vector space over Z_2 in the standard presentation. This decomposition coincides with the MOD given in [7].

Since in an MOD all Cartan subalgebras are equivalent, it makes sense to choose a basis in L that is reasonably compatible with the group $\mathrm{Aut}_{\mathrm{MOD}}(G_2)$, in whose computation we will be engaged below. In our case, a convenient basis is obtained by "disseminating" a suitable basis for one Cartan subalgebra by a cyclic subgroup

$$\langle C | C^7 = 1 \rangle \subset \mathrm{Aut}_{\mathrm{MOD}}(G_2).$$

The generator C was initially computed in the basis $\{J^d_{abc}\}$, but for the reader it is more convenient from the very beginning to work in the new basis $\{u_k, v_k | k = 1, \dots, 7\}$ written out below:

$$\begin{aligned}
&u_1 = -iJ^0_{001}, \qquad v_1 = -iJ^1_{001} \qquad (i^2 = -1),\\
&u_2 = -\frac{1}{2}(J^0_{100} - J^1_{100}), \qquad v_2 = J^1_{100},\\
&u_3 = \frac{1}{2}(J^0_{010} - J^1_{010}), \qquad v_3 = \frac{1}{2}(-J^0_{010} + 3J^1_{010}),\\
&u_4 = \frac{i}{2}(J^0_{101} - J^1_{101}), \qquad v_4 = \frac{i}{2}(-3J^0_{101} + J^1_{101}),\\
&u_5 = -J^1_{110}, \qquad v_5 = -\frac{1}{2}(J^0_{110} - 3J^1_{110}),\\
&u_6 = -\frac{i}{2}(J^0_{111} + J^1_{111}), \qquad v_6 = iJ^0_{111},\\
&u_7 = iJ^1_{011}, \qquad v_7 = -\frac{i}{2}(J^0_{011} + 3J^1_{011}).
\end{aligned}$$

The multiplication table for the algebra L in the basis $\{u_k, v_k\}$ (see Table 2) has the advantage over Table 1 that it can be specified, essentially, by just 12 entries. The remaining entries can be derived using the automorphism C:

$$C^{k-1}u_1 = u_k, \qquad C^{k-1}v_1 = v_k. \tag{1}$$

The existence of the automorphism C is however a consequence of the direct computation of Table 2, and its property described above can only be used a posteriori.

Our next subject for study is the lattice

$$\Lambda = \left\{ \sum_{k=1}^{7} (n_k u_k + m_k v_k) | n_k, m_k \in \mathbf{Z} \right\},$$

considered together with the positive definite form

$$(x|y) = -\tfrac{1}{8}K(x, y),$$

where $K(x, y)$ is the Killing form of the algebra L. We note that

$$\Lambda = \bigoplus_{k=1}^{7} \Lambda^k$$

TABLE 2

	001		100		010		101		110		111		011	
	u_1	v_1	u_2	v_2	u_3	v_3	u_4	v_4	u_5	v_5	u_6	v_6	u_7	v_7
u_1	0	0	$-u_4-v_4$	0	$-u_7-v_7$	v_7	$-2u_2-v_2$	$6u_2+3v_2$	$2u_6+v_6$	$-3u_6-v_6$	u_5+v_5	$-3u_5-2v_5$	u_3+v_3	$-v_3$
v_1	0	0	v_4	$3u_4+v_4$	$3u_7+2v_7$	$-3u_7-2v_7$	$3u_2+v_2$	$-9u_2-4v_2$	$-6u_6-3v_6$	$9u_6+4v_6$	$-v_5$	$3u_5+2v_5$	0	$-3u_3-v_3$
u_2	u_4+v_4	$-v_4$	0	0	$-u_5-v_5$	0	$-u_1-v_1$	v_1	$-2u_3-v_3$	$6u_3+3v_3$	$2u_7+v_7$	$-3u_7-v_7$	u_6+v_6	$-3u_6-2v_6$
v_2	0	$-3u_4-v_4$	0	0	v_5	$3u_5+v_5$	$3u_1+2v_1$	$-3u_1-2v_1$	$3u_3+v_3$	$-9u_3-4v_3$	$-6u_7-3v_7$	$9u_7+4v_7$	$-v_6$	$3u_6+2v_6$
u_3	u_7+v_7	$-3u_7-2v_7$	u_5+v_5	$-v_5$	0	0	$-u_6-v_6$	0	$-u_2-v_2$	v_2	$-2u_4-v_4$	$6u_4+3v_4$	$2u_1+v_1$	$-3u_1-v_1$
v_3	$-v_7$	$3u_7+2v_7$	0	$-3u_5-v_5$	0	0	v_6	$3u_6+v_6$	$3u_2+2v_2$	$-3u_2-2v_2$	$3u_4+v_4$	$-9u_4-4v_4$	$-6u_1-3v_1$	$9u_1+4v_1$
u_4	$2u_2+v_2$	$-3u_2-v_2$	u_1+v_1	$-3u_1-2v_1$	u_6+v_6	$-v_6$	0	0	$-u_7-v_7$	0	$-u_3-v_3$	v_3	$-2u_5-v_5$	$6u_5+3v_5$
v_4	$-6u_2-3v_2$	$9u_2+4v_2$	$-v_1$	$3u_1+2v_1$	0	$-3u_6-v_6$	0	0	v_7	$3u_7+v_7$	$3u_3+2v_3$	$-3u_3-2v_3$	$3u_5+v_5$	$-9u_5-4v_5$
u_5	$-2u_6-v_6$	$6u_6+3v_6$	$2u_3+v_3$	$-3u_3-v_3$	u_2+v_2	$-3u_2-2v_2$	u_7+v_7	$-v_7$	0	0	$-u_1-v_1$	0	$-u_4-v_4$	v_4
v_5	$3u_6+v_6$	$-9u_6-4v_6$	$-6u_3-3v_3$	$9u_3+4v_3$	$-v_2$	$3u_2+2v_2$	0	$-3u_7-v_7$	0	0	v_1	$3u_1+v_1$	$3u_4+2v_4$	$-3u_4-2v_4$
u_6	$-u_5-v_5$	v_5	$-2u_7-v_7$	$6u_7+3v_7$	$2u_4+v_4$	$-3u_4-v_4$	u_3+v_3	$-3u_3-2v_3$	u_1+v_1	$-v_1$	0	0	$-u_2-v_2$	0
v_6	$3u_5+2v_5$	$-3u_5-2v_5$	$3u_7+v_7$	$-9u_7-4v_7$	$-6u_4-3v_4$	$9u_4+4v_4$	$-v_3$	$3u_3+2v_3$	0	$-3u_1-v_1$	0	0	v_2	$3u_2+v_2$
u_7	$-u_3-v_3$	0	$-u_6-v_6$	v_6	$-2u_1-v_1$	$6u_1+3v_1$	$2u_5+v_5$	$-3u_5-v_5$	u_4+v_4	$-3u_4-2v_4$	u_2+v_2	$-v_2$	0	0
v_7	v_3	$3u_3+v_3$	$3u_6+2v_6$	$-3u_6-2v_6$	$3u_1-4v_1$	$-9u_1-4v_1$	$-6u_5-3v_5$	$9u_5+4v_5$	$-v_4$	$3u_4+2v_4$	0	$-3u_2-v_2$	0	0

is a decomposition of Λ_0 as the orthogonal sum of 2-dimensional isomorphic sublattices

$$\Lambda^k = \mathbf{Z}u_k + \mathbf{Z}v_k.$$

The restriction of $(x|y)$ to each of these is given by the rule

$$\left(au_k + bv_k | cu_k + dv_k\right) = 2ac + 6bd - 3(ad + bc).$$

The evenness of our lattice follows from the expression for the scalar square:

$$\left|\sum_{k=1}^{7} (a_k u_k + b_k v_k)\right|^2 = \sum_{k=1}^{7} |a_k u_k + b_k v_k|^2 = 2\sum_{k=1}^{7}\left(a_k^2 + 3b_k^2 - 3a_k b_k\right). \quad (2)$$

From (2) we immediately obtain

LEMMA 1. *All nonzero vectors $x = au_k + bv_k$ of the two shortest lengths $|x|$ in Λ^k are given below*:

$$|x|^2 = 2\colon \pm u_k,\ \pm(u_k + v_k),\ \pm(2u_k + v_k);$$

$$|x|^2 = 6\colon \pm v_k,\ \pm(3u_k + v_k),\ \pm(3u_k + 2v_k).$$

As usual, by $\operatorname{Aut}(\Lambda)$ we understand the group of Z-linear maps $\Phi\colon \Lambda \to \Lambda$ for which $|\Phi(x)| = |x|$, $x \in \Lambda$. Since $(x|x)$ is a positive definite form, we have $|\operatorname{Aut}(\Lambda)| < \infty$. In particular, $\operatorname{Aut}(\Lambda^k) = W(G_2)$, the Weyl group (a dihedral group D_6 of order 12).

THEOREM 1. *The following statements hold.*

i) *The group $G = \operatorname{Aut}_{\mathrm{MOD}}(G_2)$ is isomorphic to the (unique) nonsplit extension of an elementary abelian group E_3 of order* 8 *by the simple group* $\mathrm{GL}(W)$ *of order* 168.

ii) $G \subset \operatorname{Aut}(\Lambda)$.

PROOF. Statement ii) will follow from the arguments leading to the proof of i). As far as i) is concerned, it is convenient to assemble a certain number of automorphisms of G. In the first place, from (1) we can see that C lies both in G and in $\operatorname{Aut}(\Lambda)$. We give two more maps A, B, which are, by Lemma 1, contained in $\operatorname{Aut}(\Lambda)$:

A		B	A		B
$u_1 + v_1$	$\leftarrow\!\shortmid\ u_1 \mapsto$	$u_5 + v_5$	$-v_1$	$\leftarrow\!\shortmid\ v_1 \mapsto$	$-v_5$
$2u_5 + v_5$	$\leftarrow\!\shortmid\ u_2 \mapsto$	$-u_2 - v_2$	$-3u_5 - 2v_5$	$\leftarrow\!\shortmid\ v_2 \mapsto$	$3u_2 + 2v_2$
$-2u_7 - v_7$	$\leftarrow\!\shortmid\ u_3 \mapsto$	$-u_7$	$3u_7 + v_7$	$\leftarrow\!\shortmid\ v_3 \mapsto$	$-v_7$
$-u_6 - v_6$	$\leftarrow\!\shortmid\ u_4 \mapsto$	$-u_3$	$3u_6 + 2v_6$	$\leftarrow\!\shortmid\ v_4 \mapsto$	$3u_3 + v_3$
$-u_4 - v_4$	$\leftarrow\!\shortmid\ u_5 \mapsto$	$2u_6 + v_6$	v_4	$\leftarrow\!\shortmid\ v_5 \mapsto$	$-3u_6 - v_6$
u_2	$\leftarrow\!\shortmid\ u_6 \mapsto$	$-u_1$	v_2	$\leftarrow\!\shortmid\ v_6 \mapsto$	$3u_1 + v_1$
$-u_3 - v_3$	$\leftarrow\!\shortmid\ u_7 \mapsto$	$-u_4$	v_3	$\leftarrow\!\shortmid\ v_7 \mapsto$	$3u_4 - v_4$

A direct verification using Table 2 shows that $A, B \in G$, and also that $C = BAB^2A^3$.

Now let us proceed immediately to the computation of the whole group G. By definition, to each $\Phi \in G$ there corresponds $\varphi \in \mathrm{GL}(W)$ for which $\Phi(H_w) = H_{\varphi(w)}$ (the linearity of φ follows from the multiplicativity of the decomposition). It is not hard to show that the correspondence $\pi\colon \Phi \to \varphi$ gives an epimorphism $G \to \mathrm{GL}(W)$. Indeed, observing the numbers (abc) of the subalgebras H_w in Table 2, we find that

$$\pi(A) = \begin{Vmatrix} 1 & 0 & 0 \\ 1 & 1 & 0 \\ 0 & 1 & 1 \end{Vmatrix}, \qquad \pi(B) = \begin{Vmatrix} 1 & 0 & 1 \\ 0 & 1 & 1 \\ 0 & 1 & 0 \end{Vmatrix}$$

have orders 4 and 3 respectively. Moreover, it is obvious that the element $\pi(C)$ of the subgroup $H = \langle \pi(A), \pi(B) \rangle \subseteq \mathrm{GL}(W)$ has order 7. Thus

$$(\mathrm{GL}(W)\colon H) \leqslant \frac{168}{4 \cdot 3 \cdot 7} = 2,$$

whence $H \triangleleft \mathrm{GL}(W)$, which, in view of the simplicity of the latter, establishes the epimorphic character of π.

To complete the proof, we still have to find the kernel $\mathrm{Ker}(\pi)$, consisting of the automorphisms $\Phi \in G$ which fix each subalgebra H_w. This means that

$$\Phi(u_k) = a_k u_k + b_k v_k, \qquad \Phi(v_k) = c_k u_k + d_k v_k.$$

According to Table 2, $[\Phi(u_k), \Phi(v_{k+1})] = 0$, which leads to

$$-a_k c_{k+1} + 3b_k d_{k+1} = 0, \qquad -a_k c_{k+1} + b_k(c_{k+1} + d_{k+1}) = 0. \tag{3}$$

Moreover, the conditions for preserving the scalar product (ensured by the invariance of the Killing form) give in addition

$$\begin{aligned} a_k^2 + 3b_k^2 - 3a_k b_k &= 1, \\ c_{k+1}^2 + 3d_{k+1}^2 - 3c_{k+1}d_{k+1} &= 3, \\ 2a_k c_k + 6b_k d_k - 3a_k d_k - 3b_k c_k &= -3. \end{aligned} \tag{4}$$

We suppose at first that $c_{k+1} \neq 0$ for some k. Then from (3) it follows that $b_k \neq 0$ and $c_{k+1} = 2d_{k+1}$, $2a_k = 3d_k$. The first two equations of (4) give the explicit expressions

$$a_k = \lambda_k\sqrt{3}, \quad b_k = \lambda_k \frac{2}{\sqrt{3}}, \quad c_{k+1} = \varepsilon_{k+1} 2\sqrt{3}, \quad d_{k+1} = \varepsilon_{k+1}\sqrt{3}, \tag{5}$$

where $\lambda_k^2 = \varepsilon_{k+1}^2 = 1$.

Let us assume that $c_k = 0$. Then the third equation of (4) gives $d_k = -\lambda_k\sqrt{3}$, while the second equation, on replacing $k+1$ by k, shows that $d_k^2 = 1$, which is a contradiction.

Thus $c_k \neq 0$, and the argument given above makes it possible to assert that (5) holds for all k. However, this contradicts the relations

$$-c_1 a_2 + 3d_1 b_2 = c_4,$$

that follow from the equality $[\Phi(v_1), \Phi(u_2)] = \Phi(v_4)$ (see Table 2).

Thus we have shown that $c_{k+1} = 0$ for all k. Now from (3) and (4), one can easily deduce

$$a_k = d_k = \pm 1; \qquad b_k = c_k = 0. \tag{6}$$

The above equality $[\Phi(v_1), \Phi(u_2)] = \Phi(v_4)$ also leads to the relation $d_1 a_2 = d_4$, from which one easily obtains

$$a_k a_{k+1} = a_{k+3} \tag{7}$$

for all k.

So every automorphism $\Phi \in \operatorname{Ker}(\pi)$ is of order 2, and is determined by three parameters $a_1, a_2, a_3 \in \{\pm 1\}$. Hence $\operatorname{Ker}(\pi)$ is an elementary abelian group of order $\leqslant 8$. Since $A^4 \in \operatorname{Ker}(\pi)$ and $BA^4B^{-1} \neq A^4$, the simple group $\pi(G) = \mathrm{GL}(W)$ acts faithfully by conjugation on $\operatorname{Ker}(\pi)$, and hence $|\operatorname{Ker}(\pi)| = 8$.

The nonsplit nature of the extension follows from the easily verifiable fact that $(\Phi A)^4 \neq 1$ for any $\Phi \in \operatorname{Ker}(\pi)$. For the uniqueness of this nonsplit extension, we refer the reader to [10].

The theorem is proved.

REMARK. The arguments given at the end of the proof of Theorem 1 show also that $G = \langle A, B \rangle$.

§2. Sublattices

We intend to give a description of all the G-invariant sublattices Λ' in Λ. For this purpose, it is convenient to assemble in a separate table (see Table 3) the action of some selected automorphisms on the common vector

$$x = \sum_{i=1}^{7} (n_i u_i + m_i v_i) \in \Lambda.$$

The reader is earnestly recommended to remember the existence of Table 3 for convenience in verifying all the calculations given below. It will be proved that all the elements Λ' of interest to us are singly generated; that is, $\Lambda' = \mathbf{Z}G(x)$ for a suitable choice of x. Therefore in the first place we should study singly generated lattices. In the similarity class of sublattices $n\Lambda'$, $n \in \mathbf{Z}$, it is natural to study only the sublattice with $n = 1$. For this reason, we frequently use the condition $\mathrm{GCD}(n_i, m_i | i = 1, \dots, 7) = 1$.

We give a preliminary description of several particular sublattices which we shall meet in the course of our discussions. Our first sublattice

$$\tilde{\Lambda} = \left\{ \sum_{i=1}^{7} (n_i u_i + m_i v_i | n_i \equiv 0 \ (3) \right\}$$

is in many ways like Λ and its invariance under G can immediately be seen from Table 3.

Three sublattices are spanned by the following vectors of minimal length:

$$\Lambda_1 : \begin{cases} \pm(u_i + v_i) \pm (u_{i+1} + v_{i+1}) \pm u_{i+2}, \\ \pm u_i \pm u_{i+2} \pm (2u_{i+4} + v_{i+4}), \\ \pm(2u_i + v_i) \pm (u_{i+1} + v_{i+1}) \pm (2u_{i+4} + v_{i+4}), \\ \pm(u_i + v_i) \pm (2u_{i+2} + v_{i+2}) \pm u_{i+3}; \end{cases}$$

$$\Lambda_2 : \begin{cases} \pm u_i \pm (2u_{i+1} + v_{i+1}), \\ \pm(2u_i + v_i) \pm (u_{i+2} + v_{i+2}), \\ \pm u_i \pm (u_{i+3} + v_{i+3}); \end{cases}$$

$$\Lambda_3 : \ \pm u_i \pm (2u_{i+1} + v_{i+1}) \pm (u_{i+3} + v_{i+3}).$$

Everywhere, the combinations of signs are unrestricted, and the indices are taken modulo 7.

TABLE 3

	001		100		010		101		110		111		011	
	u_1	v_1	u_2	v_2	u_3	v_3	u_4	v_4	u_5	v_5	u_6	v_6	u_7	v_7
$\Phi_{000} = E$	n_1	m_1	n_2	m_2	n_3	m_3	n_4	m_4	n_5	m_5	n_6	m_6	n_7	m_7
Φ_{001}	$-n_1$	$-m_1$	n_2	m_2	n_3	m_3	$-n_4$	$-m_4$	n_5	m_5	$-n_6$	$-m_6$	$-n_7$	$-m_7$
Φ_{010}	n_1	m_1	n_2	m_2	$-n_3$	$-m_3$	n_4	m_4	$-n_5$	$-m_5$	$-n_6$	$-m_6$	$-n_7$	$-m_7$
Φ_{011}	$-n_1$	$-m_1$	n_2	m_2	$-n_3$	$-m_3$	$-n_4$	$-m_4$	$-n_5$	$-m_5$	n_6	m_6	n_7	m_7
Φ_{100}	n_1	m_1	$-n_2$	$-m_2$	n_3	m_3	$-n_4$	$-m_4$	$-n_5$	$-m_5$	$-n_6$	$-m_6$	n_7	m_7
Φ_{101}	$-n_1$	$-m_1$	$-n_2$	$-m_2$	n_3	m_3	n_4	m_4	$-n_5$	$-m_5$	n_6	m_6	$-n_7$	$-m_7$
Φ_{110}	n_1	m_1	$-n_2$	$-m_2$	$-n_3$	$-m_3$	$-n_4$	$-m_4$	n_5	m_5	n_6	m_6	$-n_7$	$-m_7$
Φ_{111}	$-n_1$	$-m_1$	$-n_2$	$-m_2$	$-n_3$	$-m_3$	n_4	m_4	n_5	m_5	$-n_6$	$-m_6$	n_7	m_7
A	n_1	$n_1 - m_1$	n_6	m_6	$-n_7$	$-n_7 + m_7$	$-n_5$	$-n_5 + m_5$	$2n_2 - 3m_2$	$n_2 - 2m_2$	$-n_4 + 3m_4$	$-n_4 + 2m_4$	$-2n_3 + 3m_3$	$-n_3 + m_3$
B	$-n_6 + 3m_6$	m_6	$-n_2 + 3m_2$	$-n_2 + 2m_2$	$-n_4 + 3m_4$	m_4	$-n_7 + 3m_7$	m_7	n_1	$n_1 - m_1$	$2n_5 - 3m_5$	$n_5 - m_5$	$-n_3$	$-m_3$
C	n_7	m_7	n_1	m_1	n_2	m_2	n_3	m_3	n_4	m_4	n_5	m_5	n_6	m_6
C^2B^2	$-n_1 + 3m_1$	m_1	$n_4 - 3m_4$	$-m_4$	n_5	$n_5 - m_5$	$-2n_2 + 3m_2$	$-n_2 + m_2$	n_7	m_7	$n_3 - 3m_3$	$-m_3$	$-n_6 + 3m_6$	$-n_6 + 2m_6$
C^3BA	n_1	m_1	n_2	$n_2 - m_2$	n_7	$n_7 - m_7$	$-2n_4 + 3m_4$	$-n_4 + 2m_4$	$-n_6 + 3m_6$	$-n_6 + 2m_6$	$-2n_5 + 3m_5$	$-n_5 + m_5$	$-n_3$	$-n_3 + m_3$
C^2B^2A	$2n_1 - 3m_1$	$n_1 - m_1$	$2n_5 - 3m_5$	$n_5 - m_5$	$2n_2 - 3m_2$	$n_2 - m_2$	$-2n_6 + 3m_6$	$-n_6 + m_6$	$-2n_3 + 3m_3$	$-n_3 + m_3$	$2n_7 - 3m_7$	$n_7 - m_7$	$-2n_4 + 3m_4$	$-n_4 + m_4$
C^2A^2B	$n_1 - 3m_1$	$n_1 - 2m_1$	$n_3 - 3m_3$	$-m_3$	$-n_6 + 3m_6$	m_6	n_7	$n_7 - m_7$	$-2n_4 + 3m_4$	$-n_4 + m_4$	$-n_2$	$-m_2$	$n_5 - 3m_5$	$-m_5$
C^2A^2BA	$-2n_1 + 3m_1$	$-n_1 + 2m_1$	$2n_7 - 3m_7$	$n_7 - m_7$	$-2n_4 + 3m_4$	$-n_4 + 2m_4$	$-2n_3 + 3m_3$	$-n_3 + 2m_3$	$-n_5 + 3m_5$	m_5	$-n_6$	$-m_6$	$-n_2 + 3m_2$	$-n_2 + 2m_2$

Finally, we put

$$\Lambda_4 = \Lambda_2 \cap \Lambda_3.$$

PROPOSITION 1. *The following statements hold*:

i) *Each of the sublattices* Λ_1, Λ_2, Λ_3 *is G-invariant, contains* 2Λ, *and is generated (as a* $\mathbf{Z}G$*-module) by any of its vectors of shortest length.*

ii) *Put* $V_k = \Lambda_k/2\Lambda$. *Then* $\dim V_1 = 8$, $\dim V_2 = 11$, $\dim V_3 = 6$ *and* $\dim V_4 = 3$.

iii) *The sublattices* $\Lambda_1, \dots, \Lambda_4$ *are related in the following way*:

$$\Lambda_2 \supseteq \Lambda_1; \quad \Lambda_1 \cap \Lambda_3 = 2\Lambda; \quad \Lambda_1 + \Lambda_3 = \Lambda;$$

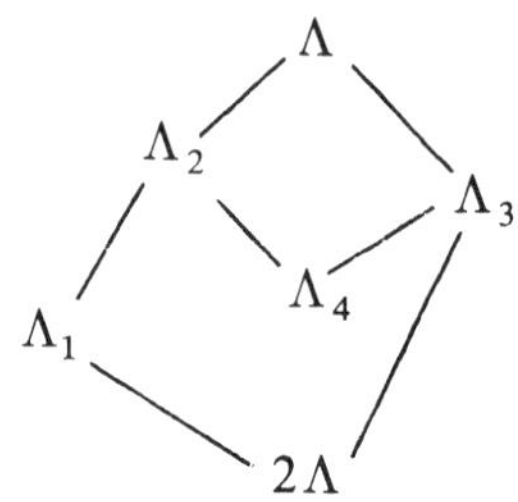

PROOF. i) The G-invariance of Λ_k is verified by acting directly with A and B. A somewhat more troublesome verification, using the automorphisms A, B, C and $\Phi_{abc} \in \operatorname{Ker}(\pi)$, shows that Λ_k is generated by any of the mentioned vectors. The fact that the signs can be chosen arbitrarily ensures the inclusion of $2\Lambda \subset \Lambda_k$.

ii), iii). A direct verification using Gaussian elimination makes it possible to write down the dimensions over Z_2 of the vector spaces V_k, and also to establish the properties $V_2 \supset V_1$, $V_1 \cap V_3 = 0$ and $V_1 + V_3 = V = \Lambda/2\Lambda$.

LEMMA 1. *The following statements hold*:

i) $\mathbf{Z}G(u_k) = \Lambda$.

ii) $\mathbf{G}(v_k) = \tilde{\Lambda}$.

iii) *If* $nu_i + mv_i \in \Lambda'$, *then*

a) $n\Lambda \subseteq \Lambda'$, *and*

b) $3m\Lambda \subseteq \Lambda'$; *in particular, if* $2\Lambda \subseteq \Lambda'$, *then* $m\Lambda \subseteq \Lambda'$.

PROOF. i) Applying C, we can suppose that $k = 1$; that is, $u_1 \in \Lambda'' = \mathbf{Z}G(u_k)$. In this case $A(u_1) = u_1 + v_1 \in \Lambda'''$ and $v_1 \in \Lambda''$. The repeated application of C completes the proof of i).

ii). On account of the G-invariance of the sublattice Λ, we have $\Lambda''' = \mathbf{Z}G(v_k) \subseteq \tilde{\Lambda}$. The action of C gives us the inclusion $v_i \subset \Lambda'''$ for all i. Since $B(v_7) - v_4 = 3u_4$, $3u_i$ also lies in Λ'''. It remains to note that $\{3u_j, v_j | j = 1, \dots, 7\}$ is a basis for $\tilde{\Lambda}$.

iii) The application of C enables us to assume that $x = n_1u_1 + m_1v_1 \in \Lambda'$, where $n_1 = n$, $m_1 = m$. Furthermore,

$$-(C^2A^2B + C^2A^2BA)x = n_1u_1 \in \Lambda',$$

$$(C^2B^2 - E)x = (-2n_1 + 3m_1)u_1 \in \Lambda' \Rightarrow 3m_1u_1 \in \Lambda'.$$

Now use i).

LEMMA 2. *Let* $\Lambda' = \mathbf{Z}G(x)$ *and* $\Lambda' \not\subseteq \tilde{\Lambda}$, *where*

$$x = \sum(n_iu_i + m_iv_i), \qquad \operatorname{GCD}(n_i, m_i | i = 1, \dots, 7) = 1.$$

Then $\Lambda' \supseteq 4\Lambda$.

PROOF. We have

$$(E + \Phi_{001} + \Phi_{010} + \Phi_{011})x = 4(n_2u_2 + m_2v_2) \in \Lambda'.$$

Applying C, we obtain $4(n_iu_2 + m_iv_2) \in \Lambda'$ for all i. By Lemma 1, iii),

$$4n_i\Lambda \subseteq \Lambda', \qquad 12m_i\Lambda \subseteq \Lambda'.$$

Hence it follows that $4d\Lambda \subseteq \Lambda'$, where $d = \mathrm{GCD}(n_i, 3m_i | i = 1, \dots, 7)$. It remains only to observe that $d = 1$, since $d | \mathrm{GCD}(3n_i, 3m_i | i = 1, \dots, 7) = 3$ and on the other hand the condition $\Lambda' \not\subseteq \tilde{\Lambda}$ shows that d is not divisible by 3. We write for brevity

$$x(a, b) = a \sum_{i=1}^{7} u_i + b \sum_{i=1}^{7} v_i.$$

LEMMA 3. i) *If* $x(a, b) \in \Lambda'$, *then* $x(2a - 3b, a - b) \in \Lambda'$.
ii) *If* $x = \sum(n_iu_i + m_iv_i) \in \Lambda'$, $a = \sum n_i$ *and* $b = \sum m_i$ *then* $x(a, b) \in \Lambda'$.

PROOF. We start with ii), which follows from the equality

$$x(a, b) = \sum_{i=1}^{7} C^j x.$$

Statement i) follows from ii) when we replace x by

$$(C^2B^2A) \cdot x(a, b) = (2a - 3b)(u_1 + u_2 + u_3 - u_4 - u_5 + u_6 - u_7) + (a - b)(v_1 + v_2 + v_3 - v_4 - v_5 + v_6 - v_7).$$

LEMMA 4. *Let* $\Lambda' \supseteq 4\Lambda$. *If for a vector* $x(a, b) \in \Lambda'$ *either* $a \not\equiv 0 \pmod 2$ *or* $b \not\equiv 0 \pmod 2$, *then* $x(1, 1) \in \Lambda'$.

PROOF. We note that the elements of type $x(c, d)$ form a **Z**-module. If we also use congruence modulo 4, then we can restrict ourselves to the cases

$$x(1, 0), \quad x(1, 2), \quad x(1, 3), \quad x(0, 1), \quad x(2, 1).$$

If $x(1, 2) \in \Lambda'$, then by Lemma 3, i), we obtain $x(-4, -1) \in \Lambda'$, whence $x(0, 1) \in \Lambda'$, and hence $x(1, 1) = x(1, 2) - x(0, 1) \in \Lambda'$. The discussion of the remaining cases is no harder.

LEMMA 5. *The vector* $x(1, 1)$ *generates the sublattice* Λ_1.

PROOF. By Lemma 3, ii), applied to $u_1 + v_1 - u_2 - v_2 + v_3$ we have $x(1, 0) \in \Lambda_1$, whence by Lemma 4 we obtain

$$\Lambda' = \mathbf{Z}G(x(1, 1)) \subseteq \Lambda_1.$$

Now we satisfy ourselves that the reverse inclusion holds. First of all, we show that $\Lambda' \supseteq 2\Lambda$. By Lemma 3, i), $x = x(1, 0) \in \Lambda_1$. Moreover, the vector $y = \Phi_{101}(C^3BA)x - x(1, 1) \in \Lambda'$ can be written in the form

$$y = 2u_1 + v_1 + 2u_2 + 2v_2 + 3u_4 + 2v_4 + 3u_6 + 2v_6.$$

Since $\Phi_{111}(y) + y = 6u_4 + 4v_4 \in \Lambda'$, we have by Lemma 2 that $2u_4 \in \Lambda'$, and by Lemma 1, i), we conclude that $2\Lambda \subseteq \Lambda'$.

So in Λ' we can work modulo 2. Applying this consideration to the vector y, we find that

$$u_4 + u_6 + 2u_1 + v_1 \in \Lambda',$$

and this is one of the vectors generating $\tilde{\Lambda}_1$.

LEMMA 6. *Let* $x = \Sigma(n_i u_i + m_i v_i) \in \Lambda'$, *in which the following conditions are satisfied*:

a) $\mathrm{GCD}(n_i, m_i | i = 1, \ldots, 7) = 1$.

b) $\Lambda' \not\subseteq \tilde{\Lambda}$.

c) *At least one of the numbers* Σn_i *or* Σm_i *is odd.*

Then $\Lambda' \supseteq \Lambda_1$.

The proof follows immediately from Lemmas 2, 3, ii), 4, and 5.

LEMMA 7. *Let* Λ' *be a sublattice containing* Λ_1 *and let the vector* $pu_1 + q(v_2 + u_4 + v_4) \in \Lambda'$. *Then*

$$p = 1 \Rightarrow \Lambda' = \Lambda;$$

$$p = 0, \qquad q = 1 \Rightarrow \Lambda' \supseteq \Lambda_2.$$

PROOF. Since $\Lambda' \supseteq \Lambda_1 \supseteq 2\Lambda$, it suffices to consider only the two possibilities $q = 0$ and $q = 1$ for q. For $q = 0$ the equality $\Lambda' = \Lambda$ follows from Lemma 1, i), and for $q = 1$ we argue as follows. By Proposition 1, i), the vector $u_1 + 2u_2 + v_2 + u_4 + v_4$, which belongs to Λ', generates Λ_3. By Proposition 1, iii), $\Lambda_1 + \Lambda_3 = \Lambda$.

For $p = 0$ it suffices to refer to Proposition 1, i).

LEMMA 8. *If* $\Lambda \supsetneqq \Lambda' \supsetneqq \Lambda_1$, *then* $\Lambda' = \Lambda_2$.

PROOF. We show first of all that $\Lambda' \supseteq \Lambda_2$. Let $x \in \Lambda'$, but $x \notin \Lambda_1$. Taking into account the fact that $\Lambda' \supseteq \Lambda_1 \supseteq 2\Lambda$, we can work modulo 2. Moreover, using the vectors which span the lattice Λ_1, we can obtain 8 ($= \dim V_2$) zero coordinates. Therefore, from the very beginning we can suppose that the vector x has the form

$$0 \neq x = au_1 + bv_1 + cu_2 + dv_2 + eu_4 + fv_4,$$

where $a, b, c, d, e, f \in \{0, 1\}$.

If $\Lambda' \not\supseteq \Lambda_2$, then by Lemma 7 the vector

$$cv_2 + (e + f)(u_4 + v_4) \equiv C^3BAx - x \pmod{2\Lambda}$$

must be zero. Hence $c = 0$ and $e = f$. Then

$$x = au_1 + bv_1 + dv_2 + f(u_4 + v_4),$$

Now an application of Lemma 7 to the vector

$$bu_1 + (d + f)(v_2 + u_4 + v_4) \equiv C^2B^2x + x \pmod{2\Lambda}$$

guarantees that $b = 0$ and $d = f$, so that the vector

$$x = au_1 + d(v_2 + u_4 + v_4)$$

falls immediately into the ambit of Lemma 7, according to which $x = 0$. However, this contradicts the choice of x.

Thus we have shown that $\Lambda' \supseteq \Lambda_2$. Now let us assume that there exists a vector $y \in \Lambda'$, but $y \notin \Lambda_2$, and reduce this to a contradiction. Using the vectors which span the lattice Λ_2, we can obtain 11 ($= \dim V_2$) zero coordinates, and suppose from the very beginning that

$$0 \neq y = au_1 + bv_1 + cu_2,$$

where $a, b, c \in \{0, 1\}$. We remark that

$$cv_2 \equiv C^3BAy - y \pmod{2\Lambda}.$$

Therefore, by Lemma 1, iii), $\Lambda' = \Lambda$, both for $c = 0$ and for $c = 1$.

LEMMA 9. *If $\Lambda' \neq \Lambda$, Λ_1, Λ_2; $\Lambda' \not\subseteq n\Lambda$ for any $n \in \mathbf{Z}$, and $\Lambda' \not\subseteq \tilde{\Lambda}$, then $\Lambda' \subseteq \Lambda_3$.*

PROOF. From Lemmas 6 and 8 it follows that for any $x = \Sigma(n_i u_i + m_i v_i)$ we have the congruences $\Sigma n_i \equiv 0$ (2) and $\Sigma m_i \equiv 0$ (2). Hence for all the images

$$\Phi(x) = \Sigma(n_i u_i + m_i v_i), \quad \Phi \in G,$$

we have the same condition $\Sigma n_i' \equiv 0$ (2) and $\Sigma m_i' \equiv 0$ (2). It is not hard to verify that the conditions of this type together determine exactly the lattice Λ_3. More precisely, Λ_3 is determined by the conditions

$$\begin{gathered} (m_i + n_i) + (m_{i+1} + n_{i+1}) + m_{i+2} \equiv 0 \quad (2) \\ m_i + m_{i+2} + n_{i+4} \equiv 0 \quad (2), \\ n_i + (m_{i+1} + n_{i+1}) + n_{i+4} \equiv 0 \quad (2), \\ (m_i + n_i) + n_{i+2} + m_{i+3} \equiv 0 \quad (2), \end{gathered} \tag{1}$$

in a certain sense dual to the basis of Λ_1.

LEMMA 10. *If $\Lambda' \not\subseteq \tilde{\Lambda}$ and $\Lambda' \not\subseteq n\Lambda$ for any $n \in \mathbf{Z}$, $n > 1$, then $\Lambda' \supseteq 2\Lambda_3$.*

PROOF. By Lemma 9, it suffices to consider the case $\Lambda' \subseteq \Lambda_3$. We show first that $\Lambda' \supseteq 2\Lambda_1$. By (1), any vector $x \in \Lambda_3$ has the form, modulo 2 in the coordinates,

$$\begin{gathered} (a, b | c, d | e, a + b + c + d | f, a + b + e | a + c + d, \\ f + c + d | a + b + d + e, b + e | a + b + f, d + f). \end{gathered} \tag{2}$$

Let $2\Lambda'' = \Lambda' \cap 2\Lambda_1$. To the lattice Λ'' we apply Lemmas 8 and 9: if $\Lambda' \not\supseteq 2\Lambda_1$, or equivalently, $\Lambda'' \not\supseteq \Lambda_1$, then $\Lambda'' \subseteq \Lambda_3$. By the assumption, in Λ' there exists a vector x of the form (2), in which not all the coordinates are even. Then in the lattice Λ'' there is a vector

$$\tfrac{1}{2}(x + \Phi_{010}x),$$

which has the form

$$(a, b | c, d | 0, 0 | f, a + b + e | 0, 0 | 0, 0 | 0, 0).$$

But we know that this vector lies in Λ_3, and hence can be determined uniquely modulo 2 by the parameters a, b, c, d, e, f (see Proposition 1, ii)). Hence it must coincide with the vector x, also lying in Λ_3. In particular $e = 0$, and $b = 0$ modulo 2. Thus we have shown that in an arbitrary vector $x \in \Lambda'$, the coordinates b and d are even. An application of the automorphism C gives the final contradiction. Thus we have proved that $\Lambda' \supseteq 2\Lambda_1$.

It remains to show that $\Lambda'' \supseteq \Lambda_2$. This, by Lemma 8, is equivalent to $\Lambda'' \neq \Lambda_1$. For the proof, we consider a vector $x \in \Lambda'$ of the form (2), not all of whose coordinates are even. We have $x \in \Lambda_3$, but $x \notin 2\Lambda$. Then obviously $x \in \Lambda''$, and since $\Lambda_3 \cap \Lambda_1 = 2\Lambda$ (see Proposition 1, ii)), we have $x \notin \Lambda_1$.

Thus all the sublattices of interest to us that are not contained in $\tilde{\Lambda}$ are between Λ_3 and $2\Lambda_2$.

LEMMA 11. *If* $\Lambda_3 \supsetneqq \Lambda' \supsetneqq 2\Lambda$, *then* $\Lambda' = \Lambda_4$.

PROOF. Putting $V' = \Lambda'/2\Lambda$, $V = \Lambda/2\Lambda$ and $V_k = \Lambda_k/2\Lambda$, we can see from Proposition 1 that

$$V = V_1 \oplus V_3, \qquad 0 \neq V' \subsetneqq V_3$$

and hence $V' + V_1 = V' \oplus V_1 \neq V, V_1$. To the subspace $V' + V_1$ there corresponds the sublattice $\Lambda' + \Lambda_1$, which by Lemma 8 coincides with Λ_2; that is, $V_2 = V' \oplus V_1$. In particular, $\dim V' = 3$ (see Proposition 1). We suppose that $V' \neq V_4$. Then the argument given above shows that $V' \cap V_4 = 0$, since $\dim V' \cap V_4 < 3$. On the other hand, $V' \subseteq V_3$, and $V_4 \subseteq V_3$, whence by dimensionality considerations it follows that $V_3 = V' \oplus V_4$. But then we have

$$V = V_1 \oplus V_3 = V_1 \oplus V' \oplus V_4 = V_2 \oplus V_4 = V_2$$

(since $V_4 \subseteq V_2$), which is a contradiction proving the lemma.

LEMMA 12. *If* $\Lambda' \not\subseteq \tilde{\Lambda}, n\Lambda, n \in \mathbf{Z}$, *and* $\Lambda' \not\supseteq 2\Lambda$, *then* $\Lambda' \subseteq \Lambda_4$.

PROOF. We put $\Lambda' \cap 2\Lambda = 2\Lambda''$. By Lemma 10 we have the inclusion $\Lambda_2 \subseteq \Lambda''$. Moreover, from the condition $\Lambda'' \neq \Lambda$ we have by Lemma 8 the equality $\Lambda'' = \Lambda_2$. Therefore $\Lambda' \subseteq \Lambda'' = \Lambda_2$. On the other hand, by Lemma 9 we can assume that $\Lambda' \subseteq \Lambda_3$. Thus $\Lambda' \subseteq \Lambda_2 \cap \Lambda_3 = \Lambda_4$.

LEMMA 13. *The only G-invariant sublattice not containing* 2Λ *but properly containing* 4Λ *can be the lattice* Λ_0 *spanned by the vectors of the form* $\varepsilon e + 2x$, *where* $x \in \Lambda_2$, $\varepsilon \in \{0, 1\}$ *and*

$$e = (1, 0|2, 2|2, 1|1, 1|1, 1|1, 0|0, 1).$$

(So far we have not claimed that Λ_0 really is G-invariant. This will be established later: see the argument before Theorem 2.)

PROOF. Lemmas 10 and 12 enable us to restrict ourselves to lattices contained in Λ_4 and containing $2\Lambda_2$. It is not hard to verify that Λ_4 is determined by the conditions

$$\begin{aligned} m_i + n_{i+1} &\equiv 0 \quad (2), \\ n_i + n_{i+2} + m_{i+2} &\equiv 0 \quad (2), \\ m_i + n_{i+3} + m_{i+3} &\equiv 0 \quad (2), \end{aligned}$$

that are in some sense dual to the basis of the lattice Λ_2, so that any vector $x \in \Lambda_4$ has the form, modulo 2 in the coordinates,

$$x = (a, b|b, c|c, a + c|a + c, a + b + c|a + b + c,$$
$$a + b|a + b, b + c|b + c, a). \tag{3}$$

Let $x \notin 2\Lambda$. The seven displayed pairs of coordinates a, b; b, c; $c, a + c$; ... have the property that the difference, coordinate by coordinate, of any two distinct pairs is, modulo 2, also a pair of this family. By the Dirichlet principle, there exist two coincident pairs, and hence also the zero pair (modulo 2). Applying the automorphism C, we can assume that x has the form, modulo 2,

$$x = (1,0|0,0|0,1|1,1|1,1|1,0|0,1).$$

Moreover, modulo $2\Lambda_2$, any vector of 2Λ can be reduced to the form

$$(0,0|2\alpha,2\beta|2\gamma,0|0,0|0,0|0,0|0,0),$$

where α, β, $\gamma \in \{0,1\}$ (see the proof of Lemma 8).

Thus the desired sublattice Λ' always contains one of the eight vectors

$$x(\alpha,\beta,\gamma) = (1,0|2\alpha,2\beta|2\gamma,1|1,1|1,1|1,0|0,1).$$

Working modulo 4, we obtain the vector

$$(2,0|2t,2t|2s,0|2,2\gamma|0,0|0,0|0,0) \equiv (C^4A - E)\cdot x(\alpha,\beta,\gamma) \in \Lambda',$$

where $s = 1 - \gamma$ and $t = \alpha - \beta$. Computing from this the vector

$$(2,0|0,0|2s,0|2,2\gamma|0,0|0,0|0,0) \in 2\Lambda_2,$$

we obtain

$$2t(u_2 + v_2) \in \Lambda'.$$

By Lemma 1, iii), we have $t = 0$; that is, $\alpha = \beta$. Again working modulo 4, we arrive at the vector

$$(0,2s|0,2\alpha|2s,0|0,0|2,0|0,0|2,0) = (C^5A^2 - E)\cdot x(\alpha,\alpha,\gamma) \in \Lambda'.$$

Subtracting from this the vector

$$(0,2s|0,2|2s,2s|0,0|2,0|0,0|2,0) \in 2\Lambda_2,$$

we obtain that

$$2(\alpha - 1)v_2 + 2sv_3 \in \Lambda'.$$

This vector can lie in Λ_2 only with $\alpha = 1$ and $s = 0$; that is, $\alpha = \beta = \gamma = 1$. We have shown that the lattice Λ' necessarily contains the vector $e = x(1,1,1)$ and that any other vector of Λ' which is not in $2\Lambda_2$ is, up to the action of the automorphism C, congruent to it modulo $2\Lambda_2$. This proves the lemma.

It remains to study the sublattices contained in $\tilde{\Lambda}$. However, there is no special necessity for this, since the lattices Λ and $\tilde{\Lambda}$ are in some sense similarly planned. Namely, there is an isomorphism φ between them inducing the isomorphism of the sublattices Λ_k and $\tilde{\Lambda}_k$. The map φ is given in the obvious way:

$$\varphi(u_i) = v_i, \quad \varphi(v_i) = 3u_i; \qquad \widetilde{(x|y)} = \tfrac{1}{3}(x|y).$$

Here obviously $\tilde{\tilde{\Lambda}} = 3\Lambda$, so that a repeated application of this process is not needed. However, we remark that the G-invariance of the sublattices $\tilde{\Lambda}_k$ for $1 \leqslant k \leqslant 4$ does not follow automatically from the isomorphism mentioned above.

By Lemma 13, any vector $y \neq 0$ of the lattice Λ_0 has the form

$$y = \varepsilon e + 2x,$$

where $\varepsilon \in \{0,1\}$ and $x \in \Lambda_2$.

If $\varepsilon = 0$, then the vector y belongs to $2\Lambda_2$ and has squared length $|y|^2 \geqslant 16$, since the minimal length of vectors of Λ_2 is 2; it is spanned by these (see the definition of

Λ_2). However, if $\varepsilon = 1$, then at least six pairs of the coordinates at u_k, v_k of the vector y are nonzero even modulo 2 (see the form of the vector e given in Lemma 13). But every nonzero vector $n_k u_k + m_k v_k \in \Lambda^k$ has by Lemma 1 of §1 a squared length 2, 6, or more. Hence $|y|^2 \geqslant 12$, and if $|y|^2 \leqslant 16$, then at most one pair (n_k, m_k) can there correspond a vector with squared length 6.

All the vectors of minimal length in Λ_0 (and the square of this turns out to be exactly 16) can be found using the following search procedure, which can be carried out either by computer or by hand. It is necessary to search through all the vectors y in the lattice Λ with $12 \leqslant |y|^2 \leqslant 16$ and such that $y - e \in 2\Lambda$. A priori, there are at most 4^7 of them, since for every pair (n_k, m_k) of coordinates of the vector y there are at most 4 possibilities. Indeed, by Lemma 1 of §1, only for one of the seven pairs are there 4 possibilities (2 for squared length 2, and 2 for squared length 6), and for the remaining pairs there remain only 2 possibilities each (the squared length is 2), and so in all there are at most $7 \cdot 4 \cdot 2^6$ possibilities. From the indicated vectors we need to select those for which $y - e \in 2\Lambda_2$. For this, we can use a simple criterion for the vector

$$x = \sum (n_k u_k + m_k v_k)$$

to belong to the lattice Λ_2:

$$m_i + n_{i+2} + m_{i+3} + n_{i+3} + n_{i+4} + m_{i+4} + m_{i+5} + n_{i+6} \equiv 0 \quad (2)$$

(it suffices to check for $i = 1, 2, 3$).

Thus in the lattice Λ_0 we obtain the large multitude M of 756 vectors of minimal length 4. Table 4 lists 54 vectors. The rest are obtained from these by negating and applying the automorphism C.

Gaussian elimination, applied to the lattice $\Lambda'_0 \subseteq \Lambda_0$ spanned by M, leads to the basis for Λ'_0 of 14 vectors $e_1, \dots, e_{14}$ of M which are listed in Table 5. It is quite easy to verify that Ae_i, $Be_i \in M$ (as usual, A and B are generators of the group G). Hence Λ'_0 is a G-invariant lattice coinciding with Λ_0. Parenthetically, we remark that only after this argument does it become clear that Λ_0 is a G-invariant lattice.

Our results concerning the exceptional lattice Λ_0 are summarized in the following theorem.

THEOREM 2. *The only G-invariant sublattice in Λ which does not contain 2Λ but properly contains 4Λ is the lattice Λ_0 with basis $\{e_1, \dots, e_{14}\}$ given in Table* 5. *All the vectors of Λ_0 of the minimal length*, 4, *form the set M of cardinality* 756, 54 *of whose vectors are listed in Table* 4; *the rest are obtained by negating and by applying the automorphism C. The basis vectors e_i, $1 \leqslant i \leqslant 14$, belong to M, so that the group* $\mathrm{Aut}(\Lambda_0)$ *of automorphisms of the lattice Λ_0 is realizable by permutations of the set M. The Gram matrix $(e_i | e_j)$ is given in Table* 6.

§3. Automorphisms

THEOREM 3. *Suppose that G, Λ and Λ_k, $1 \leqslant k \leqslant 4$ have the same meaning as before. Then the following assertions are true*:

i) $\mathrm{Aut}(\Lambda) \cong D_6 \wr S_7$, *obtained by wreathing the dihedral group D_6 of order* 12 *by the symmetric group S_7.*

ii) $\mathrm{Aut}(\Lambda_k) \subset \mathrm{Aut}(\Lambda)$, *and, moreover*, $|\mathrm{Aut}(\Lambda_k)\colon G| = Z^4$.

This theorem shows that the lattices Λ and Λ_k, $1 \leqslant k \leqslant 4$, are of little interest from the point of view of their automorphism groups. The only lattice rich in automorphism turns out to be Λ_0.

TABLE 4

0	0	2	1	1	1	1	1	1	0	0	−1	1	0
0	0	2	1	1	1	1	1	1	0	−2	−1	−3	−2
0	0	2	1	1	1	1	1	−3	−2	−2	−1	−1	0
0	0	2	1	1	1	1	1	−1	0	0	1	1	0
0	0	2	1	1	1	1	1	3	2	2	1	−1	0
0	0	2	1	1	1	1	1	−1	0	2	1	−3	−2
0	0	2	1	1	1	3	1	1	0	−2	−1	−1	0
0	0	2	1	1	1	3	1	−1	0	2	1	−1	0
0	0	2	1	1	1	−3	−1	1	0	2	1	1	0
0	0	2	1	1	1	−3	−1	−1	0	−2	−1	1	0
0	0	2	1	1	1	−1	−1	1	0	0	1	−1	0
0	0	2	1	1	1	−1	−1	1	0	2	1	3	2
0	0	2	1	1	1	−1	−1	−1	0	0	−1	−1	0
0	0	2	1	1	1	−1	−1	−1	0	−2	−1	3	2
0	0	2	1	1	1	−1	−1	−3	−2	2	1	1	0
0	0	2	1	1	1	−1	−1	3	2	−2	−1	1	0
0	0	2	1	−1	−1	1	1	1	0	0	−1	−1	0
0	0	2	1	−1	−1	1	1	1	0	−2	−1	3	2
0	0	2	1	−1	−1	1	1	−1	0	2	1	3	2
0	0	2	1	−1	−1	1	1	−1	0	0	1	−1	0
0	0	2	1	−1	−1	1	1	−3	−2	−2	−1	1	0
0	0	2	1	−1	−1	1	1	3	2	2	1	1	0
0	0	2	1	−1	−1	−1	−1	1	0	0	1	1	0
0	0	2	1	−1	−1	−1	−1	1	0	2	1	−3	−2
0	0	2	1	−1	−1	−1	−1	−1	0	0	−1	1	0
0	0	2	1	−1	−1	−1	−1	−1	0	−2	−1	−3	−2
0	0	2	1	−1	−1	−1	−1	−3	−2	2	1	−1	0
0	0	2	1	−1	−1	−1	−1	3	2	−2	−1	−1	0
0	0	2	1	−1	−1	−3	−1	1	0	2	1	−1	0
0	0	2	1	−1	−1	−3	−1	−1	0	−2	−1	−1	0
0	0	2	1	−1	−1	3	1	1	0	−2	−1	1	0
0	0	2	1	−1	−1	3	1	−1	0	2	1	1	0
0	0	2	1	−3	−1	1	1	1	0	2	1	−1	0
0	0	2	1	−3	−1	1	1	−1	0	−2	−1	−1	0
0	0	2	1	−3	−1	−1	−1	1	0	−2	−1	1	0
0	0	2	1	−3	−1	−1	−1	−1	0	2	1	1	0
0	0	2	1	3	1	1	1	1	0	2	1	1	0
0	0	2	1	3	1	1	1	−1	0	−2	−1	1	0
0	0	2	1	3	1	−1	−1	1	0	−2	−1	−1	0
0	0	2	1	3	1	−1	−1	−1	0	2	1	−1	0
0	0	0	1	1	1	1	1	1	0	2	1	−1	0
0	0	0	1	1	1	1	1	−1	0	−2	−1	−1	0
0	0	0	1	1	1	−1	−1	1	0	−2	−1	1	0
0	0	0	1	1	1	−1	−1	−1	0	2	1	1	0
0	0	0	1	−1	−1	1	1	1	0	2	1	1	0
0	0	0	1	−1	−1	1	1	−1	0	−2	−1	1	0
0	0	0	1	−1	−1	−1	−1	1	0	−2	−1	−1	0
0	0	0	1	−1	−1	−1	−1	−1	0	2	1	−1	0
2	0	4	2	0	0	0	0	0	0	0	0	0	0
−2	0	4	2	0	0	0	0	0	0	0	0	0	0
4	2	0	0	2	2	0	0	0	0	0	0	0	0
−4	−2	0	0	2	2	0	0	0	0	0	0	0	0
2	0	0	0	0	0	2	2	0	0	0	0	0	0
−2	0	0	0	0	0	2	2	0	0	0	0	0	0

TABLE 5

e_1	e_2	e_3	e_4	e_5	e_6	e_7	e_8	e_9	e_{10}	e_{11}	e_{12}	e_{13}	e_{14}
0	1	0	2	−2	4	0	0	0	0	0	0	0	0
0	0	−1	0	0	2	0	0	0	0	0	0	0	0
2	0	1	4	4	0	2	−2	4	0	0	0	0	0
1	0	0	2	2	0	0	0	2	0	0	0	0	0
1	2	0	0	0	2	4	4	0	2	−2	4	0	0
1	1	0	0	0	2	2	2	0	0	0	2	0	0
1	1	2	0	0	0	0	0	2	4	4	0	2	4
1	1	1	0	0	0	0	0	2	2	2	0	0	2
1	1	1	0	0	0	0	0	0	0	0	2	4	0
0	1	1	0	0	0	0	0	0	0	0	2	2	0
0	1	1	0	0	0	0	0	0	0	0	0	0	2
−1	0	1	0	0	0	0	0	0	0	0	0	0	2
1	0	1	0	0	0	0	0	0	0	0	0	0	0
0	−1	0	0	0	0	0	0	0	0	0	0	0	0

Note. $e_1 = 2u_2 + v_2 + u_3 + v_3 + u_4 + v_4 + u_5 - v_6 + u_7$.

TABLE 6

$$(e_i \mid e_j) = \begin{bmatrix} 16 & 8 & 0 & 4 & 4 & 4 & 4 & 0 & 8 & 0 & 4 & 0 & 0 & -4 \\ 8 & 16 & 8 & 4 & -4 & 4 & 4 & 4 & 4 & 4 & 0 & 8 & 0 & 0 \\ 0 & 8 & 16 & 8 & -4 & 0 & 4 & -4 & 4 & 4 & 4 & 4 & 4 & 8 \\ 4 & 4 & 8 & 16 & 0 & 4 & 4 & -4 & 8 & 0 & 0 & 0 & 0 & 0 \\ 4 & -4 & -4 & 0 & 16 & -4 & 4 & -4 & 8 & 0 & 0 & 0 & 0 & 0 \\ 4 & 4 & 0 & 4 & -4 & 16 & 4 & 4 & 0 & -4 & 4 & 4 & 0 & 0 \\ 4 & 4 & 4 & 4 & 4 & 4 & 16 & 0 & 4 & 4 & -4 & 8 & 0 & 0 \\ 0 & 4 & -4 & -4 & -4 & 4 & 0 & 16 & -4 & 4 & -4 & 8 & 0 & 0 \\ 8 & 4 & 4 & 8 & 8 & 0 & 4 & -4 & 16 & 4 & 4 & 0 & -4 & 4 \\ 0 & 4 & 4 & 0 & 0 & -4 & 4 & 4 & 4 & 16 & 0 & 4 & 4 & 8 \\ 4 & 0 & 4 & 0 & 0 & 4 & -4 & -4 & 4 & 0 & 16 & -4 & 4 & 8 \\ 0 & 8 & 4 & 0 & 0 & 4 & 8 & 8 & 0 & 4 & -4 & 16 & 4 & 0 \\ 0 & 0 & 4 & 0 & 0 & 0 & 0 & 0 & -4 & 4 & 4 & 4 & 16 & 4 \\ -4 & 0 & 8 & 0 & 0 & 0 & 0 & 0 & 4 & 8 & 8 & 0 & 4 & 16 \end{bmatrix}$$

THEOREM 4. *The automorphism group of the lattice* Λ_0 *is the direct product*

$$\mathrm{Aut}(\Lambda_0) = \langle T \rangle \times \Gamma,$$

where $T\colon x \to -x$ *is the central automorphism of order* 2, *and* Γ *is a simple group of order* $2^6 \cdot 3^6 \cdot 7 \cdot 13$ *isomorphic to* $G_2(3)$.

The proofs of Theorems 3 and 4 are unfortunately half of the computer type; that is, they are based on routine (for computers) calculations that are not easily adaptable to hand calculation. So we restrict ourselves to a mere outline of the ideas used.

Statement i) of Theorem 3 is very easy to prove. Since the 2-dimensional lattice Λ^k is spanned by the vectors u_k and $u_k + v_k$ of minimal length, for $\Phi \in \mathrm{Aut}(\Lambda)$ the images $\Phi(u_k)$ and $\Phi(u_k + v_k)$, each of squared length 2, must lie in Λ^s and Λ^t respectively, where $s = t$ since $(u_k | u_k + v_k) = -1 \neq 0$. Hence $\Phi(\Lambda^i) = \Lambda^{\pi(i)}$ for

TABLE 7

$\Gamma = \langle A, B, D\rangle$

$$[A]=\begin{bmatrix}
1 & 0 & 1 & 0 & 0 & 0 & -2 & -2 & 2 & 0 & 4 & -2 & 0 & 2\\
0 & 1 & 0 & 0 & 0 & 0 & 2 & 2 & 0 & 2 & -2 & 2 & 0 & 0\\
0 & -1 & -1 & 0 & 0 & 2 & 0 & 0 & -2 & -4 & 0 & 0 & 0 & -2\\
0 & 0 & 0 & -1 & 0 & -1 & -1 & 0 & 1 & 2 & 1 & -1 & 0 & 2\\
0 & 0 & 0 & 0 & -1 & 1 & 0 & 1 & -1 & -1 & 0 & 0 & 0 & 0\\
0 & 0 & 0 & 1 & -1 & 2 & 0 & 0 & -1 & -2 & 0 & 0 & 0 & -1\\
0 & 0 & -1 & 0 & 1 & -1 & 1 & 1 & 0 & 1 & -2 & 1 & 0 & -1\\
-1 & -1 & -1 & 0 & 1 & 0 & 1 & 1 & -1 & -1 & -2 & 1 & 0 & -1\\
-1 & 0 & 0 & 1 & 1 & 0 & 2 & 0 & -1 & -1 & -3 & 2 & 0 & -2\\
0 & 0 & 0 & 0 & -1 & 1 & -1 & 0 & -1 & -1 & 1 & -1 & 0 & 0\\
-1 & -1 & -1 & -1 & 0 & -1 & 0 & 1 & -1 & 1 & -1 & 0 & 0 & 0\\
0 & 0 & 1 & -1 & -1 & -1 & -2 & -2 & 1 & 1 & 3 & -2 & 0 & 2\\
0 & 0 & 0 & 1 & 1 & 0 & 2 & 0 & 0 & 0 & -2 & -1 & 0 & -1\\
1 & 1 & 1 & 0 & 0 & -1 & -1 & -1 & 3 & 2 & 2 & -1 & -1 & 2
\end{bmatrix}$$

$$[B]=\begin{bmatrix}
0 & -2 & -1 & 0 & 0 & -2 & -2 & -2 & 0 & 0 & 0 & -4 & 0 & 0\\
1 & 1 & 0 & 0 & 0 & 2 & 2 & 2 & 0 & 0 & 0 & 2 & 0 & 0\\
-1 & 0 & 1 & 0 & 0 & 0 & -2 & -2 & 0 & -2 & 2 & 0 & 0 & 0\\
0 & -1 & -1 & 0 & 0 & -1 & 0 & 1 & 0 & 1 & -1 & -1 & 0 & 0\\
0 & 0 & 0 & 0 & 0 & 0 & -1 & 0 & 0 & -1 & 1 & 0 & 0 & 0\\
-1 & 0 & 1 & 0 & 0 & 0 & -1 & -1 & 0 & -1 & 1 & 0 & 0 & 1\\
1 & 1 & 0 & 0 & 1 & 1 & 2 & 1 & 1 & 1 & 0 & 2 & 0 & 0\\
0 & 1 & 0 & -1 & 0 & 1 & 0 & 1 & 0 & 0 & 1 & 2 & 0 & 0\\
0 & 2 & 1 & 0 & 0 & 2 & 1 & 1 & 0 & 0 & 0 & 3 & 0 & 0\\
-1 & -1 & 0 & 0 & 0 & -1 & -1 & -1 & 0 & -1 & 0 & -1 & -1 & 0\\
0 & 0 & 0 & 0 & 0 & 0 & 1 & 1 & 0 & 1 & 0 & 1 & 0 & 0\\
0 & -1 & 0 & 1 & -1 & -2 & -1 & -1 & 0 & 1 & -1 & -3 & 0 & 0\\
0 & 1 & 0 & 0 & 0 & 2 & 1 & 1 & 0 & 0 & 0 & 2 & 0 & 0\\
1 & -1 & -1 & 0 & 0 & -1 & 0 & 0 & 0 & 1 & -1 & -2 & 1 & 0
\end{bmatrix}$$

$$[D]=\begin{bmatrix}
1 & -1 & -1 & 1 & 1 & -1 & 0 & -2 & 2 & 1 & 0 & -1 & 0 & -1\\
0 & 0 & 0 & -2 & 0 & 0 & 0 & 1 & -1 & 0 & 1 & 0 & 0 & 1\\
0 & 2 & 0 & 2 & -1 & 1 & 0 & 2 & 0 & 0 & -2 & 1 & -1 & -1\\
0 & -2 & -1 & -1 & 1 & -1 & -1 & -1 & 0 & 0 & 1 & -1 & 1 & 0\\
0 & 0 & -1 & 0 & 0 & 0 & 0 & 1 & -1 & 0 & -1 & 0 & 0 & -1\\
0 & 1 & 0 & 1 & 0 & 0 & 1 & 1 & 0 & 0 & -2 & 1 & -1 & -1\\
0 & 1 & 1 & -1 & -1 & 1 & 0 & 1 & -1 & 0 & 1 & 1 & 1 & 1\\
0 & 1 & 0 & 0 & -1 & 2 & 0 & 2 & -1 & -1 & 0 & 1 & 0 & 0\\
0 & 2 & 2 & 1 & -1 & 2 & 1 & 1 & 0 & -1 & 0 & 1 & -1 & 1\\
0 & 0 & 0 & 1 & 0 & -1 & 0 & 0 & 0 & 0 & -1 & 0 & 0 & -1\\
0 & 0 & 0 & -1 & 0 & 0 & -1 & 0 & -1 & -1 & 1 & 0 & 1 & 0\\
0 & -2 & -1 & 0 & 1 & -2 & -1 & -3 & 1 & 0 & 0 & -2 & 0 & -1\\
0 & 1 & 1 & 0 & 0 & 1 & 1 & 1 & 0 & 0 & 0 & 1 & 0 & 1\\
0 & -2 & -1 & -1 & 1 & -1 & 0 & -2 & 1 & 1 & 1 & -1 & 0 & 0
\end{bmatrix}$$

Note. $[A]$, $[B]$, $[D]$ are the matrices of the automorphisms A, B, D of the lattice in the basis $\{e_1, \ldots, e_{14}\}$.

some permutation $\pi \in S_7$, and we can take any permutation of S_7 for π. Since the subgroup corresponding to the identity permutation is isomorphic to

$$\underbrace{D_6 \times D_6 \times \cdots \times D_6}_{7}$$

(see the remarks after Lemma 1 in §1), we conclude that $\operatorname{Aut}(\Lambda) \cong D_6 \wr S_7$. ∎

ii) We start with the lattice Λ_4. Formula (3) of §2 and the ensuing arguments ensure that any vector $x \in \Lambda_4$ which does not lie in 2Λ has squared length

$|x|^2 \geqslant 12$. Hence the vectors $2u_i$ and $2(u_i + v_i)$, with squared length 8, under any $\Phi \in \operatorname{Aut}\Lambda_4$ must go into 2Λ; that is, $\Phi(2\Lambda) \subset 2\Lambda$. Therefore Φ extends to an automorphism of the lattice Λ, so that $\operatorname{Aut}(\Lambda_4) \subset \operatorname{Aut}(\Lambda)$.

For the groups $\operatorname{Aut}(\Lambda_k)$, $k = 1, 2, 3$, one fails to draw the analogous conclusions immediately from the properties of Λ_k. This, as well as the computation of the orders $|\operatorname{Aut}(\Lambda_k)|$, $1 \leqslant k \leqslant 4$, was carried out by computer.

Proceeding to the proof of Theorem 4, we note first of all that the group $\operatorname{Aut}(\Lambda_0)$ acts transitively on the set M of 756 vectors of length 4 mentioned in Theorem 2. The stabilizer of a point is isomorphic to Aut(SL(3, 3)). The group $\operatorname{Aut}(\Lambda_0)$ also acts transitively and faithfullly on the set of $52 \cdot 756$ ordered pairs (P, Q), where $P, Q \in M$ and $(P|Q) = 8$. The stabilizer of such a pair has order 216. Hence $|\operatorname{Aut}(\Lambda_0)| = 2^7 \cdot 3^6 \cdot 7 \cdot 13$. The cyclic group $\langle T \rangle$ is a direct summand of $\operatorname{Aut}(\Lambda_0)$ since the Sylow 2-subgroup can be represented as $\langle T \rangle \times G_2$, where G_2 is a Sylow 2-subgroup of $G = \operatorname{Aut}_{\mathrm{MOD}}(G_2)$ (see §1).

The identification of the group $\Gamma \subset \operatorname{Aut}(\Lambda_0)$ of order $2^6 \cdot 3^6 \cdot 7 \cdot 13$ with $G_2(3)$ can be achieved in two ways. First, the transitive representations mentioned above and the explicit description of the corresponding stabilizers enable us to prove the simplicity of Γ. Hence by the equality of the orders $|\Gamma|$ and $|G_2(3)|$ it follows that $\Gamma = G_2(3)$.

As an intermediate step we wrote down by computer the most general automorphisms both as permutations and as matrices. In particular it can easily be verified that

$$\Gamma = \langle A, B, D \rangle.$$

With respect to the basis $\{e_1, \dots, e_{14}\}$ the automorphisms A, B and D have the form given in Table 7. We recall that $\langle A, B \rangle = G$. The automorphism D of order 13 is taken from the stabilizer of the point (vector) e_1.

The other method of finding the isomorphism $\Gamma \cong G_2(3)$ is based on the description of the centraliser of an involution in Γ and the use of Janko's theorem [11].

REMARK. A faithful integral representation of degree 14 for $G_2(3)$ is apparently indicated here for the first time, even though the existence of such a representation is of course not new. It suffices, as in [8], to use the ideas of Conway when calculating the Schur index over Q of the complex irreducible character θ_{10} of the group $G_2(3)$, $\theta_{10}(1) = 14$, whose values are given in [12].

BIBLIOGRAPHY

1. J. W. S. Cassels, *Rational quadratic forms*, Academic Press, 1978.
2. Jean-Pierre Serre, *Cours d'arithmétique*, Presses Univ. France, Paris, 1970.
3. J. H. Conway, *A group of order* 8,315,553,613,086,720,000, Bull. London Math. Soc. **1** (1969), 79–88.
4. ______, *A characterization of Leech's lattice*, Invent. Math. **7** (1969), 137–142.
5. Robert L. Griess, Jr., *A construction of F_1 as automorphisms of a* 196,883-*dimensional algebra*, Proc. Nat. Acad. Sci. U.S.A. **78** (1981), 686–691.
6. A. I. Kostrikin, I. A. Kostrikin and V. A. Ufnarovskiĭ, *Orthogonal decompositions of simple Lie algebras*, Dokl. Akad. Nauk SSSR **260** (1981), 526–530; English transl. in Soviet Math. Dokl. **24** (1981).
7. ______, *Multiplicative decompositions of simple Lie algebras*, Dokl. Akad. Nauk SSSR **262** (1982), 29–33; English transl. in Soviet Math. Dokl. **25** (1982).

8. P. E. Smith, Ph.D. Thesis, Cambridge, England, 1975.
9. J. G. Thompson, *A conjugacy theorem for* E_8, J. Algebra **38** (1976), 525–530.
10. Günter Sandlöbes, *Perfect groups of order less than* 10^4, Comm. Algebra **9** (1981), 477–490.
11. Zvonimir Janko, *A characterization of the simple group* $G_2(3)$, J. Algebra **12** (1969), 360–371.
12. Hikoe Enomoto, *The characters of the finite Chevalley group* $G_2(q)$, $q = 3^f$, Japan J. Math. (N.S.) **2** (1976), 191–248.

Translated by L. QUEEN

Труды Матем. Инст. Стеклов.
165(1984)

Proceedings of the Steklov
Institute of Mathematics
1985, Issue 3

NEW EXACT SOLUTIONS AND COHOMOLOGY ANALYSIS OF ORDINARY AND SUPERSYMMETRIC YANG-MILLS EQUATIONS

UDC 530.145

YU. I. MANIN

ABSTRACT. A method is developed of constructing locally free sheaves on incidence quadrics in the projective superspace and the dual space, which yields a new class of solutions of the standard and supersymmetric Yang-Mills equations. It is shown that the set of cohomology invariants of such sheaves corresponds to field components.

Bibliography: 15 titles.

§1. Introduction

The construction of self-dual solutions of the Yang-Mills equations based on the Penrose-Ward transformation reduces to the construction of holomorphic vector bundles on three-dimensional projective space P^3 or its domains (see [1] and [2]). The requirement that the solutions of instanton and monopole types should be real together with the boundary or asymptotic conditions render the resultant geometric problem so stringent that the corresponding class of bundles can be described quite explicitly (see [3] and [4]).

A generalization of the Penrose transformation to the complete system of Yang-Mills (source-free) equations was described in the remarkable papers of Witten [5] and Isenberg, Yasskin and Green [6]. To obtain holomorphic solutions, it is necessary to construct bundles on an analytic five-dimensional space of null-lines, i.e., third-order infinitesimal neighborhood $L^{(3)}$ of the incidence quadric $L \subset P^3 \times \hat{P}^3$, or on its domains covered with two-dimensional quadrics. It was subsequently noted by Khenkin and Manin (see [7]–[10]) that a systematic study of extensions of bundles and cohomology classes from L to $L^{(3)}$ leads to other physically important systems of equations, in particular, to Yang-Mills-Dirac and Higgs equations.

However, such a reduction has not been exploited, since the construction of bundles on $L^{(3)}$ with the required properties was not available. There is a natural construction (i.e., it is possible to combine self-dual and anti-self-dual bundles by means of direct sums and direct products) which leads to bundles over all of $P^3 \times \hat{P}^3$.

1980 *Mathematics Subject Classification*. Primary 81E10; Secondary 14F05.

V. Gal′perin showed in 1981 (work carried out at the Faculty of Mechanics and Mathematics, Moscow State University) that the exterior direct sum of a complex instanton with a complex anti-instanton admits nontrivial deformations on L. It is true that such deformations do not extend to $L^{(1)}$.

After a summary of basic concepts in §2, we shall show in §3 that a modification of Gal′perin's idea can be used to construct nontrivial bundles of Witten-Isenberg-Yasskin-Green type on arbitrary infinitesimal neighborhoods L^m. In particular, such a construction leads to a new class of exact solutions of the vacuum Yang-Mills equations for $m \geqslant 3$.

In the second part of his paper [5], Witten showed that certain bundles on $L^{5|2N} \subset P^{3|N} \times \hat{P}^{3|N}$ correspond for $N = 3$ to the solutions of the supersymmetric system of Yang-Mills equations, where $P^{3|N}$ is a projective superspace of dimension $3|N$. Transition to infinitesimal neighborhoods is not required since they are replaced by anticommuting coordinates. Our construction described in §3 is applicable to this case without any modifications, and we thus obtain exact solutions of the supersymmetric Yang-Mills equations. More details concerning these solutions are given in §4.

We aim in this paper to describe examples of new bundles. However, this does not exhaust all the possibilities of our method; it is clear that m-complexes of arbitrary length could be considered instead of m-monads; supermonopoles and deformations of monopole-antimonopole combinations are of interest; and there are many other obvious applications.

The aim of §5 is to reproduce the component structure of supersymmetric Yang-Mills fields and equations in the cohomology language directly in terms of the superspace $L^{5|6}$. We shall show that the formalism of the theory of extensions and obstructions which was applied in [7]–[10] to discuss the standard field equations by the twistor method is just as effective in the case of supergeometry.

The present paper can be regarded as the second part of [11], where, among other things, all the required elements of the theory of analytic superspaces were explained.

§2. Yang-Mills sheaves and Yang-Mills fields

1. Principal spaces. We operate with analytic superspaces over C and consider analytic coherent sheaves on them. The dimension of an irreducible supermanifold is denoted by $p|q$ if it admits an open covering by superdomains in $C^{p|q}$ with p even and q odd coordinates. We use an analogous notation to denote the rank of a free supermodule, etc. We shall denote by T a fixed linear space (of supertwistors) of dimension $4|N$, $N \geqslant 0$; T^* is the corresponding dual superspace. These spaces are then used to construct a number of flag superspaces [11]. The following double fibration plays central role in our construction of the supertwistor Radon-Penrose transformation:

$$L = F(1|0,3|N;T) \overset{\pi_1}{\leftarrow} F = F(1|0,2|0,2|N,3|N;T) \overset{\pi_2}{\to} M$$
$$= F(2|0,2|N;T)$$

(of type A_N in the notation of [11]). Here, M is a compact complex model of the Minkowski superspace of dimension $4|4N$; L is the superspace of complex light-like lines in M (each line has dimension $1|2N$); and F is a space of incidence pairs connecting L and M. More detailed description of the present fibration and similar double fibrations can be found in [11].

We further set $P = F(1|0; T)$ and $\hat{P} = F(3|N; T) = F(1|0; T^*)$. The projections $\rho\colon L \to P$ and $\hat{\rho}\colon L \to \hat{P}$ induce a closed embedding $L \subset P \times \hat{P}$. We denote by $L^{(m)}$ the mth order infinitesimal neighborhood of L corresponding to this embedding which is determined by a sheaf of ideals I^{m+1}, where I is the sheaf of ideals of L. Since there are tautological sheaves $\mathcal{O}(1)$ on P and $\hat{P}$, we can define sheaves $\mathcal{O}(a, b) = \rho^*(\mathcal{O}_p(a)) \otimes \hat{\rho}^*(\mathcal{O}_{\hat{p}}(b))$ on $P \times \hat{P}$ and $L^{(m)}$. The space L in $P \times \hat{P}$ is determined by the equation $s = 0$, where $s \in H^0(P \times \hat{P}, \mathcal{O}(1, 1)) = T^* \otimes T$ is the standard section corresponding to id_T. When N is required in the notation, we can write L_N, F_N, etc.

To each point $x \in M$, we assign a quadric $L(x) = \pi_1\pi_2^{-1}(x) = P^1(x) \times \hat{P}^1(x) \subset L$, where $P^1(x) \subset P$ and $\hat{P}^1(x) \subset \hat{P}$. To domains $U \subset M$ there correspond domains $L(U) \subset L$, $P(U) \subset P$, and $\hat{P}(U) \subset \hat{P}$ covered by such quadrics and lines. We regard such domains as open supermanifolds in L, P, and $\hat{P}$; the same applies to $L(U)^{(m)}$.

2. Yang-Mills sheaves and Yang-Mills files. By a *Yang-Mills sheaf* we mean a pair $(L_0, \mathcal{E}_L)$, where $L_0 \subset L$ is an open subspace containing $L(U)$ for some nonempty $U \subset M$ and $\mathcal{E}_L$ is a locally free sheaf on L_0 whose restriction to an arbitrary quadric $L(x)$, $x \in U$, is free, i.e., isomorphic to $\mathcal{O}^{p|q}_{L(x)} = \mathcal{O}^p_{L(x)} \oplus \Pi\mathcal{L}^q_{L(x)}$ (Π is the functor of change of parity). We shall call the last condition *U-triviality* of the sheaf $\mathcal{E}_L$.

The pairs $(\mathcal{E}, \nabla)$ are said to be *Yang-Mills fields* on U, where $\mathcal{E}$ is a locally free sheaf on U and $\nabla\colon \mathcal{E} \to \mathcal{E} \otimes \Omega^1 U$ is a superanalytic connection defined by its covariant derivative.

For $N = 0$, the equation $\nabla(\Phi_{\pm}) = 0$ is said to be the *Yang-Mills (source-free) equation*, where $\Phi \in H^0(U, \operatorname{End}\mathcal{E} \otimes \Omega^2 U)$ is the curvature form of ∇ and $\Phi_{\pm}$ are its self-dual and anti-self-dual parts. They can be defined in terms of the spinor decomposition $\Omega^1 M = \mathcal{S} \otimes \mathcal{S}_-$, where $\mathcal{S}_{\pm}$ are tautological sheaves on $M = G(2; T^{4|0})$, $\mathcal{S}_+ \subset \mathcal{O}_M \otimes T$, $\mathcal{S}_- = \mathcal{S}_+^{\perp} \subset \mathcal{O}_M \otimes T^*$ and $\Phi_{\pm} \in \operatorname{End}\mathcal{E} \otimes S^2\mathcal{S}_{\pm} \otimes \Lambda^2\mathcal{S}_{\mp}$.

For $N = 3$, by the *supersymmetric Yang-Mills equation* we mean the condition of integrability of $(\mathcal{E}, \nabla)$ along complex light-like lines (more precisely, the integrability of $(\pi_2^*(\mathcal{E}), \nabla_{F/L})$ along fibers of the projection $\pi_1\colon F \to L$). Integrability conditions can be naturally imposed for arbitrary N, but they represent a complete system of equations which is physically meaningful [5] only for $N = 3$. For $N = 0$, they are satisfied automatically.

To any U-trivial Yang-Mills sheaf $\mathcal{E}_L$, we can assign a Yang-Mills field $(\mathcal{E}, \nabla)$ on U (the Penrose-Ward-Witten-Yasskin-Isenberg-Green transformation of $\mathcal{E}_L$). In fact, we have $\mathcal{E} = \pi_{2*}\pi_1^*\mathcal{E}_L|U$ and $\nabla = \pi_{2*}(\nabla_{F/L})$, where $\nabla_{F/L}\colon \pi_1^*\mathcal{E}_L \to \pi_1^*\mathcal{E}_L \otimes \Omega^1 F/L$ is a relative connection defined by $\operatorname{Ker}\nabla_{F/L} = \pi^{-1}(\mathcal{E}_L)$. By definition, the field $(\mathcal{E}, \nabla)$ is integrable along the fibers of π_1 and, therefore, it is a solution of the supersymmetric Yang-Mills equation for $N = 3$. For $N = 0$ and a connected U, the above construction yields an arbitrary field with a trivial monodromy along the intersection of light-like lines with U. Such a field satisfies the Yang-Mills equation provided $\mathscr{E}_L$ can be extended to $L(U)^{(3)}$ (see [5] and [6] and, for details, [7], [10], and [12]).

3. Self-dual and combined Yang-Mills sheaves. A Yang-Mills sheaf $(L_0, \mathscr{E}_L)$ is said to be *self-dual* if $L_0 = \rho^{-1}(P_0)$, where $P_0 \subset P$ is an open set containing some $P(U)$ and $\mathscr{E}_L = \rho^*(\mathscr{E}_P)$, where $\mathscr{E}_p$ is a locally free sheaf on P_0 whose restriction to an arbitrary line $P^1(x)$, $x \in U$, is free. Anti-self-dual sheaves $\hat{\rho}^*(\mathscr{E}_{\hat{P}})$ have analogous properties. We shall apply the term *U-triviality* also to $\mathscr{E}_P$ and $\mathscr{E}_{\hat{P}}$.

A Yang-Mills sheaf is said to be *combined* if it belongs to the tensor algebra generated by self-dual and anti-self-dual sheaves, i.e., it is isomorphic to a direct sum $\oplus_i \rho^*(\mathscr{E}_P^{(i)}) \otimes \hat{\rho}^*(\mathscr{E}_{\hat{P}}^{(i)})$. Combined sheaves can be clearly extended to $P(U) \times \hat{P}(U)$ and, therefore, the corresponding fields for $N = 0$ are also solutions of the standard Yang-Mills equations. We take the view that self-dual and anti-self-dual fields have been well (although not exhaustively) investigated and, therefore, combined sheaves are not independent objects. In fact, this is not quite true. In particular, self-dual solutions of instanton type for supersymmetric equations have only recently been considered by A. Semikhatov. Although their definition based on the monad theory is almost self-evident, their standard component analysis or discussion in the spirit of §5 has not been achieved.

Nevertheless, it is our aim to demonstrate that many combined sheaves admit deformations on $L(U)^{(m)}$ which do not reduce to deformations of their components, and we shall also give explicit construction of a series of such deformations. It is essential in the present problem that a deformation of a U-trivial sheaf remains U'-trivial (possibly for a smaller $U' \subset U$), since free sheaves on a quadric do not admit deformations.

If a Yang-Mills field $(\mathscr{E}, \nabla)$ corresponds to a Yang-Mills sheaf $\mathscr{E}_L$, most of the information about the field can be obtained directly from $\mathscr{E}_L$ without the Penrose transformation. We shall briefly discuss this assertion.

4. Strengths, structure group, and finiteness of the Euclidean action. a) The curvature $\Phi(x)$ of the connection ∇ at a point x corresponds to a cohomology class on $L(x)$ classifying $\mathscr{E}_L | L(x)^{(2)}$, where $L(x)^{(2)}$ is a second-order neighborhood of $L(x)$ in L. This applies strictly to $x \in U$, where U is the maximum domain of triviality of $\mathscr{E}_L$. Outside U, the Yang-Mills field has nonremovable singularities (although the sheaf $\mathscr{E}$ itself may remain locally free; see §4). For proof, see [7].

b) Let $S^4 \subset M_{\mathrm{red}} = G(2; \mathbf{C}^4)$ be a compact real form M. It is realized as the set of real points of the Grassmannian $G(2; T_0)$ relative to an antiholomorphic involution j induced by the quaternion structure on T_0. If the domain of U-triviality of the sheaf $\mathscr{E}_L$ contains S^4, its Euclidean action is finite. Any supersymmetric action is finite provided its coefficients are entire functions of the coefficients of the curvature form Φ.

c) In general, the action is complex since we consider sheaves $\mathscr{E}_L$ and $\mathscr{E}$ with a structure group GL. If there is real or quaternion structure on $\mathscr{E}_L$ extending $(j, \hat{j})$, then it is possible to reduce GL to a real subgroup. If there is also a nondegenerate holomorphic inner product on L, a reduction to one of the simple classical Lie groups is possible (we consider a restriction of $\mathscr{E}$ to a real subspace of M, i.e., S^4).

If a Yang-Mills sheaf which is U-trivial for $U \supset S^4$ and admits reduction to a compact simple structure group is self-dual (or anti-self-dual), then it is called an instanton (or an anti-instanton).

The cohomological properties of standard instantons are such that the exterior direct sum of an instanton and an anti-instanton admits a nontrivial deformation on L but such a deformation does not extend to $L^{(3)}$ and, therefore, it is not a vacuum Yang-Mills field (although it may admit an interesting physical interpretation). However, in the supersymmetric case, even a deformation on L yields a solution of the equations which describe specific interaction of an instanton with an anti-instanton in an internal space.

As in the case of Euclidean structures, we can consider a real structure on M corresponding to the standard Minkowski space and the associated real structure on $\mathcal{E}_L$ and $\mathcal{E}$. The monad constructions of §3 are suitable for investigation of real structures, but we shall not be concerned with them in this paper.

§3. Principal constructions

1. Extensions, deformations, and cohomology. Our first observation is that the existence of nontrivial deformations of combined sheaves on $L^{(m)}$ is related to the fact that certain cohomology groups of self-dual and anti-self-dual sheaves do not vanish. The group $H^1(L(U)^{(m)}, \rho^*\mathcal{E}_P \otimes \hat{\rho}^*\mathcal{E}_{\hat{P}})$ plays a prominent role (more standard notation for this group is $H^1(L(U), j_m^*(\rho^*\mathcal{E}_P \otimes \hat{\rho}^*\mathcal{E}_{\hat{P}}))$, where j_m: $L(U^{(m)}) \to P(U) \times \hat{P}(U)$ is a closed embedding). It appears in the following context.

a) The extensions $0 \to \rho^*\mathcal{E}_{\hat{P}} \to \mathcal{E}_L \to \rho^*\mathcal{E}_P^* \to 0$ can be classified by means of its elements. (In the supersymmetric case, the even elements of H^1 classify such extensions and the odd elements classify extensions of $\rho^*\Pi\mathcal{E}_P^*$ by $\hat{\rho}\mathcal{E}_{\hat{P}}$.) The classification of the sheaves $\mathcal{E}_L$ corresponding to different elements of H^1 modulo isomorphisms is clearly a separate problem; but, at least for $H^1 \neq 0$, sheaves $\mathcal{E}_L$ not isomorphic to $\rho^*\mathcal{E}_P^* \oplus \hat{\rho}^*\mathcal{E}_{\hat{P}}$ exist. They are U-trivial Yang-Mills sheaves for any U for which $\mathcal{E}_P$ and $\mathcal{E}_{\hat{P}}$ are U-trivial, since an arbitrary extension of free sheaves on $L(x) = P^1 \times \hat{P}^1$ is free.

b) The elements of the group $H^1(L^{(m)}(U), \rho^*\mathcal{E}_P^* \otimes \hat{\rho}^*\mathcal{E}_{\hat{P}} \oplus \rho^*\mathcal{E}_{\hat{P}} \otimes \hat{\rho}^*\mathcal{E}_P^*)$ classify infinitesimal first-order deformations of the direct sum $\rho^*\mathcal{E}_P \oplus \hat{\rho}^*\mathcal{E}_{\hat{P}}$ on $L^{(m)}(U)$ such that $\rho^*\mathcal{E}_P$ and $\rho^*\mathcal{E}_{\hat{P}}$ are not deformed. In fact, an infinitesimal deformation of a sheaf $\mathcal{F}$ is classified by the group $H^1(\mathcal{F}^* \otimes \mathcal{F})$ which has the property that its even cohomology classes correspond in the supersymmetric case to deformations with an even parameter and the odd classes correspond to deformations with an odd parameter. The problem of whether an infinitesimal deformation can be extended to a global deformation for an even parameter requires special investigation. As discussed above, mixed orientations of such deformations correspond to a superposition of the extension of $\rho^*\mathcal{E}_P$ by $\hat{\rho}^*\mathcal{E}_{\hat{P}}$ and of $\hat{\rho}^*\mathcal{E}_{\hat{P}}$ by $\rho^*\mathcal{E}_P$. Since the square of odd constants vanishes, there are no such problems in the corresponding extensions.

c) Infinitesimal deformations of the combined sheaf $\rho^*\mathcal{E}_P \otimes \hat{\rho}^*\mathcal{E}_{\hat{P}}$ are classified by 1-cohomologies of $\rho^* \operatorname{End} \mathcal{E}_P \otimes \hat{\rho}^* \operatorname{End} \mathcal{E}_{\hat{P}}$ (End, Hom, $\mathcal{F}^*$, etc. in the supersymmetric case refer to sheaves of interior morphisms which are Z_2-graded and coherent). The sheaves of endomorphisms of (anti-)self-dual sheaves belong themselves to the same class.

Our second observation is that the main contribution to the group $H^1(L(U)^{(m)}, \rho^*\mathscr{E}_P \otimes \hat\rho^*\mathscr{E}_{\hat P})$ is due to groups

$$H^1(P(U), \mathscr{E}_P(-m-1)) \quad \text{and} \quad H^1(\hat P(U), \mathscr{E}_{\hat P}(-m-1)).$$

This is the simplest situation.

2. PROPOSITION. *Assume that* $H^0(P(U), \mathscr{E}_P) = H^1(P(U), \mathscr{E}_P) = 0$ *and the similar relation holds for* $\mathscr{E}_{\hat P}$. *Then*

$$H^1\big(L(U)^{(m)}, \rho^*\mathscr{E}_P \otimes \hat\rho^*\mathscr{E}_{\hat P}\big) = H^1(P(U), \mathscr{E}_P(-m-1)) \otimes H^1(\hat P(U), \mathscr{E}_{\hat P}(-m-1)).$$

PROOF. We write on $P(U) \times \hat P(U)$ the following exact sequence:

$$0 \to \mathscr{O}_{P\times\hat P}(-m-1, -m-1) \overset{s^{m+1}}{\to} \mathscr{O}_{P\times\hat P} \to j_m^*\mathscr{O}_{L^{(m)}} \to 0$$

and take its tensor product with $\rho^*\mathscr{E}_P \otimes \hat\rho^*\mathscr{E}_{\hat P}$. The boundary differential δ: $H^1 \to H^2$ yields the required isomorphism using the Künneth formula. (For noncompact $P(U)$ and $\hat P(U)$, the corresponding tensor products are completed when working in the analytic category.)

We shall now quote some results about the cohomology of self-dual sheaves. We shall consider separately the two cases when U is either small or so large that $P(U) = P$.

3. PROPOSITION. *Let* $U \subset M$ *be a connected Stein manifold and assume that any nonempty intersection of* U *with a complex light-like line is connected and simply connected. Let also* $N = 0$, *let* $\mathscr{E}_P$ *be a U-trivial sheaf on* $P(U)$, *let* $(\mathscr{E}, \nabla)$ *be the corresponding Yang-Mills field on* U, *and let* $\mathscr{S}_\pm$ *tautological* (*spinor*) *bundles on* U. *Then there are the following isomorphisms*:

a) $H^0(P(U), \mathscr{E}_P) = H^0(U, \operatorname{Ker}\nabla)$. *Here* $H^1(P(U), \mathscr{E}_P)$ *is isomorphic to the* (*central*) *homology group of the complex*

$$H^0(U, \mathscr{E}) \overset{\nabla}{\to} H^0(U, \mathscr{E} \otimes \Omega^1 U) \overset{\nabla_+}{\to} H^0\big(U, \mathscr{E} \otimes S^2\mathscr{S}_+ \otimes \Lambda\mathscr{S}_-\big),$$

where ∇_+ *is the composite of* ∇ *and the projection* $\Omega^2 \to \Omega^2_+$.

b) $H^1(P(U), \mathscr{E}_P(-i)) = \operatorname{Ker} H^0(D_i)$. *Here the* D_i *are the following differential operators on* U:

$$D_1\colon \mathscr{E} \otimes \mathscr{S}_- \otimes \Lambda^2\mathscr{S}_+ \to \mathscr{E} \otimes \mathscr{S}_+ \otimes \big(\Lambda^2\mathscr{S}_+\big)^2 \otimes \Lambda^2\mathscr{S}_-$$

(*the Dirac operator on the background field* $(\mathscr{E}, \nabla)$);

$$D_2\colon \mathscr{E} \otimes \Lambda^2\mathscr{S}_+ \to \mathscr{E} \otimes \big(\Lambda^2\mathscr{S}_+\big)^2 \otimes \Lambda^2\mathscr{S}_-$$

(*the Klein-Gordon operator on the background field* $(\mathscr{E}, \nabla)$; *it is a second-order operator*); *and*

$$i \geqslant 3\colon \qquad D_3\colon \mathscr{E} \otimes S^{i-2}\mathscr{S}_+ \otimes \Lambda^2\mathscr{S}_+ \to \mathscr{E} \otimes S^{i-3}\mathscr{S}_+ \otimes \mathscr{S}_- \otimes \big(\Lambda^2\mathscr{S}_+\big)^2$$

(*the operator of massless fields of helicity* $(i-2)/2$ *on the background* $(\mathscr{E}, \nabla)$).

Analogous results hold for $\mathscr{E}_{\hat P}$ *with the substitution of* $\mathscr{S}_\mp$ *for* $\mathscr{S}_\pm$.

For proofs, see [13] (where $\mathscr{E} = \mathscr{O}_U$ in a major part of the text), [15] (where the general semi-flat case is considered) and [9] (where analogous calculations were carried out for $L(U)$).

This result shows that the elements of $H^2(L(U)^{(i-1)}, \rho^*\mathscr{E}_P \otimes \hat{\rho}^*\mathscr{E}_{\hat{P}})$ can be constructed from the solution of the equations $D_i\psi = 0$ and $\hat{D}_i\hat{\psi} = 0$. For $H^1(\mathscr{E}_P) \neq 0$, it is necessary to evaluate $\mathrm{Ker}(H^2(s^{i+1}))$ in $\mathrm{Ker}\, D_i \otimes \mathrm{Ker}\, \hat{D}_i$ (if $\alpha\colon \mathscr{F} \to \mathscr{G}$ is a sheaf morphism, then $H^i(\alpha)\colon H^i(\mathscr{F}) \to H^i(\mathscr{G})$ is the corresponding morphism of cohomology.

We shall now consider the case $P(U) = P$ and the possibility of pure algebraic constructions. We shall further assume that $N \geqslant 0$.

4. Cohomology of the projective superspace. The projective superspace $P = P(T_0 \otimes T_1)$ (i.e., the Grassmannian $1|0$ of subspaces of T) has the following structure: $P_{\mathrm{red}} = P(T_0)$, and $\mathscr{O}_P = S(T_1 \otimes \mathscr{O}_{P(T_0)}(-1))$ (a symmetric algebra will be naturally considered in the superalgebra; since T_1 is odd, $\mathscr{O}_P$ is a sheaf of Grassmann algebras). Therefore, in addition to the standard closed inclusion $P_{\mathrm{red}} \subset P$, there is a projection $P \to P_{\mathrm{red}}$. The tautological sheaf on P is denoted by $\mathscr{O}_P(-1)$; it induces an analogous sheaf on P_{red}. The following statements are straightforward generalizations of standard results (we assume that $\dim T_0 = 4$).

a) The dualizing sheaf $\mathrm{Ber}\, \Omega^1 P$ is isomorphic to $\Pi^N \mathscr{O}_P(-4 + N)$.

b) $H^0(P, \mathscr{O}_P(n)) = S^n(T^*)$ for $N \geqslant 0$ and 0 for $N < 0$; $H^i(P, \mathscr{O}(n)) = 0$ for all n, $i = 1, 2$; and $H^3(P, \mathscr{O}(n)) = S^{-n+N-4}(T) \otimes \mathrm{Bar}\, T$ for $n \leqslant N - 4$ and $= 0$ for $n > N - 4$.

c) For any locally free sheaf $\mathscr{E} = \mathscr{E}_P$ on P, the qualitative behavior of the group $H^i(P, \mathscr{E}(n))$ is the same as for $\mathscr{O}_P$: the groups $H^0(P, \mathscr{E}(n))$ are nonzero only for $n \geqslant n_0$, $H^3(P, \mathscr{E}(n)) \neq 0$ for $n \leqslant n_3$, and $H^{1,2}(P, \mathscr{E}(n)) \neq 0$ for $n^-_{1,2} \leqslant n \leqslant n^+_{1,2}$.

d) Invertible sheaves on P of rank $1|0$ (or $0|1$) are up to isomorphism exhausted by the sheaves $\mathscr{O}(n)$ (or $\Pi\mathscr{O}(n)$).

5. Monads. A *monad* (on an arbitrary superspace) is a complex of locally free sheaves $\mathscr{F}_\cdot : \mathscr{F}_{-1} \to^\alpha \mathscr{F}_0 \to^\beta \mathscr{F}_1$ in which α is locally a direct inclusion, β is a surjection, and $\beta \circ \alpha = 0$. We shall denote by $\mathscr{E}(\mathscr{F}_\cdot) = \mathrm{Ker}\beta/\mathrm{Im}\,\alpha$ the sheaf of zero cohomology groups of the monad. We shall construct U-trivial sheaves of the form $\mathscr{E}(\mathscr{F}_\cdot)$, where all the $\mathscr{F}_i$ are direct sums of invertible sheaves. The corresponding fields include as special cases instantons and superinstantons. We shall write $\alpha(i)$ instead of $\alpha \otimes \mathrm{id}_{\mathscr{O}(i)}\colon \mathscr{F}_{-1}(i) \to \mathscr{F}_0(i)$, etc.

The fundamental theory of monads on P^n can be found in [14].

6. Proposition. *Let $\mathscr{F}_\cdot$ be a monad consisting of direct sums of invertible sheaves on P, and put $\mathscr{E} = \mathscr{E}(\mathscr{F}_\cdot)$. For all i,*

$$H^0(\mathscr{E}(i)) = \mathrm{Ker}\, H^0(\beta(i))/\mathrm{Im}\, H^0(\alpha(i));$$
$$H^1(\mathscr{E}(i)) = \mathrm{Coker}\, H^0(\beta(i));$$
$$H^2(\mathscr{E}(i)) = \mathrm{Ker}\, H^3(\alpha'(i));\ \alpha'\colon \mathscr{F}_{-1} \to \mathrm{Ker}\beta;$$
$$H^3(\mathscr{E}(i)) = \mathrm{Ker}\, H^3(\beta(i))/\mathrm{Im}\, H^3(\alpha'(i)).$$

Proof. We set $\mathscr{K}(i) = \mathrm{Ker}\beta(i)$ and write the two exact sequences $0 \to \mathscr{K}(i) \to \mathscr{F}_0(i) \to^\beta \mathscr{F}_1(i) \to 0$ and $0 \to \mathscr{F}_{-1}(i) \to \mathscr{K}(i) \to \mathscr{E}(i) \to 0$. We then calculate the cohomology $\mathscr{K}(i)$ from the first sequence and $\mathscr{E}(i)$ from the second sequence using that $H^k(\mathscr{F}_a(i)) = 0$ for $k = 1, 2$.

To verify U-triviality of the sheaf defined by a monad, we shall use the following criterion.

7. PROPOSITION. *Let $\mathcal{F}.$ be a monad on P^1 with the properties that $\mathcal{F}_0$ is free and $H^0(\mathcal{F}_{-1}) = H^0(\mathcal{F}_1^*) = 0$. Then one can define a map $\varepsilon = \varepsilon(\mathcal{F}.)$: $H^1(\mathcal{F}_{-1}(-1)) \to H^0(\mathcal{F}_{-1}(-1))$ which is an isomorphism if and only if $\mathcal{E}(\mathcal{F}.)$ is free on P^1.*

PROOF. As before, we set $\mathcal{K} = \operatorname{Ker}\beta$. From $0 \to \mathcal{K}(-1) \to \mathcal{F}_0(-1) \to \mathcal{F}_1(-1) \to 0$, we determine $H^0(\mathcal{F}_1(-1)) = H^1(\mathcal{K}(-1))$ and $H^0(\mathcal{K}(-1)) = 0$. From $0 \to \mathcal{F}_{-1}(-1) \to \mathcal{K}(-1) \to \mathcal{E}(-1) \to 0$ we obtain a morphism $H^1(\mathcal{F}_{-1}(-1)) \to H^1(\mathcal{K}(-1)) \to H^0(\mathcal{F}_1(-1))$ which is by definition ε. Its kernel and cokernel are, respectively, $H^0(\mathcal{E}(-1))$ and $H^1(\mathcal{E}(-1))$. However, these two groups vanish if and only if $\mathcal{E}$ is free.

8. ***m*-monads on *P*.** We define an *m-monad* ($m \geqslant 0$) to be a complex

$$\mathcal{F}.: F_{-1} \otimes \mathcal{O}_P(-m-1) \xrightarrow{\alpha} F_0 \otimes \mathcal{O}_P \xrightarrow{\beta} \mathcal{C}_1 \otimes \mathcal{O}_P(m+1),$$

where the F_i are linear (super)spaces. We set $\mathcal{E}_P = \mathcal{E}(F.)$. For $0 \leqslant i \leqslant m$ it follows from Proposition 6 that

$$H^1(P, \mathcal{E}_P(-i-1)) = F_1 \otimes S^{m-i}(T^*).$$

We thus have enough one-dimensional cohomology groups to construct nontrivial extensions and deformations. It will be shown in §4 that the sheaf $\mathcal{E}(F.)$ can be U-trivial for nonempty U.

The conditions of Proposition 2 can be verified for a monad as follows:

$$\left.\begin{array}{l} H^0(P, \mathcal{E}_P) = 0 \Leftrightarrow \operatorname{Ker} H^0(\beta) = 0 \\ H^1(P, \mathcal{E}_P) = 0 \Leftrightarrow \operatorname{Coker} H^0(\beta) = 0 \end{array}\right\} \Leftrightarrow F_0 \overset{H^0(\beta)}{\tilde{\to}} F_1 \otimes S^{m+1}(T^*).$$

The same conditions are satisfied for the sheaf $\mathcal{E}_P^*$ which is defined by the dual monad if the map

$$F_0^* \to^{H^0(\alpha^*)} F_{-1}^* \otimes S^{m+1}(T^*)$$

is also an isomorphism. A monad having these two properties is said to be *maximal*.

9. ***m*-monads on $L^{(i)}$.** We shall consider the following diagram of sheaves on $P \times \hat{P}$:

$$\mathcal{F}.: F_{-1}^+ \otimes \mathcal{O}(-m-1, 0) \oplus F_{-1}^- \otimes \mathcal{O}(, -m-1) \xrightarrow{\alpha} (F_0^+ \oplus F_0^-)$$
$$\otimes \mathcal{O} \xrightarrow{\beta} F_1^+ \otimes \mathcal{O}(m+1, 0) \oplus F_1^- \otimes \mathcal{O}(0, m+1),$$

where

$$\alpha = \begin{Vmatrix} \alpha^+ & 0 \\ 0 & \alpha \end{Vmatrix} \quad \text{and} \quad \beta = \begin{Vmatrix} \beta^+ & \gamma^+ \\ \gamma^- & \beta^- \end{Vmatrix}.$$

When writing morphisms in the form of a block matrix, we shall always assume that it is multiplied from the left by a column of components. In particular, we have

$$\gamma^+: F_0^- \otimes \mathcal{O} \to F_1^+ \otimes \mathcal{O}(m+1, 0), \qquad \gamma^-: F_0^+ \otimes \mathcal{O} \to F_1^- \otimes \mathcal{O}(0, m+1).$$

We shall assume that the following conditions are satisfied:

$$\beta^+\alpha^+ = 0, \quad \beta^-\alpha^- = 0, \quad \gamma^+\alpha^- = s^{i+1}\gamma', \quad \gamma^-\alpha^+ = s^{i+1}\gamma''.$$

Moreover, let α be a locally direct inclusion and β a surjection. Then the restriction $\mathcal{F}.|L^{(i)}$ is a monad and, hence, it determines a sheaf on $L^{(i)}$. Such monads will be called *standard*.

A monad $\mathcal{F}.$ is a deformation of the direct sum of m-monads lifted, respectively, from P and $\hat{P}$. In fact, Let $\mathcal{F}.(t_+, t_-)$ be the complex which is obtained from $\mathcal{F}.$ by replacing $\gamma^\pm$ by $t_\pm\gamma^\pm$. For $t_\pm = 0$, it is the direct sum $\mathcal{F}.(0,0) = \mathcal{F}.^+ \oplus \mathcal{F}.^-$. If, for example, $t_- \neq 0$ and $t_0 = 0$, there is a natural inclusion of complexes $\mathcal{F}.^- \subset \mathcal{F}.(0, t_-)$ with the corresponding extension $0 \to \mathcal{E}(\mathcal{F}.^-) \to \mathcal{E}(\mathcal{F}.(0,t)) \to \mathcal{E}(\mathcal{F}.^+) \to 0$. The class of this extension $e(t_-)$ depending on t_- lies in the group

$$H^1\big(L^{(i)}, \mathcal{E}^*(\mathcal{F}.^+) \otimes \mathcal{E}(\mathcal{F}.^-)\big) = H^1\big(P, \mathcal{E}^*_P(-i-1)\big) \otimes H^1\big(\hat{P}, \mathcal{E}_{\hat{P}}(-i-1)\big)$$
$$= \big(F^+_{-1}\big)^* \otimes S^{m-i}(T^*) \otimes F^-_1 \otimes S^{m-i}(T)$$

(we assume that the monads $\mathcal{F}.^\pm$ are maximal). On the other hand, we have on $P \times \hat{P}$ the morphism

$$t_-\gamma^-\alpha^+ = t_-s^{i+1}\gamma'' \colon F^+_{-1} \otimes \mathcal{O}(-m-1, 0) \to F^-_1 \otimes \mathcal{O}(0, m+1)$$

and it is clear that $t_-\gamma''$ can be regarded as an element of the same space

$$\operatorname{Hom}(F^+_{-1}, F^-_1) \otimes H^0\big(P \times \overline{P}, \mathcal{O}(m-i, m-i)\big)$$

that contains $e(t_-)$. It is easy to verify that $e(t_-) = \text{const} \cdot t_-\gamma''$ under such an identification.

To verify that the monads $\mathcal{F}.$ for all possible $t_\pm\gamma^\pm$ where $t^2_\pm = 0$ realize all infinitesimal deformations of $\mathcal{F}.^+ \oplus \mathcal{F}.^-$, it is sufficient to show that the corresponding pairs (γ', γ'') can assume all admissible values. We restrict ourselves to the case of maximal $\mathcal{F}.^\pm$. It is then possible to write the monad $\mathcal{F}.$ in the following form, which will be referred to as *standard*:

$$F^+_{-1} \otimes \mathcal{O}(-m-1, 0) \oplus F^-_{-1} \otimes \mathcal{O}(0, -m-1)$$
$$\downarrow \begin{Vmatrix} a^+ & 0 \\ 0 & a^- \end{Vmatrix}$$
$$\mathcal{O} \otimes \big[F^+_{-1} \otimes S^{m+1}(T) \oplus F^-_{-1} \otimes S^{m+1}(T^*)\big]$$
$$\xrightarrow{\begin{Vmatrix} B^+ & C^+ \\ C^- & B^- \end{Vmatrix}} \mathcal{O} \otimes \big[F^+_1 \otimes S^{m+1}(T^*) \oplus F^-_1 \otimes S^{m+1}(T)\big]$$
$$\downarrow \begin{Vmatrix} b^+ & 0 \\ b & b^- \end{Vmatrix}$$
$$F^+_1 \otimes \mathcal{O}(m+1, 0) \oplus F^-_1 \otimes \mathcal{O}(0, m+1).$$

In this notation, the morphisms $a^\pm$ and $b^\pm$ are defined once and for all, and the elements of an invertible constant operator B represent parameters of the monad. More precisely,

$$a^+(f^+ \otimes t) = f^+ \otimes h(ts^{m+1}),$$

where $f^+ \in F^+_{-1}$, t is a section of $\mathcal{O}(-m-1, 0)$ and $h \colon \mathcal{O}(0, m+1) \to S^{m+1}(T) \otimes \mathcal{O}$ is a canonical morphism; a^- can be similarly expressed. Furthermore, b^+ is determined from the requirement that $H^0(b^+)$ is an identity map on $F^+_1 \otimes S^{m+1}(T^*)$. The matrix B describes an identity isomorphism between two natural realizations of the monad sheaf $\mathcal{F}_0$ in which α and β assume, respectively, canonical form.

To prove the existence of a monad with chosen γ' and γ'', it is necessary to make use of the freedom of choice of $C^\pm$ ($B^\pm$ are determined by the monads $\mathcal{F}.^\pm$). We may choose C^- to be an arbitrary element in $\operatorname{Hom}(F^+_{-1}, F^-_1) \otimes S^{m+1}(T^*) \otimes S^{m+1}(T)$, for example, $\gamma' s^{i+1}$, $i \leqslant m$.

The following result shows that all the monads belonging to some natural class are standard.

10. THEOREM. *Let the sheaf of cohomology groups of a monad*

$$\mathcal{F}.: \bigoplus_{a+b=m+1} F_{-1}^{(a,b)} \otimes \mathcal{O}(-a,-b) \xrightarrow{\alpha} F_0 \otimes \mathcal{O} \xrightarrow{\beta} \bigoplus_{a+b=m+1} F_1^{(a,b)} \otimes \mathcal{O}(a,b),$$

where the $F_j^{(a,b)}$ $(a, b \geqslant 0)$ *are linear superspaces, be defined on* $L^{(i)}$ *and be U-free for a nonempty U. Then the following statements hold:*

a) *The monad* $\mathcal{F}.$ *is isomorphic to a standard monad; in particular, it determines two monads* $\mathcal{F}.^{\pm}$ *lifted, respectively, from P and* $\hat{P}$.

b) *The sheaf* $\mathcal{E}(\mathcal{F}.)|L(x)$ *is free if and only if* $\mathcal{E}(\mathcal{F}.^{+})|P^1(x)$ *and* $\mathcal{E}(\mathcal{F}.^{-})|\hat{P}^1(x)$ *are free.*

PROOF. We first note that $F_{-1}^{(-a,-b)} = \{0\}$ for $ab > 0$. To show that this result holds, we prove the following statement: if $F_{-1}^{(a,b)} \neq \{0\}$ for a pair (a, b), $ab > 0$, then $H^1(\mathcal{E}(\mathcal{F}.)(-1,-1))|L(x) \neq 0$ for an arbitrary quadric $L(x) \subset L$ and, therefore, $\mathcal{E}(\mathcal{F}.)|L(x)$ cannot be free. We shall write $\mathcal{F}_i(x)$ instead of $\mathcal{F}_1|L(x)$, etc.; $0 \to \mathcal{K}(x) \to \mathcal{F}_0(x) \to \mathcal{F}_1(x) \to 0$ implies that $H^2(\mathcal{K}(x)(-1,-1)) = 0$; it can be seen from $0 \to \mathcal{F}_{-1}(x) \to \mathcal{K}_1(x) \to \mathcal{E}(x) \to 0$ that δ: $H^1(\mathcal{E}(x)(-1,-1)) \to H^2(\mathcal{F}_{-1}(-1,-1))$ is a surjection. Finally, we obtain

$$H^2\big(F_{-1} \otimes \mathcal{O}_{L(x)}(-a-1,-b-1)\big) \neq 0 \quad \text{for } a, b > 0.$$

Applying this result to the dual monad, we find that U-triviality of $\mathcal{E}(\mathcal{F}.^*)$ also implies that $F_1^{(a,b)} = \{0\}$ for $a, b > 0$. We thus find that $\mathcal{F}.$ has the following form on $L^{(i)}$:

$$\begin{gathered} F_{-1}^{+} \otimes \mathcal{O}(-m-1,0) \oplus F_{-1}^{-} \otimes \mathcal{O}(0,-m-1) \\ \xrightarrow{(\alpha^+,\alpha^-)} F_0 \otimes \mathcal{O} \xrightarrow{(\beta^+,\beta^-)} F_1^{+} \otimes \mathcal{O}(m+1,0) \oplus F_1^{-} \otimes \mathcal{O}(0,m+1). \end{gathered}$$

We have $\beta^+\alpha^+ + \beta^+\alpha^- = 0$ and $\beta^-\alpha^+ + \beta^-\alpha^- = 0$. However, $\beta^+\alpha^+$ is product with a section of the sheaf

$$\operatorname{Hom}(F_{-1}^{+}, F_1^{+}) \otimes \mathcal{O}(2m+2,0)$$

and $\beta^+\alpha^-$ product with a section of

$$\operatorname{Hom}(F_{-1}^{+}, F_1^{-}) \otimes \mathcal{O}(m+1,m+1).$$

It follows that $\beta^+\alpha^+ = 0$ and $\beta^+\alpha^- = 0$ on $L^{(i)}$; similarly, $\beta^-\alpha^+ = 0$ and $\beta^-\alpha^- = 0$.

Let F_0^+ (or F_0^-) be a minimal subspace in F_0 such that the α^+ (α^-) image is contained in $F_0^+ \otimes \mathcal{O}$ (or $F_0^- \otimes \mathcal{O}$). Since (α^+, α^-) is a locally direct inclusion, we obtain $F_0^+ \cap F_0^- = \{0\}$. We may assume that $F_0 = F_0^+ \oplus F_0^-$ (alternatively, we can redefine, for example, F_0^- by adding to it the missing complement in F_0). The monad then assumes the standard form

$$F_0 = F_0^+ \oplus F_0^-, \; \alpha = \begin{Vmatrix} \alpha^+ & 0 \\ 0 & \alpha^- \end{Vmatrix}, \quad \beta = \begin{Vmatrix} \beta^+ & \gamma^+ \\ \gamma^- & \beta^- \end{Vmatrix}.$$

All its sheaves and morphisms can be extended uniquely from $L^{(i)}$ to $P \times \hat{P}$, and only relations such as $\gamma^+\alpha^- = 0$ on $L^{(i)}$ are transformed to $\gamma^+\alpha^- = s^{i+1}\gamma'$ on $P \times \hat{P}$ since the extension of $\gamma^+\alpha^-$ is multiplication by a section of the sheaf

$$\operatorname{Hom}(F_{-1}^{+}, F_1^{+}) \otimes \mathcal{O}(m+1,m+1).$$

We shall now clarify for which $L(x)$ the sheaf $\mathscr{E}(x)$ is free. We shall use the following criterion to decide whether such a sheaf is free: $H^i(\mathscr{E}(x)(\varepsilon,\eta)) = 0$ for all i and $(\varepsilon,\eta) = (-1,0)$, $(0,-1)$, $(-1,-1)$. Its necessity is obvious. To prove its sufficiency, we consider the Koszul resolution of the geometric fiber of $\mathscr{E}(x)$ at an arbitrary point $(u,v) \in L(x)$ and show that it leads to a canonical trivialization of $\mathscr{E}(x)$, which establishes the isomorphism of $H^0(\mathscr{E}(x))$ with this fiber.

As usual, we write the monad $\mathscr{F}_\cdot(\varepsilon,\eta)$ restricted to $L(x)$:

$$0 \to F_{-1}^+ \otimes \mathscr{O}_{L(x)}(-m-1+\varepsilon,\eta) \oplus F_{-1}^- \otimes \mathscr{O}_{L(x)}(\varepsilon,-m-1+\eta) \overset{\tilde\alpha(\varepsilon,\eta)}{\to} \mathscr{K}(x)(\varepsilon,\eta) \to \mathscr{E}(x)(\varepsilon,\eta) \to 0,$$

$$0 \to \mathscr{K}(x)(\varepsilon,\eta) \to (F_0^+ \oplus F_0^-) \otimes \mathscr{O}_{L(x)}(\varepsilon,\eta) \to F_1^+ \otimes \mathscr{O}_{L(x)}(m+1+\varepsilon,\eta) \oplus F_1^- \otimes \mathscr{O}_{L(x)}(\varepsilon,m+1+\eta) \to 0.$$

We obtain the following isomorphisms from the second triplet:

$$\delta^+(\beta)\colon H^1(\mathscr{K}(x)(-1,0)) \tilde\to F_1^+ \otimes H^0(\mathscr{O}_{L(x)}(m,0)),$$

$$\delta^-(\beta)\colon H^1(\mathscr{K}(x)(0,-1)) \tilde\to F_1^- \otimes H^0(\mathscr{O}_{L(x)}(0,m)),$$

and all the remaining groups $H^i(\mathscr{K}(x)(\varepsilon,\eta))$ are zero. Consequently, $H^i(\mathscr{E}(x)(-1,-1)) = 0$ for all i, and the other groups are zero if and only if the following maps are isomorphisms:

$$\delta^+(\beta)\circ H^1(\tilde\alpha(-1,0))\colon F_{-1}^+ \otimes H^1(\mathscr{O}_{L(x)}(-m-2,0)) \to F_1^+ \otimes H^0(\mathscr{O}_{L(x)}(m,0)),$$
$$\delta^-(\beta)\circ H^1(\tilde\alpha(0,-1))\colon F_{-1}^- \otimes H^1(\mathscr{O}_{L(x)}(0,-m-2)) \to F_-^{-1} \otimes H^0(\mathscr{O}_{L(x)}(0,m)).$$

For $\gamma^\pm = 0$, i.e., for

$$\beta = \begin{Vmatrix} \beta^+ & 0 \\ 0 & \beta^- \end{Vmatrix},$$

the last two maps are clearly $\varepsilon(\mathscr{F}_\cdot^+(x))$ and $\varepsilon(\mathscr{F}_\cdot^+(x))$ in the notation of Proposition 7. We shall now show that they are independent of $\gamma^\pm$, since the requirement that $\mathscr{E}(\mathscr{F}_\cdot(x))$ should be free is equivalent to $\mathscr{E}(\mathscr{F}_\cdot^+(x))$ and $\mathscr{E}(\mathscr{F}_\cdot^-(x))$ being free simultaneously.

In fact, we define a cohomology class from $H^1(\mathscr{F}_{-1}(x)(-1,0))$, for example, by a cocyle in the usual two-element covering (lifted from $P^1(x)$ to $P^1(x) \times \hat P^1(x)$). To apply $\delta^+(\beta)\circ H^1(\tilde\alpha(-1,0))$ to it, we have to split this cocycle in $\mathscr{F}_0(x)(-1,0)$ and then apply the morphism $H^0(\beta(x)(-1,0))$ to one of the two terms. However, only the component of the morphism $\beta(x)(-1,0)$ that lies in $F_1^- \otimes \mathscr{O}_{L(x)}(-1,m+1)$ depends on $\gamma^\pm$, and H^0 of this component is zero. For details, see the calculations of §4.2, below.

11. Remarks. a) It is essential that deformations of combined sheaves which have been constructed on $L^{(i)}$ remain nontrivial after restriction to $L^{(0)} = L$. In fact, the proof of Proposition 2 and the identifications made in subsection 9 indicate that the map of the restriction of groups H^2 from $L^{(i)}$ to L reduces to multiplication by s^i: $S^{m-i}(T \oplus T^*) \to S^m(T \oplus T^*)$.

b) It is probable that the sheaf $\mathscr{E}_L$ on $L^{(m)}$ associated with the m-monad with maximal $\mathscr{F}_\cdot^{\pm}$ and $(\gamma', \gamma'') \neq (0,0)$ cannot be excluded to $L^{(m+1)}$. To verify this, it would be necessary to find the obstruction to such an extension which lies in $H^2(L, \mathrm{End}, \mathscr{E}_L(-m-1, -m-1))$.

§4. Calculations

1. Coordinate expressions for simplest *m*-monads on *P*. Let $(z_1, \ldots, z_4; \zeta_1, \ldots, \zeta_N)$ be a basis in T^* and let (Z^a) be a basis in $S^{m+1}(T^*)$ consisting of monomials of degree $m+1$, i.e., $Z^a = z_1^{a_1} \cdots z_4^{a_4} \zeta_1^{\alpha_1} \cdots \zeta_N^{\alpha_N}$, where $a_i \geqslant 0$, $\alpha_i = 0$ or 1, and $\Sigma a_i + \Sigma \alpha_j = m+1$. We set $A(Z) = (Z^a)^t$ (column of monomials) and $B(Z) = (\Sigma_b \beta_b^a Z^b)$ (row of polynomials), where the constants β_b^a satisfy the requirement that $B(Z)$ should be even in the sense of superalgebra together with the following two additional conditions: a) $\Sigma_{a+b=c} \beta_b^a = 0$ for any c, and b) the elements $B(Z)$ generate $S^{m+1}(T^*)$. The diagram

$$\mathscr{F}_\cdot: \mathscr{O}_P(-m-1) \xrightarrow{A(Z)} \mathscr{O}_P^{r(m)|s(m)} \xrightarrow{B(Z)} \mathscr{O}_P(m+1),$$

where $r(m)|s(m) = \dim S^{m+1}(T^*)$, then represents a maximal monad on P, and all the maximal monads with $\dim F_{-1} = 1$ can be obtained in this way.

Replacing the constants β_b^a by the matrices of maps $F_{-1}^+ \to F_1^+$, we obtain the general case.

2. Criterion of *U*-triviality. We shall consider the line P^1: $\{z_3 = z_4 = 0; \zeta_1 = \cdots = \zeta_N = 0\}$ and express the condition for the restriction of $\mathscr{E}(\mathscr{F}_\cdot)$ to this line to be free in terms of (β_b^a). We shall use Proposition 3.7, but first we must simplify $\mathscr{F}_\cdot | P^1$. We have $\mathscr{F}_0 = \mathscr{F}_0' \oplus \mathscr{F}_0''$, where $\mathscr{F}_0'$ is generated by the monomials $z_1^{a_1} z_2^{a_2}$, $a_1 + a_2 = m+1$, and $\mathscr{F}_0''$ is generated by all the other monomials. Hence,

$$A(Z) = \begin{pmatrix} A'(Z) \\ A''(Z) \end{pmatrix}, \qquad B(Z) = (B'(Z) | B''(Z)),$$

where $A''(Z)|P^1 = 0$ and $B''(Z)|P^1 = 0$. It follows that $\mathscr{E}(\mathscr{F}_\cdot|P^1)$ contains the sheaf $\mathscr{F}_0''|P^1$ as a direct summand and it is sufficient to determine when the sheaf of cohomology groups of a monad on P^1

$$\mathscr{G}_\cdot: \mathscr{O}_{P^1}(-m-1) \xrightarrow{A'(Z)} \mathscr{O}_{P^1}^{m+2} \xrightarrow{B'(Z)} \mathscr{O}_P(m+1)$$

is free. This depends only on $(\beta_{b_1 b_2}^{a_1 a_2})$ with $a_1 + a_2 = b_1 + b_2 = m+1$. We shall determine the map $\varepsilon(\mathscr{G}_\cdot)$: $H^1(\mathscr{G}_{-1}(-1)) \to H^0(\mathscr{G}_1(-1))$ from Proposition 3.7. The basis $H^1(\mathscr{G}_{-1}(-1)) = H^1(\mathscr{O}(-m-2))$ of the Čech complex for a covering U_i: $\{z_i \neq 0\}$ is represented by cocycle classes

$$h_i(U_{12}) = z_1^{-i} z_2^{i-m-2}, \qquad i = 1, \ldots, m+1.$$

They should be regarded as cochains with values in $\mathscr{K}(-1)$, and it is necessary to split explicitly their images in $\mathscr{F}_0(-1)$ after the application of $A'(Z)$. The required splitting is as follows:

$$z_1^{-i} z_2^{i-m-2} \xrightarrow{A'(Z)} \left(z_1^{a_1 - i} z_2^{i - a_1 - 1}\right)^t_{a_1 = 0, \ldots, m+1}$$

$$= \left(\cdots z_1^{a_1-i} z_2^{i-a_1-1} \cdots \Big|_{a_1 \leqslant i-1} 0 \cdots 0\right)^t + \left(0 \cdots 0 \Big| \cdots z_1^{a_1-i} z_2^{i-a_1-1} \cdots\right)^t_{a_1 \geqslant i}.$$

The first term on the right is regular on U_1 and the second on U_2. Finally, we obtain

$$\varepsilon(h_i) = B'(Z)h_i^2 = -B'(Z)h_i^1 = -\sum_{a_1=0}^{i-1}\sum_{b_1=0}^{m+1}\beta^{a_1,m+1-a_1}_{b_1,m+1-b_1}z_1^{a_1+b_1-i}z_2^{m+i-a_1-b_1}.$$

The sheaf $\mathscr{E}(\mathscr{F}_\cdot)|P^1$ is free if and only if all the polynomials $\varepsilon(h_i)$, $i = 1,\ldots, m+1$, are linearly independent. The set of such β is open in the sense of Zariski. To verify that the set in question is nonempty, we impose the following additional conditions: $\beta^{a_1,a_2}_{b_1,b_2} = 0$ if $(a_1+b_1, a_2+b_2) \neq (m+1, m+1)$; $\beta^{a_1a_2}_{b_1b_2} = \beta(a_1) - \gamma$ otherwise. Let also

$$\sum_{a_1=0}^{m+1}\beta(a_1) = (m+2)\gamma$$

which guarantees that $\sum_{a+b=c}\beta^a_b = 0$ for all a, b of the form $(a_1a_20\cdots 0)$. The condition that the aforementioned sheaf is free reduces to a system of inequalities: for all $i \leqslant m+1$,

$$\sum_{a_1=0}^{i-1}(\beta(a_1)-\gamma) \neq 0$$

since

$$\varepsilon(h_i) = -\left(\sum_{a_1=0}^{i-1}(\beta(a_1)-\gamma)\right)z_1^{m+1-i}z_2^{i-1}, \qquad i = 1,\ldots, m+1.$$

To complete such a submatrix (β^a_b) to the full matrix of a monad, we may set $\beta^a_b = 0$ if a or b are of the form $(a_1a_20\cdots 0)$ an b or a is not of this form. The elements of $B(Z)$ are then polynomials in z_1 and z_2 alone or only the remaining coordinates, and the corresponding requirements on the second part of the polynomials can be satisfied independently.

3. Simple m-monads on $L^{(m)}$ in coordinates. Let W^a be a dual basis of monomials in $S^{m+1}(T)$ of elements $(w_1,\ldots, w_4, \nu_1, \ldots, \nu_N)$. Let the equation of L be

$$s = \sum z_a w_a + \sum \zeta_n \nu_a = 0.$$

We set $s^{m+1} = \sum_a C(a,m)Z^aW^a$ in $S(T \oplus T^*)$. Let also $A_+(Z) = (Z^a)^t$, $A_-(W) = (W^a)^t$, $B_+(Z) = (\sum_b \beta^a_{b+}Z^b)$ and $B_-(W) = (\sum_b\beta^a_{b-}W^b)$. Finally, we have

$$\Gamma_+(Z) = t_+(C(a,m)Z^a), \qquad \Gamma_-(W) = t_-((-1)^{\tilde{a}}C(a,m)W^a),$$

where $t_\pm$ are constants and $\tilde{a}$ is the parity of W^a, i.e., $\sum\alpha_i$. If $B_\pm$ satisfy the conditions of §1, then the following diagram on $P \times \hat{P}$:

$$\mathscr{F}_\cdot\colon \mathscr{O}(-m-1,0)\oplus\mathscr{O}(0,-m-1)$$
$$\xrightarrow{\alpha} \mathscr{O}\otimes\left[S^{m+1}(T)\oplus S^{m+1}(T^*)\right] \xrightarrow{\beta} \mathscr{O}(m+1,0)\oplus\mathscr{O}(0,m+1),$$

where

$$\alpha = \begin{pmatrix} A_+(Z), & 0 \\ 0 & A_-(W)\end{pmatrix}, \qquad \beta = \begin{pmatrix} B_+(Z), & \Gamma_+(Z) \\ \Gamma_-(W), & B_-(W)\end{pmatrix},$$

determines a monad on $L^{(m)}$ since $\Gamma_+ A_- = t_+ s^{m+1}$ and $\Gamma_- A_+ = t_- s^{m+1}$ (apart from exceptional values of $t_\pm$ for which β might not be surjective). Such a monad is U-trivial by §2 and Theorem 3.10.

4. Ranks of simple monads. The simplest solution of the vacuum Yang-Mills equations as far as the structure of the monad is concerned is obtained for $m = 3$ and $N = 0$. Since $\dim S^4(C^4) = 35$, the rank of $\mathscr{E}_L$ is 66.

A gauge group of more acceptable size can be obtained for the supersymmetric case: $m = 0$ and $N = 3$. It is the group $\mathrm{GL}(4|6; C)$.

In fact, quite accurate information not only about the rank but also about the sheaf $\mathscr{E} = \pi_{2*}\pi_1^*\mathscr{E}_L$, where $\mathscr{E}_L$ is the cohomology sheaf of a standard m-monad, can be obtained on the basis of calculations described in §2.

We set $\pi_1^*(\mathscr{F}_i) = \mathscr{G}_i$ and write α and β instead of $\pi_1^*(\alpha)$ and $\pi_1^*(\beta)$. We then have $\mathscr{E} = \pi_{2*}\mathscr{E}_F$, where $\mathscr{E}_F = \mathscr{E}(\mathscr{G}_\cdot)$.

We shall denote by $\mathscr{G}_0' \subset \mathscr{G}_0$ the minimal locally free sheaf containing the image of α. If α is realized in the standard form (§3.9), then

$$\mathscr{G}_0' = F_{-1}^+ \otimes S^{m+1}(\mathscr{S}_{+F}) \oplus F_{-1}^- \otimes S^{m+1}(\mathscr{S}_{-F})$$

where $\mathscr{S}_{\pm F} = \pi_2^*(\mathscr{S}_\pm)$ and the inclusion $\mathscr{G}_0' \subset \mathscr{G}_0$ is induced by the inclusions $\mathscr{S}_{+F} \subset \mathscr{O}_F \otimes T$ and $\mathscr{S}_{-F} \subset \mathscr{O}_F \otimes T^*$. We set $\beta' = \beta | \mathscr{G}_0'$.

5. PROPOSITION. *The Yang-Mills sheaf $\mathscr{E}$ on M, the carrier of the connection ∇, has the following structure. There is an exact sequence $0 \to \mathscr{E}''' \to \mathscr{E} \to \mathscr{E}' \to 0$, where*

$$\mathscr{E}' = F_{-1}^+ \otimes S^{m-1}(\mathscr{S}_+) \otimes \Lambda^2\mathscr{S}_+ \oplus F_{-1}^- \otimes S^{m-1}(\mathscr{S}_-) \otimes \Lambda^2(\mathscr{S}_-) \text{ for } m \geqslant 1;$$

$$\mathscr{E}' = 0 \textit{ for } m = 0;$$

$$\mathscr{E}'' = F_1^+ \otimes \mathscr{S}_+^\perp \cdot S^m(T^* \otimes \mathscr{O}_M) \oplus F_1^- \otimes \mathscr{S}_-^\perp \cdot S^m(T \otimes \mathscr{O}_M).$$

Assume further that the morphism β': $\mathscr{G}_0' \to \mathscr{G}_1$ over an open subset $\pi_2^{-1}(U)$, where $U \subset M$, is surjective. Then this exact sequence splits canonically on U; $\mathscr{E}|U = \mathscr{E}'|U \oplus \mathscr{E}'''|U$.

Special case. The simplest non-self-dual solution of the supersymmetric ($N = 3$) Yang-Mills equations is a connection on the sheaf $\mathscr{S}_+^\perp \oplus \mathscr{S}_-^\perp$, i.e., an "interacting" sum of a superinstanton and an antisuperinstanton.

PROOF. The exact sequence $0 \to \mathscr{K} = \operatorname{Ker}\beta \to \mathscr{G}_0 \to^{\beta} \mathscr{G}_1 \to 0$ on F yields $\pi_{2*}\mathscr{K} = \operatorname{Ker}\pi_{2*}(\beta)$ and $R^1\pi_{2*}\mathscr{K} = 0$. The kernel of the morphism $\pi_{2*}(\beta)$ can be naturally identified with $\mathscr{E}''$ if $\mathscr{F}_\cdot$ is written in the standard form and $\mathscr{G}_0$ is realized as

$$F_1^+ \otimes S^{m+1}(T^*) \otimes \mathscr{O}_F \oplus F_1^- \otimes S^{m+1}(T) \otimes \mathscr{O}_F$$

and β as a morphism such that $H^0(\beta)$ on L is an identity map.

We then obtain from the exact sequence $0 \to \mathscr{G}_{-1} \to \mathscr{K} \to \mathscr{E}_F \to 0$

$$0 \to \pi_{2*}\mathscr{K} = \mathscr{E}'' \to \mathscr{E} \to R^1\pi_{2*}\mathscr{G}_{-1} \to 0.$$

Using the standard isomorphism

$$R^1\pi_{2*}\mathscr{O}_F(-m-1, 0) = S^{m-1}(\mathscr{S}_+) \otimes \Lambda^2\mathscr{S}_+$$

and the analogous isomorphism for $\mathscr{O}(0, -m-1)$, we can identify $R^1\pi_{2+}\mathscr{G}_{-1}$ with $\mathscr{E}'$. It remains to show that this exact sequence splits canonically on U.

We shall consider the monad $\mathcal{G}'_{\cdot}$: $0 \to \mathcal{G}_{-1} \to^{\alpha} \mathcal{G}'_0 \to^{\beta'} \mathcal{G}_1 \to 0$ over $\pi_2^{-1}(U)$. We set $\mathcal{K}' = \operatorname{Ker}\beta'$; repeating our calculations, we obtain $\pi_{2*}\mathcal{K}' = 0$ and $\pi_{2*}(\mathcal{E}(\mathcal{G}'_{\cdot})) = R^1\pi_{2*}\mathcal{G}_{-1} = \mathcal{E}'$. The inclusion of monads $\mathcal{G}'_{\cdot} \subset \mathcal{G}_{\cdot}$ determines the inclusion $\mathcal{E}(\mathcal{G}'_{\cdot}) \subset \mathcal{E}(\mathcal{G}_{\cdot})$ and then $\mathcal{E}' = \pi_{2*}(\mathcal{E}(\mathcal{G}')) \subset \pi_{2*}(\mathcal{E}(\mathcal{G})) = \mathcal{E}$, which splits our exact triplet.

§5. Cohomological component analysis

1. It is our aim to demonstrate in this section that the supergeometric version of the formalism of sheaf extension to infinitesimal extensions applied to $\mathcal{E}_L$ can reproduce the results of the component analysis of supersymmetric Yang-Mills equations ($N = 3$). The component analysis of fields (and equations) which is used in physical papers is based on the idea that all the objects are expanded in a local system of coordinates $(x^a, \Theta^{\alpha i}, \Theta_{\beta i})$ on M in terms of monomials of the odd coordinates on Θ and are replaced by the set of coefficients of such an expansion. The coefficients are functions of x and are regarded as the standard fields in space-time. For example, the scalar superfield on $M = G(2|0, 2|N; C^{4|N})$ (not restricted by connections), i.e., a section of the structure sheaf, is represented as a system consisting of one scalar field on $M_{\text{red}} = G(2|0, T^{4|0})$, N fields of spin $1/2$, $\binom{N}{2}$ fields of spin 1, etc.

In this approach, it is necessary to search for the transformation laws of all the coefficients that are required in transformations of coordinates and gauge transformations of the fields, which obscures the geometric interpretation of the structures under study. On the other hand, this approach preserves the advantages of supersymmetric field theories over classical field theories and is mathematically just as well founded as any coordinate calculations.

However, there is yet another aspect of this approach that can be treated within coordinate-free description. From the mathematical point of view, all the sheaves of $\mathcal{O}_M$-modules are filtered by powers of the ideal $J = \mathcal{O}_{M,1} + \mathcal{O}_{M,1}^2$, and the component analysis of sheaves, connections, etc. includes completely invariant calculation of the corresponding graded objects, of spectral sequences of filtered complexes, etc.

In the example studied here, it is natural to regard certain cohomology classes describing successive extensions of $\mathcal{E}_L/J\mathcal{E}_L$ from L_{red} to L as "components" of the sheaf $\mathcal{E}_L$. Such classes transform to fields on M_{red} after Penrose-Radon-type transformation (see [9]). Equations for such fields originate from the conditions that there should be no obstructions to the extension. We shall summarize the main results of the extension formalism in the next subsection. The rest of the paper comprises comments on the table (see below) which contains the principal results of this section, i.e., formulation of the calculations of [9] in the supersymmetric case.

2. Extensions and obstructions. Let $(Y, \mathcal{O}_Y)$ be an analytic supermanifold, $J = \mathcal{O}_{Y,1} + \mathcal{O}_{Y,1}^2$, and $Y^{(n)} = (Y, \mathcal{O}_Y/J^{n+1})$. The superspace $Y^{(n)}$ is a closed subspace of $Y^{(n+1)}$, $Y^{(0)} = Y_{\text{red}}$, and $Y^{(n)} = Y$ for $n \gg 0$. Let G be a connected analytic supergroup over C and ρ a finite-dimensional analytic representation $G \to \operatorname{GL}(p|q)$. The pair (e, ρ), where $e \in H^1(Y, G(\mathcal{O}_Y))$, can be regarded as a locally free sheaf on Y whose structure group reduces to G in the representation ρ. Here, $G(\mathcal{O}_Y)$ denotes the sheaf of noncommutative groups, i.e., of sections $Y \times G$ over Y. In particular, for $G = \operatorname{GL}(p|q)$ and if ρ is the fundamental representation, the elements of $H^1(Y, \operatorname{GL}(p|q; \mathcal{O}_Y))$ classify locally free sheaves of rank $p|q$ on Y.

Now let $e^{(n)} \in H^1(Y, G(\mathcal{O}/J^{n+1}))$, and let $\mathcal{G}$ be the locally free sheaf on $Y^{(0)} = Y_{\text{red}}$ corresponding to the pair $(e^{(0)}, \text{Ad})$, where $e^{(0)}$ is the reduction of $e^{(n)} \bmod J$ and Ad is the adjoint representation. The following results hold:

a) There is a map

$$\omega\colon H^1\big(Y, G\big(\mathcal{O}_Y/J^{n+1}\big)\big) \to H^1\big(Y, \big(\mathcal{G} \otimes J^{n+1}/J^{n+2}\big)_0\big)$$

such that $\omega(e^{(n)}) = 0$ if and only if $e^{(n)}$ can be extended to a class $e^{(n+1)} \in H^1(Y, G(\mathcal{O}_Y/J^{n+2}))$.

The map ω can be constructed as follows. There is a commutative diagram of exact triplets

$$\begin{array}{ccccccccc}
0 \longrightarrow & \big(1 + J^{n+1}/J^{n+2}\big)^* & \longrightarrow & \mathcal{O}_{Y^{(n+1)}} & \longrightarrow & \mathcal{O}_{Y^{(n)}} & \longrightarrow & 0, \\
& \downarrow d \otimes 1 & & \downarrow d \otimes 1 & & \downarrow d & & \\
0 \longrightarrow & J^{n+1}/J^{n+2} & \longrightarrow & \Omega^1 Y^{(n+1)} \otimes & & \mathcal{O}_{Y^{(n)}} & \longrightarrow & \Omega^1 Y^{(n)} \to 0.
\end{array}$$

This diagram yields the following commutative diagram of sheaves:

$$\begin{array}{ccccccc}
0 \to \big(\mathcal{G} \otimes J^{n+1}/J^{n+2}\big)_0 & \xrightarrow{\exp} & G\big(\mathcal{O}_{Y^{(n+1)}}\big) & \longrightarrow & G\big(\mathcal{O}_{Y^{(n)}}\big) \to 1, \\
\| & & \downarrow D \otimes 1 & & \downarrow D \\
0 \to \big(\mathcal{G} \otimes J^{n+1}/J^{n+2}\big)_0 & \longrightarrow & \big(\mathcal{G} \otimes \Omega^1 Y^{(n+1)} \otimes \mathcal{O}_{Y^{(n)}}\big)_0 & \longrightarrow & \big(\mathcal{G} \otimes \Omega^1 Y^{(n)}\big)_0 \to 0,
\end{array}$$

where $Dg = g^{-1} dg$ in the matrix realization. The map of obstruction ω is the composition

$$\omega\colon H^1\big(Y, G\big(\mathcal{O}_{Y^{(n)}}\big)\big) \overset{H^1(D)}{\to} H^1\big(Y, \big(\mathcal{G} \otimes \Omega^1 Y^{(n)}\big)_0\big) \overset{\delta}{\to} H^2\big(Y, \big(\mathcal{G} \otimes J^{n+1}/J^{n+2}\big)_0\big).$$

The subscript 0 refers to Z_2-grading.

b) If $\omega(e^{(n)}) = 0$, then the action of the group $H^1(Y, (\mathcal{G} \otimes J^{n+1}/J^{n+2})_0)$ on the set of extensions $\{e^{(n+1)}\}$ of the class $e^{(n)}$ is transitive.

c) The action is effective it, for one of the extensions, the map $H^0(Y, \mathcal{G}^{(n+1)}) \to H^0(Y, \mathcal{G}^{(n)})$ is a surjection, where $\mathcal{G}^{(n+1)}$ and $\mathcal{G}^{(n)}$ are sheaves associated with $(e^{(n+1)}, \text{Ad})$ and $(e^{(n)}, \text{Ad})$.

(Even if the action of the group is not effective, we can define canonically the difference $e_1^{(n+1)} - e_2^{(n+1)} \in H^1(Y, (\mathcal{G} \otimes J^{n+1}/J^{n+2})_0)$.)

3. Analysis of $\mathcal{E}_L$. We recall that $L = F(1|0, 3|N; T)$, $T = C^{4|N}$, and $L_{\text{red}} = F(1, 3; T_0)$. It can easily be verified that

$$J^{n+1}/J^{n+2} = S^{n+1}\big(J/J^2\big) = S^{n+1}\big(T_1 \otimes \mathcal{O}(-1,0) \oplus T_1^* \otimes \mathcal{O}(0,-1)\big),$$

where the corresponding symmetric (i.e., Grassmann) algebra is constructed over $\mathcal{O} = \mathcal{O}_{L\,\text{red}}$; $J^{n+1} = 0$ for $n \geqslant 2N$.

Let $e \in H^1(L(U), G(\mathcal{O}))$, and let $e^{(n)}$ be the image of e in $H^1(L(U), G(\mathcal{O}/J^{n+1}))$. We may assume that e is constructed successively from $e^{(n)}$ to $e^{(n+1)}$. The pair $(e^{(0)}, \rho)$ represents a Yang-Mills sheaf L_{red} with a structure group reduced to G. It corresponds to an ordinary Yang-Mills field on $U \subset G(2; T_0)$ (the supergroup G plays the role of gauge group).

Assuming that $e^{(n)}$ has been constructed and $\omega(e^{(n)}) = 0$, we can choose an extension $\tilde{e}^{(n+1)}$ and parametrize other extensions by the elements $h^{(n+1)}$ of the corresponding cohomology group and write $\omega(\tilde{e}^{(n+1)} + h^{(n+1)}) = 0$. The elements

$h^{(n+1)}$, just like the elements of groups containing obstructions ω, are transformed to fields on U by the Radon-Penrose transformation. The operator ω itself determines the form of the equations. We shall not discuss the equations; their structure should be clear from [9] and [10], where an infinitesimal neighborhood $L^{(3)}$ was considered instead of L_3.

TABLE OF COHOMOLOGY ($N = 3$)

n	H^1	H^2
1	$\mathscr{G}_1 \otimes (\mathscr{S}_+ \otimes \Lambda^2\mathscr{S}_- \otimes T^{(1)} \oplus \mathscr{S}_- \otimes \Lambda^2\mathscr{S}_+ \otimes T^{(-1)})$	0
2	$\mathscr{G}_0 \otimes (\Lambda^2\mathscr{S}_+ \otimes T^{(-2)} \oplus \Lambda^2\mathscr{S}_- \otimes t^{(2)} \oplus \Lambda^2\mathscr{E}_+ \otimes \Lambda^2\mathscr{S}_- \otimes T^{(1)} \otimes T^{(-1)})$	0
3	$\operatorname{Ker} D_{\nabla,1}$,	$\operatorname{Coker} D_{\Delta,1}$
	where $D_{\nabla,1}\colon \mathscr{G}_1 \otimes (\mathscr{S}_+ \otimes \Lambda^2\mathscr{S}_+ \otimes T^{(-3)} \oplus \mathscr{S}_- \otimes \Lambda^2\mathscr{S}_- \otimes T^{(3)}) \to$	$\to \mathscr{G}_1 \otimes (\mathscr{S}_- \otimes (\Lambda^2\mathscr{S}_+)^2 \otimes T^{(-3)} \oplus \mathscr{S}_+ \otimes (\Lambda^2\mathscr{S}_-)^2 \otimes T^{(3)})$
4	0	$\mathscr{G}_0 \otimes ((\Lambda^2\mathscr{S}_+)^2 \otimes \Lambda^2\mathscr{S}_- \otimes t^{(-3)} \otimes T^{(1)} \oplus \Lambda^2\mathscr{S}_+ \otimes (\Lambda^2\mathscr{S}_-)^2 \otimes T^{(3)} \otimes T^{(-1)} \oplus \Lambda^2\mathscr{S}_+ \otimes \Lambda^2\mathscr{S}_- \otimes T^{(2)} \otimes T^{(-2)})$
5	0	$\mathscr{G}_1 \otimes (\mathscr{S}_+ \otimes \Lambda^2\mathscr{S}_+ \otimes \Lambda^2\mathscr{S}_- \otimes T^{(-3)} \otimes T^{(2)} \oplus \mathscr{S}_- \otimes \Lambda^2\mathscr{S}_- \otimes \Lambda^2\mathscr{S}_+ \otimes T^{(3)} \otimes T^{(-2)})$
6	0	$\operatorname{Ker} \nabla_3$, where $\nabla_3\colon \mathscr{G} \otimes \Omega^3 U \to \mathscr{G} \otimes \Omega^4 U$ (covariant differential)

The table shows for $N = 3$ all the cohomology groups of interest. The same conditions on U are imposed as in Proposition 3.3. Each entry (n, H^i) contains a sheaf $\mathscr{F}(n, i)$ on U such that

$$H^i\big(L(U), (\mathscr{G}_L \otimes J^n/J^{n+1})_0\big) \simeq H^0(U, \mathscr{F}(n, i)).$$

The form of such sheaves can be deduced from [9]. We use the following notation: $(\mathscr{G}, \nabla)$ is a sheaf with a connection on U associated with $(e^{(0)}, A\alpha \circ \rho)$ on $L(U)$; $T^{(i)} = S^i(T_1^*)$ (for $i \geqslant 0$), or $S^{-i}(T_1)$ (for $i < 0$); D_∇ is the Dirac operator on the background ∇; the tautological sheaves $e_\pm$ are constructed for $G(2, T_0) = M_{\text{red}}$.

4. Comments. The most important observation about the table is that, as we move up in stages $L(U)^{(n)}$, we first obtain fields (up to $n \leqslant 3$) and then equations ($6 \geqslant n \geqslant 3$). The first equation corresponding to vanishing of the obstruction $\omega(e^{(2)})$ is the Dirac equation: its origin was demonstrated by Witten [5] for the gauge group $\mathrm{GL}(1|0)$ in the coordinate formulation. The last obstruction is the current for the ordinary Yang-Mills field, i.e., the zeroth component of the supersymmetric field (see [7], [10], and [12]).

To list the remaining equations would be a lengthy business.

It should be noted that the concept of a "locally free sheaf of Yang-Mills type on $L(U)$" is far more concise than the concept of a "solution of the supersymmetric $N = 3$ system of Yang-Mills equations". However, to construct quantum field theory on L, we lack fundamental objects, i.e., the Lagrangian and the corresponding version of functional integration.

BIBLIOGRAPHY

1. M. F. Atiyah, *Geometry on Yang-Mills fields*, Scuola Norm. Sup., Pisa, 1979.

2. A. Douady and J.-L. Verdier, editors, *Les équations de Yang-Mills* (Sém., Ecole Norm. Sup., Paris, 1977–78), Astérisque No. 71–72, Soc. Math. France, Paris, 1980.

3. V. G. Drinfel′ and Yu. I. Manin, *Instantons and sheaves in* CP^3, Funktsional. Anal. i Prilozhen. **13** (1979), no. 2, 59–74; English transl. in Functional Anal. Appl. **13** (1979).

4. N. J. Hitchin, *Monopoles and geodesics*, Comm. Math. Phys. **83** (1982), 579–602.

5. Edward Witten, *An interpretation of classical Yang-Mills theory*, Phys. Lett. **77B** (1978), 394–398.

6. James Isenberg, Philip B. Yasskin and Paul S. Green, *Non-self-dual gauge fields*, Phys. Lett. **78B** (1978), 462–464.

7. Yu. I. Manin, *Gauge fields and holomorphic geometry*, Itogi Nauki: Sovremennye Problemy Mat., vol. 17, VINITI, Moscow, 1981, pp. 3–55; English transl. in J. Soviet Math. **21** (1983), no. 4.

8. G. M. Kenkin [Khenkin] and Yu. I. Manin, *Twistor description of classical Yang-Mills-Dirac fields*, Phys. Lett. **95B** (1980), 405–408.

9. ______, *On the cohomology of twistor flag spaces*, Composito Math. **44** (1981), 103–111.

10. Yu. I. Manin and G. M. Khenkin, *Yang-Mills-Dirac equations as Cauchy-Riemann equations on twistor space*, Yadernaya Fiz. **35** (1982), 1610–1626; English transl. in Soviet J. Nuclear Phys. **35** (1982).

11. Yu. I. Manin, *Flag superspaces and supersymmetric Yang-Mills equations*, Problems of High-Energy Physics and Quantum Field Theory, Serpukhov, 1982; English version in Arithmetic and Geometry (I. R. Shafarevich Sixtieth Birthday Vols.), Vol. II, Progress in Math., vol. 36, Birkhäuser, 1983, pp. 175–198.

12. N. P. Buchdahl, *Analysis on analytic spaces and non-self-dual Yang-Mills fields*, Preprint, Oxford, 1982.

13. Michael C. Eastwoood, Roger Penrose and R. O. Wells, Jr., *Cohomology and massless fields*, Comm. Math. Phys. **78** (1980/81), 305–351.

14. Wolf Barth and Klaus Hulek, *Monads and moduli of vector bundles*, Manuscripta Math. **25** (1978), 323–347.

15. N. J. Hitchin, *Linear field equations on self-dual spaces*, Proc. Roy. Soc. London Ser. A **370** (1980), 173–191.

Translated by D. and J. MATHON

Труды Матем. Инст. Стеклов.
165(1984)

Proceedings of the Steklov
Institute of Mathematics
1985, Issue 3

THE GROUP K_2 FOR A LOCAL FIELD

UDC 512.54

A. S. MERKUR'EV

ABSTRACT. It is proved that the group of the title is the direct sum of the roots of 1 contained in the field and a complete torsion-free group.
Bibliography: 6 titles.

This note is devoted to a proof of Tate's conjecture on the structure of the group $K_2(F)$ for a local field F [6]. Let n be the number of roots of unity contained in F; the Hilbert symbol of order n defines the epimorphism $\Gamma\colon K_2(F) \to \mu_n$ taking the generators $\{a, b\}$ of the group $K_2(F)$ to the element $(a, b)_n$ of the group μ_n of roots of unity of order n [2].

Moore [5] has shown that the kernel of the homomorphism Γ coincides with the group $nK_2(F)$, which is divisible. In particular, the homomorphism Γ splits, and the group $K_2(F)$ is isomorphic to the direct sum of the groups μ_n and $nK_2(F)$. Subsequently the structure of the divisible group $nK_2(F)$ has been studied; in [4] it is shown that this group has no l-torsion as long as l is a prime number different from the characteristic of the residue class field.

Tate has stated the conjecture (see, for example [6]) that the group $nK_2(F)$ is torsion-free, i.e. is uniquely divisible.

The main result of this work is the following

THEOREM. *Let F be a local field, and n the number of roots of unity contained in F. Then the group $nK_2(F)$ is uniquely divisible; in particular, $K_2(F)$ is isomorphic to the direct sum of the group μ_n and a uniquely divisible group.*

COROLLARY. *The torsion subgroup $K_2(F)_{\mathrm{tors}}$ is a cyclic group of order n.*

For local fields of nonzero characteristic this theorem was proved by Tate in [6]. Therefore we shall assume that all local fields under consideration have characteristic zero.

We introduce the following notation.

F is a local field of characteristic zero.

v is an exponent of the field F.

$\mathcal{O}$ is the ring of integers in F.

p is the characteristic of the residue class field of F.

q is the number of elements of the residue class field.

e is the ramification index of F over $\mathbf{Q}_p$, $e_0 = e/(p-1)$.

1980 *Mathematics Subject Classification*. Primary 12B22.

ξ_{p^r} is a primitive root of unity of order p^r.
$F_r = F(\xi_p r)$.
s is the largest integer for which $\xi_{p^s} \in F$.
p^t is the highest power of p dividing pe.
U_i are the groups of the standard filtration of the group U of units of the field F [2].

In order to indicate the dependence on the field F we will occasionally write v_F, $\mathcal{O}_F$, U_F, etc.

The Hilbert symbol of order p^s defines the homomorphism $K_2(F) \to \mu_{p^s}$ taking $\{a, b\}$ to $(a, b) = (a, b)_{p^s} \in \mu_{p^s}$. Nonzero elements a and b of F will be called *orthogonal* (written $a \perp b$) if $(a, b) = 1$.

The theorem is a consequence of two propositions.

PROPOSITION 1. *Let the field F satisfy the following conditions*:
1) *The extension F_r/F is totally ramified for any natural number r.*
2) $2s \geqslant t + 1$ (*in particular*, $\xi_p \in F$).
3) $\xi_{p^s} \notin U^p U_{e+1}$.
Then, for any element $x \in F^*$ orthogonal to ξ_p, $\{\xi_p, x\} = 0 \in K_2(F)$.

PROPOSITION 2. *Any local field can be embedded in a local field satisfying conditions* 1)–3) *of Proposition* 1.

We show first how the theorem follows from the propositions. In [6] it is shown that if Tate's conjecture hold in some local field, then the conjecture is true in any local subfield. By this result and Proposition 2 we may assume that F satisfies 1)–3) of Proposition 1. As noted above, it is enough to prove that the group $nK_2(F)$ has no p-torsion. An arbitrary element of order p in $K_2(F)$ has the form $\{\xi_p, x\}$, where $x \in F^*$ [6]. If $\{\xi_p, x\} \in nK_2(F)$, then $(\xi_p, x) = 1$, i.e. $\xi_p \perp x$, and, by Proposition 1, $\{\xi_p, x\} = 0 \in K_2(F)$, as was to be proved.

We now pass to the proof of Proposition 1. We assume that the field F satisfies conditions 1)–3) of that proposition (in particular, $\xi_p \in F$).

We select a subset $\Omega \subset \mu_{q-1}$ whose images in the residue class field form a basis over the prime field. Denote by S the set of natural numbers relatively prime to p and smaller than pe_0. Fix a prime element π and choose $\delta \in U_{pe_0}$ not a pth power [3]. Considering the behavior of the filtration of the group of units under the operation of raising to the pth power [3], it is not hard to prove the following well-known assertion:

LEMMA 1. *The elements $1 - \theta\pi^a$ ($\theta \in \Omega$, $a \in S$) and δ generate U_1 as a $\mathbf{Z}_p$-module.*

LEMMA 2. *The orders of the elements $\{1 - \theta\pi^a, \pi\}$ ($\theta \in \Omega$, $a \in S$) and $\{(1 - \eta\pi^{pe_0})^{p^t}, \pi\}$ ($\eta \in \mu_{q-1}$) in the group $K_2(F)$ are finite and relatively prime to p. In particular, $1 - \theta\pi^a \perp \pi$.*

PROOF. We have

$$a(q-1)\{1 - \theta\pi^a, \pi\} = (q-1)\{1 - \theta\pi^a, \pi^a\}$$
$$= -(q-1)\{1 - \theta\pi^a, \theta\} = -\{1 - \theta\pi^a, \theta^{q-1}\} = 0,$$

and the numbers $a(q-1)$ and p are relatively prime. Analogously, if $pe_0 = p^t m$, where $(m, p) = 1$, then

$$m(q-1)\left\{(1-\eta\pi^{pe_0})^{p^t}, \pi\right\} = 0.$$

We now take the element δ in the form $\delta = 1 - \eta\pi^{pe_0}$, where $\eta \in \mu_{p-1}$ [3]. Since $\xi_{p^s} \in U_1$, by Lemma 1 there exist $\alpha_{\theta,a}$, $\beta \in \mathbf{Z}_p$ $(\theta \in \Omega,\ a \in S)$, so that

$$\xi_{p^s} = (1-\eta\pi^{pe_0})^{\beta} \prod_{\substack{\theta\in\Omega \\ a\in S}} (1-\vartheta\pi^a)^{\alpha_{\theta,a}}.$$

Since the element (δ, π) generates the group μ_{p^s} [3], from Lemma 2 we find that if $\xi_{p^s} \perp \pi$, then β is divisible by p^s.

We set

$$\Delta = \beta p e_0 \frac{\eta\pi^{pe_0}}{1-\eta\pi^{pe_0}} + \sum_{\substack{\theta\in\Omega \\ a\in S}} \alpha_{\theta,a} a \frac{\theta\pi^a}{1-\theta\pi^a}.$$

LEMMA 3. $\Delta \neq 0$.

PROOF. Suppose that $\Delta = 0$. Since $pe_0 \geqslant e+1$ and $\xi_{p^s} \notin U^p U_{e+1}$ (condition 3) of Proposition 1), there is an $a \in S$ such that $a \leqslant e$ and $\alpha_{\theta,a}$ is not divisible by p for some $\theta \in \Omega$. Choose the least such a_0 and consider the equation $\Delta = 0$ modulo π^{a_0+1}. Because for $a < a_0$ the element $\alpha_{\theta,a}\pi^a$ is divisible by $p\pi^a$ (and consequently by π^{a_0+1}) we see that the element

$$\sum_{\theta\in\Omega} \alpha_{\theta,a_0}\theta\pi^{a_0}$$

is divisible by π^{a_0+1}, i.e. the residue of $\sum_{\theta\in\Omega}\alpha_{\theta,a}\theta$ equals $\bar{0}$. However this contradicts the choice of Ω and a_0.

For arbitrary natural numbers $r_{\theta,a}$ $(\theta \in \Omega,\ a \in S)$ and r consider the polynomial

$$f(X) = \xi_{p^s} - (1-\eta X^{pe_0})^r \prod_{\substack{\theta\in\Omega \\ a\in S}} (1-\theta X^a)^{r_{\theta,a}}.$$

If the numbers $r_{\theta,a}$ are close (in the p-adic topology) to the respective $\alpha_{\theta,a}$, and r is close to β, then $f(\pi)$ is close to zero, and $f'(\pi)$ is close to the element $\xi_{p^s}\pi^{-1}\Delta$, which is different from zero by Lemma 3.

By the theorem on the roots of polynomials over a local field [1], the polynomial f has a root π_1 sufficiently close to π, so that, for example, $\pi/\pi_1 \in U_1^p$. In this way we have proved

LEMMA 4. *Let π be an arbitrary prime element of the field F. Then there exist natural numbers $r_{\theta,a}$ $(\theta \in \Omega,\ a \in S)$, r and a prime element π_1 so that $\pi/\pi_1 \in U_1^p$ and*

$$\xi_{p^s} = (1-\eta\pi_1^{pe_0})^r \prod_{\substack{\theta\in\Omega \\ a\in S}} (1-\theta\pi_1^a)^{r_{\theta,a}}. \tag{1}$$

If $\xi_{p^s} \perp \pi$, then we may take r to be divisible by p^s.

With this we proceed directly to the proof of Proposition 1. It is necessary to show that if the element $x \in F^*$ is orthogonal to ξ_p, then $\{\xi_p, x\} = 0 \in K_2(F)$.

First let $x = \pi$ be a prime element and $\xi_{p^s} \perp \pi$. By Lemma 4 find a prime element π_1 satisfying (1), so that $\pi = \pi_1 u^p$, for some unit $u \in U$. We raise both sides of (1) to the p^{s-1} power and multiply by π_1. Since r is divisible by p^s, $p^{s-1}r$ is divisible by p^t (condition 2) of Proposition 1). Therefore, by Lemma 2, applied to the element π_1, the order of the element $\{\xi_p, \pi_1\}$ is relatively prime to p, i.e. $\{\xi_p, \pi_1\} = 0$, and consequently $\{\xi_p, \pi\} = \{\xi_p, \pi_1\} + \{\xi_p, u^p\} = 0$.

Now let the element x be arbitrary and assume as before that $\xi_{p^s} \perp x$. By condition 1) of Proposition 1 the field

$$F_{2s} = F\left(\sqrt[p^s]{\xi_{p^s}}\right)$$

is totally ramified over F; therefore the norm π_0 of a prime element of F_{2s} is a prime in F. By one of the properties of the Hilbert symbol [2], $\xi_{p^s} \perp \pi_0$. But for some integer k the element $\pi_1 = x\pi_0^k$ is prime and orthogonal to ξ_{p^s}; and by what was shown above $\{\xi_p, x\} = \{\xi_p, \pi_1\} - k\{\xi_p, \pi_0\} = 0$.

We consider, finally, the general case. Since $\xi_p \perp x$, there is a $y \in F^*$ for which $\xi_{p^s} \perp xy^p$. Therefore, by the above, $\{\xi_p, x\} = \{\xi_p, xy^p\} = 0 \in K_2(F)$, and this completes the proof of the proposition.

We now prove Proposition 2. We will show that for sufficiently large k the field F_k satisfies conditions 1)–3) of Proposition 1. This is clear for 1) and 2), and so it remains to consider 3).

LEMMA 5.

$$\lim_{\substack{k \to \infty \\ x \in \mathcal{O}_{F_k}}} v\left(\mathrm{Tr}_{F_k/F}(x)\right) = \infty$$

($\mathrm{Tr}_{F_k/F}$ *is the trace map*).

PROOF. For the field $F = \mathbf{Q}_p$ the lemma follows from a description of the integral elements of F [1]. Consider now the general case. Let x be an integral element of F_k, and choose $y \in \mathcal{O}_F$ so that

1) $t = y\mathrm{Tr}_{F_k/F}(x) = \mathrm{Tr}_{F_k/F}(xy) \in \mathbf{Q}_p$;
2) $v_F(y) < r = (F\colon \mathbf{Q}_p)$.

We denote the field $\mathbf{Q}_p(\xi_{p^k})$ by E_k; then

$$rt = \mathrm{Tr}_{F_k/\mathbf{Q}_p}(xy) = \mathrm{Tr}_{E_k/\mathbf{Q}_p}(z),$$

where $z = \mathrm{Tr}_{F_k/E_k}(xy)$ is an integer in E_k. It follows that

$$v_F\left(\mathrm{Tr}_{F_k/F}(x)\right) = v_F(rt) - v_F(y) - v_F(r) > v_F\left(\mathrm{Tr}_{E_k/\mathbf{Q}_p}(z)\right) - r - v_F(r) \to \infty$$

as $k \to \infty$, by the first part of the proof.

LEMMA 6. *For sufficiently large k the norm in the extension F_k/F of the group U_{e+1} in the field F_k belongs to U_F^p.*

PROOF. Let $u \in U_{e+1} \subset F_k$; then $u = 1 + px$, where $v_{F_k}(x) \geqslant 1$. Since $2e \geqslant pe_0$,

$$N_{F_k/F}(u) \equiv 1 + p\mathrm{Tr}_{F_k/F}(x) \pmod{U_{pe_0+1}}.$$

By Lemma 5 it is clear that for sufficiently large k the right side of this congruence lies in the group U_{pe_0+1}, i.e. $N_{F_k/F}(u) \in U_{pe_0+1} \subset U^p$ [3], and this finishes the proof.

To complete the proof of Proposition 2 we must show that for sufficiently large k the field F_k satisfies condition 3) of Proposition 1. Let r be any natural number. With the help of Lemma 6 we find an m with the property that for $k \geqslant m$

$$N_{F_{k+r}/F_r}(U_{e+1}) \subset U^p_{F_r}.$$

We shall assume that $\xi_{p^{k+r}} \in U^p U_{e+1} \subset F_{k+r}$ and derive a contradiction. Applying the norm map F_{F_{k+r}/F_r} to the element $\xi_{p^{k+r}}$, we obtain

$$\xi_{p^r} \in N_{F_{k+r}/F_r}(U^p U_{e+1}) \subset U^p_{F_r},$$

but this contradicts the fact that $\xi_{p^{r+1}} \notin F_r$ for large r.

This proves Proposition 2.

BIBLIOGRAPHY

1. Z. I. Borevich and I. R. Shafarevich, *Number theory*, 2nd rev. ed., "Nauka", Moscow, 1972; English transl. of 1st ed., Academic Press, 1966.

2. J. W. S. Cassels and A. Fröhlich, *Algebraic number theory* (Proc. Instructional Conf., Brighton, 1965), Academic Press, London, and Thompson, Washington, D. C., 1967.

3. John Milnor, *Introduction to algebraic K-theory*, Princeton Univ. Press, Princeton, N. J., and Univ. of Tokyo Press, Tokyo, 1971.

4. Joseph E. Carroll, *On the torsion in K_2 of local fields*, Algebraic K-theory II (Proc. Conf., Seattle, 1973), Lecture Notes in Math., vol. 342, Springer-Verlag, 1973, pp. 464–473.

5. Calvin C. Moore, *Group extensions of p-adic and adelic linear groups*, Inst. Hautes Études Sci. Publ. Math. No. 35 (1969), 157–222.

6. J. Tate, *On the torsion in K_2 of fields*, Algebraic Number Theory (Proc. Internat. Sympos., Kyoto, 1976), S. Iyanaga editor, Proc. Taneguchi Internat. Sympos., Division of Math., No. 2, Japan Soc. Promotion Sci., Tokyo, 1977, pp. 243–261.

Translated by P. MORTON

Труды Матем. Инст. Стеклов.
165(1984)

Proceedings of the Steklov
Institute of Mathematics
1985, Issue 3

SURFACES OF TYPE K3 WITH A FINITE AUTOMORPHISM GROUP AND A PICARD GROUP OF RANK THREE

UDC 512.7 + 511.3

V. V. NIKULIN

ABSTRACT. Hyperbolic quadratic forms over **Z** of rank three whose groups of automorphisms are generated by reflections relative to the elements with square (-2) are classified up to finite index. The analogous problem was solved earlier by the author for forms of rank $\geqslant 5$ (see [7] and [8]) and by È. B. Vinberg for rank 4 (not published); and it is very simple for rank $\leqslant 2$ (see [6]). Thanks to the global Torelli theorem for surfaces of type K3 (Piatetski-Shapiro and Shafarevich [6]) this completes the classification of algebraic surfaces of type K3 with a finite group of automorphisms. The author notes that the same question of the finiteness of the group of birational automorphisms of an algebraic surface can be reduced to Enriques surfaces (for them it is open).

An interesting application is cited, relating to the arithmetic decomposition of a convex polygon into smaller polygons.

Figures: 10. Tables: 3. Bibliography: 14 titles.

A classical consequence of the theory of algebraic curves and Riemann surfaces, developed in the preceding century, asserts that the group of birational automorphisms of an algebraic curve (or the group of automorphisms of a compact Riemann surface) is finite if the genus $g > 1$, and infinite if $g \leqslant 1$ (see [1], and also [13] and [14]).

The analogous question is undoubtedly of interest also in the case of algebraic surfaces: When is the group of birational automorphisms of a projective algebraic surface finite? Thanks to the classification of algebraic surfaces in the classic works of the Italian school of algebraic geometry at the beginning of our century (see [4], and also [5]) this question remained open only for two classes of algebraic surfaces: K3 surfaces and Enriques surfaces (see [2], [3], and also [5], [9] and [10]). At that time only examples of K3 surfaces both with finite and with infinite groups of automorphisms were known, as well as examples of Enriques surfaces with infinite groups of automorphisms [12].

Progress was achieved in 1971 by Piatetski-Shapiro and Shafarevich [6], who obtained, as a consequence of a global Torelli theorem, that the group of automorphisms of a K3 surface X over an algebraically closed field of characteristic zero is isomorphic, up to a finite group, to the factor group $O(S)/W^{(2)}(S)$, where $O(S)$ is the group of automorphisms of the Picard lattice S of the surface X (i.e. S is the Picard group of X, considered together with the intersection index form), and

1980 *Mathematics Subject Classification*. Primary 14J25; Secondary 14E05, 14J05, 52A10.

$W^{(2)}(S)$ is the subgroup generated by all reflections relative to elements with square (-2) of the lattice S. Since Enriques surfaces have as universal cover (two-sheeted) K3 surfaces the analogous account (we do not go through it) holds also for their groups of automorphisms. Since the Picard lattice of the surface is hyperbolic, in particular, the question of a description of the K3 surfaces with a finite group of automorphisms was posed by the author [6] in terms of the description of hyperbolic lattices S (over $\mathbf{Z}$) for which the factor group $O(S)/W^{(2)}(S)$ is finite. For $\operatorname{rk} S = 1$ this question is trivial, and for $\operatorname{rk} S = 2$ it is easily solved (that was also done in [6], §7) thanks to the known simple description of the groups of automorphisms of a hyperbolic lattice of rank two.

For $\operatorname{rk} S \geqslant 3$ the question turns out to be much more difficult. For $\operatorname{rk} S \geqslant 5$ the answer was obtained in [8] and [9]: it turns out that there are only finitely many such lattices S, and all of them were found. For $\operatorname{rk} S = 4$, the answer, also finite, was later obtained by È. B. Vinberg (regrettably, this result has not yet been published, although it was mentioned back in 1979 [8]). After this the description of the lattices S of rank 3 was obtained by the author. Our goal in this paper is to present this description (Theorems 1.3 and 2.5). Thus we complete the description of K3 surfaces defined over an algebraically closed field of characteristic zero which have a finite group of automorphisms.

We remark that in [10] and [11] assertions of finiteness were generalized to hyperbolic varieties over arbitrary rings of algebraic integers whose groups of automorphisms are generated by reflections up to finite index. Therefore, there is in principal a general algorithm to describe lattices of rank 3 of interest to us. But this algorithm, to all appearances, is very unwieldy in this case. In this paper we succeed in carrying out the classification in relatively few pages thanks only to drawing in other, rather subtle, considerations.

The question of a description of the projective algebraic surfaces with a finite group of birational automorphisms remains open only for Enriques surfaces.([1])

At the end of §2 we cite an interesting (in our opinion) topological application connected with the arithmetic subdivision of convex polygons into triangles.

In this paper we adhere to the notation of [9], §1. There the reader will find all preliminary information necessary to us, and which we will use without special mention.

§1. The compact case

Let S be a hyperbolic lattice of rank 3, having a finite factor group $A(S) = O'(S)/W^{(2)}(S)$ and a fundamental polygon $\mathscr{M}$ for $W^{(2)}(S)$ in the Lobachevsky plane $\mathscr{L}(S)$ which is bounded. Our goal is to describe all such lattices S. As usual, in considering the semidirect product $O'(S) = A(S) \cdot W^{(2)}(S)$, we will consider elements of $A(S)$ as automorphisms of S carrying $\mathscr{M}$ into itself.

1°. The group $A(S)$ and the vertices of $\mathscr{M}$. Since $A(S)$ is a finite group, being a group of automorphisms of S over $\mathbf{Z}$, it is finite crystallographic of dimension two ([9], §7.2°). Therefore $A(S)$ is one of the groups C_n or D_n, $n = 1, 2, 3, 4, 6$, where C_n

---

[1] The first example of Enriques surfaces with finite group of automorphisms was constructed by I. V. Dolgachev. The full description of Enriques surfaces with finite group of automorphisms was recently obtained by the author.

is the group of rotations and D_n is that of all symmetries of the regular n-gon.

Let v be a vertex of $\mathcal{M}$, and let $h \in S$ be the primitive element of S defining v, i.e. $v = h \cdot \mathbf{R}^+$. Let r_1 and r_2 be sides of $\mathcal{M}$ going out from v, and δ_1 and δ_2 the elements orthogonal to them and pointing outward with square (-2) of S. Then either $\delta_1 \cdot \delta_2 = 0$ (if v is a vertex of type $\mathbf{A}_1^2$), or $\delta_1 \cdot \delta_2 = 1$ (if v is a vertex of type $\mathbf{A}_2$). In both of these cases the lattice $K = [\delta_1, \delta_2]$ spanned by δ_1 and δ_2 is a maximal even lattice; therefore the imbeddings $K \subset S$ are primitive, and the sublattices $[h]$ (spanned by h) and $K = [\delta_1, \delta_2]$ are orthogonal complements of each other in S. On the strength of the connection between orthogonal complements for primitive imbeddings, the super-lattice $[h] \oplus K \subset S$ is determined by subgroups $\Gamma_h \subset \mathfrak{A}_h$ and $\Gamma_K \subset \mathfrak{A}_K$ of the discriminant groups $\mathfrak{A}_h = [h]^*/[h]$ and $\mathfrak{A}_K = K^*/K$ and an isomorphism

$$\gamma_S\colon\ q_h|\Gamma_h \tilde{\to} -q_K|\Gamma_K \tag{1.1}$$

of the restrictions to these subgroups of the discriminant forms of the quadratic lattices $[h]$ and K (by definition, the graph of γ_s is $\Gamma_S = S/([h] \oplus K) \subset \mathfrak{A}_h \oplus \mathfrak{A}_K$). From this we see that the discriminant form of S is equal to

$$q_S = \left(q_h \oplus q_K|(\Gamma_S)^{\perp}_{q_h \oplus q_K}\right)/\Gamma_S \tag{1.2}$$

(we remark that the subgroup Γ_S is isotropic relative to $q_h \oplus q_K$).

The group $\mathfrak{A}_h = [h]^*/[h] \cong \mathbf{Z}/t\mathbf{Z}$, where $t = h^2$ is even and $t > 0$. Therefore on the strength of the isomorphism (1.1) $\Gamma_h \cong \Gamma_K$ are cyclic groups. The subgroup $\Gamma_h \subset \mathfrak{A}_h$ has order $m = |\Gamma_h| = |\Gamma_K|$, with $m|t$. On the other hand,

$$\mathfrak{A}_K \simeq \begin{cases} (\mathbf{Z}/2\mathbf{Z})^2 & \text{for type } \mathbf{A}_1^2, \\ \mathbf{Z}/3\mathbf{Z} & \text{for type } \mathbf{A}_2. \end{cases} \tag{1.3}$$

Therefore, since $\Gamma_K \subset \mathfrak{A}_K$, we obtain $m = 1$ or 2 (for type $\mathbf{A}_1^2$), or 3 (for type $\mathbf{A}_2$). In this way, every vertex v in $\mathcal{M}$ has for invariants

$$(\text{type}, m), \tag{1.4}$$

where $m = 1$ or 2 (for type $\mathbf{A}_1^2$) or 3 (for type $\mathbf{A}_2$).

We denote, further, by $A(S)_v$ the stabilizer subgroup of v in $\mathcal{M}$ relative to $A(S)$. Evidently, either it is trivial (equal to C_1), or $A(S)_v \cong D_1$ has order two; it interchanges the sides r_1 and r_2 issuing from v (or the elements δ_1 and δ_2) of the polygon $\mathcal{M}$.

FUNDAMENTAL THEOREM 1.1. *The invariants* (1.4) *determine the conjugacy class of the vertex of the fundamental* (*for* $W^{(2)}(S)$) *polygon* $\mathcal{M}$ *relative to* $A(S)$. *In this case, for* $A(S)$, *the vertices of* $\mathcal{M}$ *and their stabilizer subgroups have only those possibilities which are presented in Table* 1.

PROOF. Evidently (by (1.2) and (1.3)), for $m = 1$ we have

$$\mathfrak{A}_S \simeq \begin{cases} \mathbf{Z}/t\mathbf{Z} \oplus (\mathbf{Z}/2\mathbf{Z})^2 & \text{for type } \mathbf{A}_1^2, \\ \mathbf{Z}/t\mathbf{Z} \oplus \mathbf{Z}/3\mathbf{Z} & \text{for type } \mathbf{A}_2. \end{cases} \tag{1.5}$$

TABLE 1

Case	vertices of $\mathcal{M}$	stabilizer subgroups of the vertices of $\mathcal{M}$	the group $A(S)$	Case	vertices of $\mathcal{M}$	stabilizer subgroups of the vertices of $\mathcal{M}$	the group $A(S)$
I	$(\mathbf{A}_1^2, 1)$	D_1	D_6	V	$(\mathbf{A}_1^2, 2)$	C_1	D_2
II	$(\mathbf{A}_2, 1)$	D_1	D_4, D_6		$(\mathbf{A}_2, 1)$	D_1	
III	$(\mathbf{A}_2, 3)$	C_1	D_2	VI	$(\mathbf{A}_1^2, 2)$	D_1	D_2
IV	$(\mathbf{A}_1^2, 2)$	D_1	D_2		$(\mathbf{A}_2, 3)$	C_1	
	$(\mathbf{A}_2, 1)$	D_1		VII	$(\mathbf{A}_1^2, 2)$	C_1	C_2, D_1, D_2
					$(\mathbf{A}_2, 3)$	C_1	

From (1.1)–(1.3), as direct calculations show, it follows that for $m \neq 1$

$$\mathfrak{A}_S \simeq \begin{cases} \mathbf{Z}/t\mathbf{Z} & \text{for type } \mathbf{A}_1^2, \\ \mathbf{Z}/t\mathbf{Z}/3 & \text{for type } \mathbf{A}_2 \end{cases} \tag{1.6}$$

(in particular, the group $\mathfrak{A}_S$ is cyclic). Hence as candidates for invariants (1.4) of the vertices of $\mathcal{M}$ we obtain the possibilities presented in Table 1 and one additional possibility: $(\mathbf{A}_1^2, 2)$. Below we will rule out the last possibility.

We show that the invariants (1.4) determine the conjugacy class of the vertex v and find its stabilizer subgroup (relative to $A(S)$). The Formulae (1.5) and (1.6) indicate that for the given invariants (type, m) of the vertex v the number

$$t = \begin{cases} |\operatorname{discr} S| \cdot 2^{-2} & \text{for } (\mathbf{A}_1^2, 1), \\ |\operatorname{discr} S| \cdot 3^{-1} & \text{for } (\mathbf{A}_2, 1), \\ |\operatorname{discr} S| & \text{for } (\mathbf{A}_1^2, 2), \\ |\operatorname{discr} S| \cdot 3^{-1} & \text{for } (\mathbf{A}_2, 3) \end{cases}$$

(we recall that $|\operatorname{discr} S| = |\mathfrak{A}_S|$).

Thus the number $t = h^2$ is determined by the invariants (1.4). For type $\mathbf{A}_2$ they clearly determine also the subgroups $\Gamma_h \subseteq \mathfrak{A}_h$ and $\Gamma_k \subseteq \mathfrak{A}_K$. We remark, further, that the automorphism of K transposing δ_1 and δ_2 operates on $\mathfrak{A}_K \simeq \mathbf{Z}/3\mathbf{Z}$ as a permutation. Therefore, up to a transposition of δ_1 and δ_2, they determine also the isomorphism γ_S, and the collection

$$(t, \Gamma_h, \Gamma_K, \gamma_S) \tag{1.7}$$

determines the isomorphism class of the vertex v ([9], §7.3°). If the $\Gamma_h \simeq \Gamma_K$ are trivial, then the automorphism of K transposing δ_1 and δ_2 clearly extends to an automorphism of S which is the identity on h; and that is why it belongs to $A(S)$. But if $\Gamma_h \simeq \Gamma_K \simeq \mathbf{Z}/3\mathbf{Z}$, then this extension does not exist, since it does not agree with γ_S. This shows that for type $\mathbf{A}_2$ the invariants (1.4) determine the isomorphism class of the vertex v, with

$$A(S)_v \simeq \begin{cases} D_1, & \text{if } m = 1, \\ C_1, & \text{if } m = 3. \end{cases} \tag{1.8}$$

All that was said above obviously carries over to type $\mathbf{A}_1^2$, if $m = 1$. In this case $A(S)_v \simeq D_1$. Let the invariants of the vertex v be $(\mathbf{A}_1^2, 2)$. Then for the group Γ_K there are two possibilities (up to transposition of δ_1 and δ_2): $\Gamma_K = [z]$, where $z = \delta_1/2 \pmod K$, or $z = (\delta_1 + \delta_2)/2 \pmod K$. In the first case $q_K(z) = -\frac{1}{2} \pmod 2$, and in the second case $q_K(z) = 1 \pmod 2$.

Let w be a generating element of the subgroup of order 2 of the cyclic group $\mathfrak{A}_h \simeq \mathbf{Z}/t\mathbf{Z}$. On the strength of the isomorphism (1.1) we get that $q_h(w)$ must be equal to $\frac{1}{2} \pmod 2$ in the first case and $1 \pmod 2$ in the second. Since $q_h(w)$ is determined by the number t, for the same reason the invariants (1.4) determine the subgroup Γ_K. Since $\Gamma_h \simeq \Gamma_K \simeq \mathbf{Z}/2\mathbf{Z}$, clearly the isomorphism γ_S: $\Gamma_h \simeq \Gamma_K$ is uniquely determined. In the same way the invariants (1.4) determine the collection $(t, \Gamma_h, \Gamma_K, \gamma_S)$, and it determines the isomorphism class of the vertex v. It is easy to see (we will not need it) that in the case $(\mathbf{A}_1^2, 2)$, in addition,

$$A(S)_v \simeq \begin{cases} C_1, & \text{if } z = \delta_1/2 \pmod K, \\ D_1, & \text{if } z = (\delta_1 + \delta_2)/2 \pmod K. \end{cases}$$

The proven assertion about the conjugacy of the vertices of $\mathcal{M}$ and (1.8) show that the vertices of $\mathcal{M}$ (and not only their invariants (1.4)) and their stabilizer subgroups are either as stated in Table 1, or $(\mathbf{A}_1^2, 2)$ with stabilizer subgroup C_1 or D_1.

In order to exclude the last two possibilities and to describe the possible groups $A(S)$, we will need one more consideration: the canonical isomorphism

$$A(S) \to O(q_S), \tag{1.9}$$

is injective, where $O(q_S)$ is the group of automorphism of the discriminant form ([9], Theorem 7.2.1). Thus, if among the vertices of $\mathcal{M}$ there are vertices with invariants $(\mathbf{A}_1^2, 2)$ or $(\mathbf{A}_1^2, 3)$, then the group $\mathfrak{A}_S$ is cyclic (see (1.6)), whence it follows easily that

$$O(q_S) \simeq \prod_p O(q_{S_p}) \simeq (\mathbf{Z}/2\mathbf{Z})^a$$

is a 2-elementary group, where q_{S_p} is the p-component of q_S. Therefore on the strength of the injectivity of the homomorphism (1.9) the group $A(S)$ is also 2-elementary, which for two-dimensional crystallographic groups is true only for the groups $A(S) \simeq C_1$, C_2, D_1, or D_2. In this way, if among the vertices of $\mathcal{M}$ there are vertices with invariants $(\mathbf{A}_1^2, 2)$ or $(\mathbf{A}_2, 3)$, the $A(S) \simeq C_1$, C_2, D_1, or D_2. This proves that in cases 3–7 from Table 1 the group $A(S)$ is as shown on the right of the table, if we take into account that $\mathcal{M}$ has at least three vertices. This also shows the impossibility of the case of vertices of $\mathcal{M}$ with invariants $(\mathbf{A}_1^2, 2)$ and stabilizer subgroup D_1. In the case $(\mathbf{A}_1^2, 2)$ with stabilizer subgroup C_1 it is necessary, in addition, to make use of the fact that the sum of the angles of an n-gon in the Lobachevsky plane is less than $\pi(n-2)$. This then proves that in cases 1 and 2 from the table the groups $A(S)$ are as indicated.

The theorem is proved.

2°. Classification of the fundamental polygons $\mathcal{M}$ of the lattices S. The fundamental polygons $\mathcal{M}$ for the groups $W^{(2)}(S)$ of the lattices S are determined up to congruence (as is any polygon in the Lobachevsky plane) by their Gram matrix

$$(a_{ij}) = (\delta_i \cdot \delta_j), \qquad 1 \leqslant i,\ j \leqslant k, \tag{2.1}$$

where $\{\delta_1, \ldots, \delta_m\} = P(\mathcal{M})$ is a set of elements with square (-2), orthogonal to the boundaries of $\mathcal{M}$ and pointing outward. This matric clearly determines a sublattice of finite index in S, generated by elements of square (-2) of S. We will consider the elements $\delta_1, \ldots, \delta_k$ to be ordered corresponding to the sides around $\mathcal{M}$. We note that, since S is hyperbolic,

$$\operatorname{rk}(a_{ij}) = 2, \qquad \operatorname{sign}(a_{ij}) = (1, 2, k-3), \tag{2.2}$$

where $\operatorname{sign}(a_{ij})$ is the collection of "numbers of positive, negative, zero squares" respectively of the symmetric matrix a_{ij}. We note also the geometric meaning of the coefficients a_{ij}:

$$a_{ij} = \delta_i \cdot \delta_j = 2\cos\varphi, \tag{2.3}$$

if δ_i and δ_j are orthogonal to lines meeting at an angle φ; and

$$a_{ij} = \delta_i \cdot \delta_j = 2\cosh\rho,$$

if δ_i and δ_j are orthogonal to nonintersecting lines with distance ρ between them.

THEOREM 1.2. *The Gram matrices of the fundamental polygons $\mathcal{M}$ for the groups $W^{(2)}(S)$ of the lattices S are the following and only the following (the polygons $\mathcal{M}$ corresponding to them are depicted in Figure* 1):

$$\begin{matrix} U_1 = \\ (\textit{Figure } 1,\ I\ D_6) \end{matrix} \begin{Vmatrix} -2 & 0 & 4 & 6 & 4 & 0 \\ 0 & -2 & 0 & 4 & 6 & 4 \\ 4 & 0 & -2 & 0 & 4 & 6 \\ 6 & 4 & 0 & -2 & 0 & 4 \\ 4 & 6 & 4 & 0 & -2 & 0 \\ 0 & 4 & 6 & 4 & 0 & -2 \end{Vmatrix},$$

$$\begin{matrix} U_2 = \\ (\textit{Figure } 1,\ II\ D_6) \end{matrix} \begin{Vmatrix} -2 & 1 & 7 & 10 & 7 & 1 \\ 1 & -2 & 1 & 7 & 10 & 7 \\ 7 & 1 & -2 & 1 & 7 & 10 \\ 10 & 7 & 1 & -2 & 1 & 7 \\ 7 & 10 & 7 & 1 & -2 & 1 \\ 1 & 7 & 10 & 7 & 1 & -2 \end{Vmatrix},$$

$$\begin{matrix} U_3 = \\ (\textit{Figure } 1,\ III\ D_4) \end{matrix} \begin{Vmatrix} -2 & 1 & 4 & 1 \\ 1 & -2 & 1 & 4 \\ 4 & 1 & -2 & 1 \\ 1 & 4 & 1 & -2 \end{Vmatrix}, \qquad \begin{matrix} U_4 = \\ (\textit{Figure } 1,\ III\ D_2) \end{matrix} \begin{Vmatrix} -2 & 1 & 3 & 1 \\ 1 & -2 & 1 & 6 \\ 3 & 1 & -2 & 1 \\ 1 & 6 & 1 & -2 \end{Vmatrix},$$

$$\begin{matrix} U_5 = \\ (\textit{Figure } 1,\ IV\ D_2) \end{matrix} \begin{Vmatrix} -2 & 1 & 3 & 0 \\ 1 & -2 & 0 & 3 \\ 3 & 0 & -2 & 1 \\ 0 & 3 & 1 & -2 \end{Vmatrix}, \qquad \begin{matrix} U_6 = \\ (\textit{Figure } 1,\ IV\ D_2) \end{matrix} \begin{Vmatrix} -2 & 1 & 5 & 6 & 5 & 1 \\ 1 & -2 & 0 & 5 & 11 & 9 \\ 5 & 0 & -2 & 1 & 9 & 11 \\ 6 & 5 & 1 & -2 & 1 & 5 \\ 5 & 11 & 9 & 1 & -2 & 0 \\ 1 & 9 & 11 & 5 & 0 & -2 \end{Vmatrix}.$$

PROOF. By Theorem 1.1, the fundamental polygons and their Gram matrices have the form given by Figure 1, $I\ D_y$–$VII\ D_2$ (the numbering in Figure 1 is given to correspond with Table 1). The Gram matrices are easily written out from conditions (2.2) and (2.3). Accordingly, cases $I\ D_6$, $II\ D_6$, $III\ D_4$, and $IV\ D_2$ give the Gram matrices U_1, U_2, U_3, and U_5, respectively. In the remaining cases it is necessary, in addition, to make use of the fact that the Gram matrices a_{ij} must be integral. The equations and inequalities on the parameters indicated in Figure 1 allow all possible integral Gram matrices to be easily calculated in the remaining cases, except case *VII* D_2. For these, case *III* D_2 gives matrices U_2 and U_4, case *VI* D_2 gives the matrix U_6, and in the remaining cases the Gram matrices cannot be integral. In case *VII* D_2 the Gram matrix also cannot be integral, but to verify this is much more involved. We did this, having made calculations both by hand and on a computer. Here we limit ourselves to proving the finiteness of the number of integral Gram matrices in case *VII* D_2, so that it will remain to sort out the finite number of possibilities. Utilizing considerations of divisibility, this is not hard to do.

On the strength of the figure, for *VII* D_2 all such matrices correspond to solutions of the system of equations

$$(w^2 - 4)(s^2 - 4) = 4u^2, \tag{2.4}$$

$$(w^2 - 4)(t^2 - 4) = 4v^2, \tag{2.5}$$

$$u^2v^2 - 4u^2 - 4v^2 - 2uvw - 3w^2 + 12 = 0, \tag{2.6}$$

where u^2, v, w^2, uw, s^2, t^2 and ts are integers and u, v, w, s and t are greater than 2. (We remark that (2.4), (2.5) and (2.6) are the conditions for the determinants of the Gram matrices with elements $(e, f, \delta_1, \delta_2)$, $(\delta_1, f, e, \delta_3)$ and $(\delta_1, \delta_2, \delta_3, e)$ respectively to be equal to zero.)

Since $s > 2$, and $s^2 \in \mathbf{Z}$, then $s^2 \geqslant 5$. From (2.4) it follows that then

$$4u^2 \geqslant w^2 - 4. \tag{2.7}$$

From (2.6) and the inequalities $u > 2$ and $v > 2$ we obtain that

$$u = \frac{vw}{v^2 - 4} + \sqrt{\frac{v^2w^2}{v^2 - 4} + \frac{4v^2 + 3w^2 - 12}{v^2 - 4}}.$$

Hence by (2.7)

$$\begin{aligned} u &\leqslant \frac{v\sqrt{4v^2 + 4}}{v^2 - 4} + \sqrt{\frac{20v^4 - 60v^2}{(v^2 - 4)^2}} = \sqrt{4 + \frac{36}{v^2 - 4} + \frac{80}{(v^2 - 4)^2}} \\ &+ \sqrt{20 + \frac{100}{v^2 - 4} + \frac{80}{(v^2 - 4)^2}} \leqslant \sqrt{4 + \frac{36}{5} + \frac{80}{5^2}} \\ &+ \sqrt{20 + \frac{100}{5} + \frac{80}{5^2}} < \sqrt{108}, \end{aligned}$$

since $v \geqslant 3$. Since u^2 is an integer and $u \geqslant 2$, then $u = \sqrt{m}$, where m is an integer and $5 \leqslant m \leqslant 107$. In exactly the same way it follows that $3 \leqslant v \leqslant 25$, where v is an integer. From this, in an obvious way, we get the finiteness of the number of solutions of the system (2.4)–(2.6).

In this way, the Gram matrices $U_1, \ldots, U_6$ exhaust the Gram matrices of the polygon $\mathscr{M}$. That they really are Gram matrices of a polygon for groups $W^{(2)}(S)$ of lattices S is shown in the following subsection.

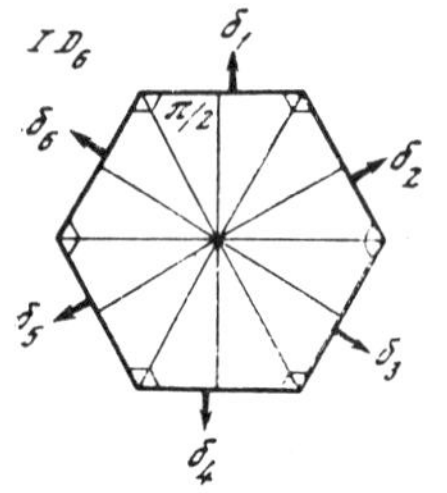

−2	0	4	6	4	0
0	−2	0	4	6	4
4	0	−2	0	4	6
6	4	0	−2	0	4
4	6	4	0	−2	0
0	4	6	4	0	−2

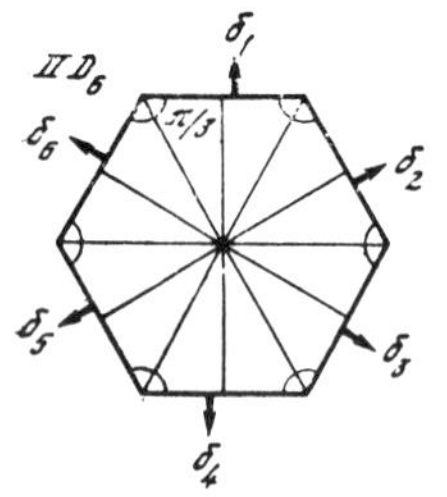

−2	1	7	10	7	1
1	−2	1	7	10	7
7	1	−2	1	7	10
10	7	1	−2	1	7
7	10	7	1	−2	1
1	7	10	7	1	−2

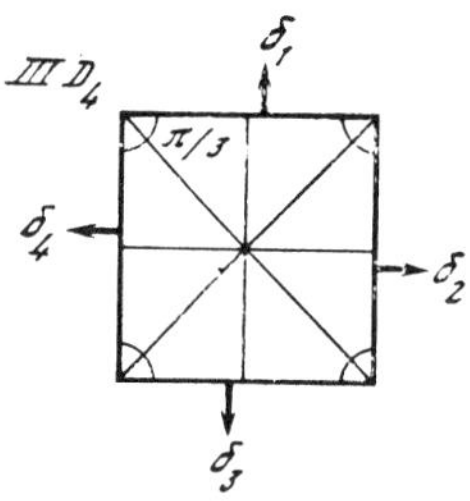

−2	1	4	1
1	−2	1	4
4	1	−2	1
1	4	1	−2

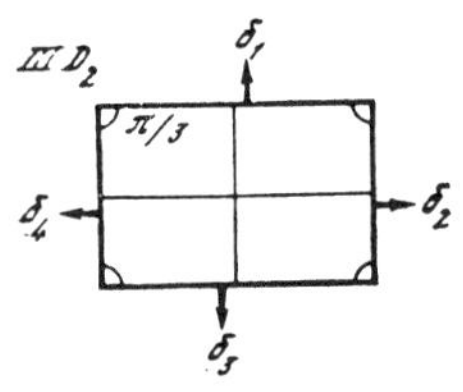

−2	1	a_{13}	1
1	−2	1	a_{24}
a_{13}	1	−2	1
1	a_{24}	1	−2

$a_{13} = \delta_1 \cdot \delta_3$, $a_{24} = \delta_2 \cdot \delta_4$, where $(a_{13} - 2)(a_{24} - 2) = = 4$, $a_{13} > 2$, $a_{24} > 2$.

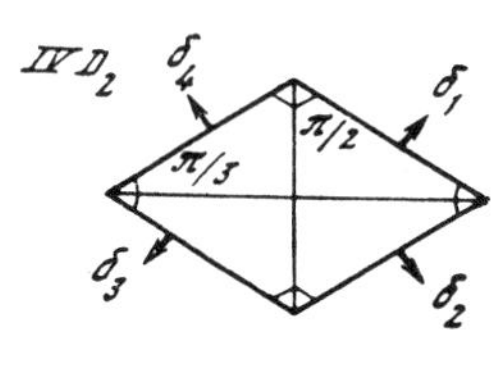

−2	1	3	0
1	−2	0	3
3	0	−2	1
0	3	1	−2

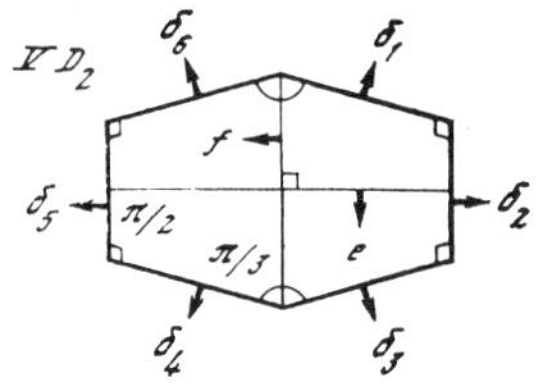

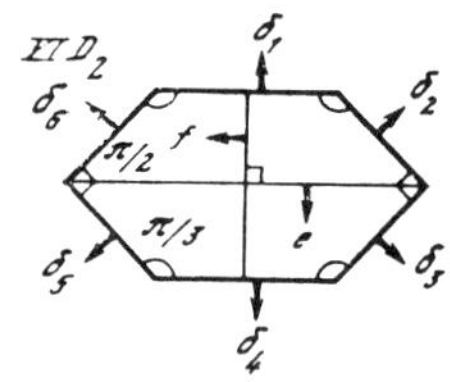

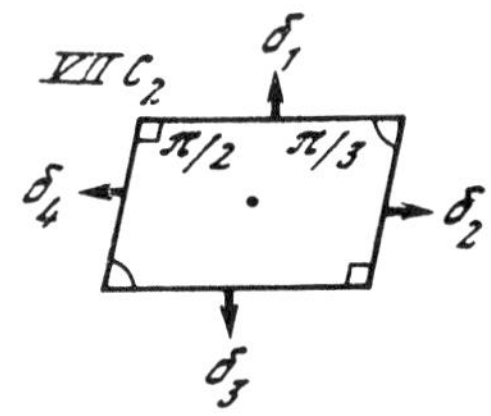

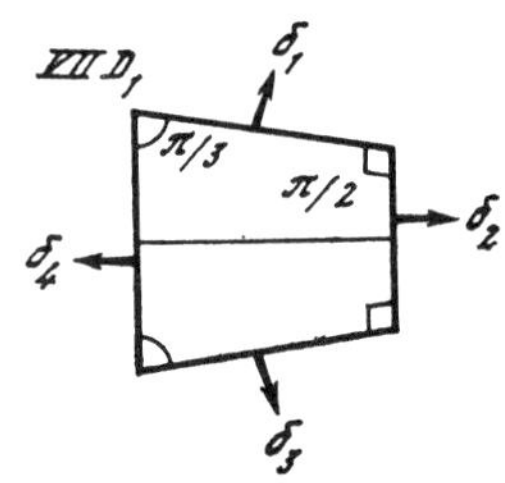

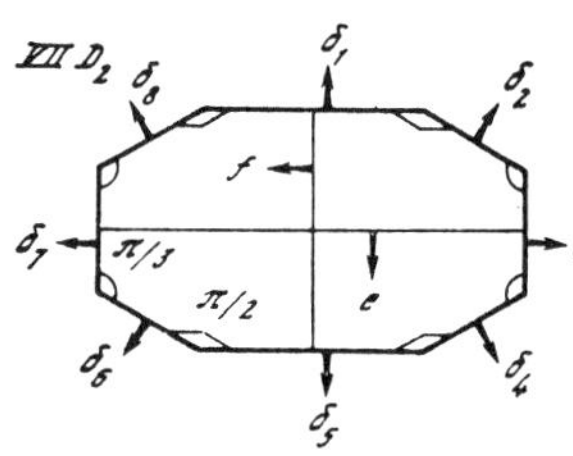

FIG. 1

$$\begin{Vmatrix} -2 & 0 & -2+u^2 & 1+u^2 & \sqrt{3}\,v & 1 \\ 0 & -2 & 0 & \sqrt{3}\,v & -2+v^2 & \sqrt{3}\,v \\ -2+u^2 & 0 & -2 & 1 & \sqrt{3}\,v & 1+u^2 \\ 1+u^2 & \sqrt{3}\,v & 1 & -2 & 0 & -2+u^2 \\ \sqrt{3}\,v & -2+v^2 & \sqrt{3}\,v & 0 & -2 & 0 \\ 1 & \sqrt{3}\,v & 1+u^2 & -2+u^2 & 0 & -2 \end{Vmatrix}$$

$u = \delta_1 \cdot e$, $v = \delta_2 \cdot f$, where $u > 2$, $v > 2$, $(4-u^2)(4-v^2) = 12$.

$$\begin{Vmatrix} -2 & 1 & 1+\sqrt{2}\,u & -2+u^2 & 1+\sqrt{2}\,u & 1 \\ 1 & -2 & 0 & 1+\sqrt{2}\,u & v^2 & -2+v^2 \\ 1+\sqrt{2}\,u & 0 & -2 & 1 & -2+v^2 & v^2 \\ -2+u^2 & 1+\sqrt{2}\,u & 1 & -2 & 1 & 1+\sqrt{2}\,u \\ 1+\sqrt{2}\,u & v^2 & -2+v^2 & 1 & -2 & 0 \\ 1 & -2+v^2 & v^2 & 1+\sqrt{2}\,u & 0 & -2 \end{Vmatrix}$$

$u = \delta_1 \cdot e$, $v = \delta_2 \cdot f$, where $u > 2$, $v > 2$, $4u^2 + 4v^2 + 4\sqrt{2}\,u - u^2 \cdot v^2 - 4 = 0$.

$$\begin{Vmatrix} -2 & 1 & a_{13} & 0 \\ 1 & -2 & 0 & a_{24} \\ a_{13} & 0 & -2 & 1 \\ 0 & a_{24} & 1 & -2 \end{Vmatrix}$$

$a_{13} = \delta_1 \cdot \delta_3$, $a_{24} = \delta_2 \cdot \delta_4$, where $(a_{13} - 2)(a_{24} - 2) = 1$, $a_{13} > 2$, $a_{24} > 2$.

$$\begin{Vmatrix} -2 & 0 & a_{13} & 1 \\ 0 & -2 & 0 & a_{24} \\ a_{13} & 0 & -2 & 1 \\ 1 & a_{24} & 1 & -2 \end{Vmatrix}$$

$a_{13} = \delta_1 \cdot \delta_3$, $a_{24} = \delta_2 \cdot \delta_4$, where $(a_{13} - 2)(a_{24}^2 - 4) = 4$, $a_{13} > 2$, $a_{24} > 2$.

$$\begin{Vmatrix} -2 & 0 & v & uw & -2+w^2 & uw & v & 0 \\ 0 & -2 & 1 & -2+u^2 & uw & -2+u^2+s^2 & 1+ts & -2+s^2 \\ v & 1 & -2 & 1 & v & 1+ts & -2+t^2 & 1+ts \\ uw & -2+u^2 & 1 & -2 & 0 & -2+s^2 & 1+ts & -2+u^2+s^2 \\ -2+w^2 & uw & v & 0 & -2 & 0 & v & uw \\ uw & -2+u^2+s^2 & 1+ts & -2+s^2 & 0 & -2 & 1 & -2+u^2 \\ v & 1+ts & -2+t^2 & 1+ts & v & 1 & -2 & 1 \\ 0 & -2+s^2 & 1+ts & -2+u^2+s^2 & uw & -2+u^2 & 1 & -2 \end{Vmatrix}$$

$u = \delta_2 \cdot e$, $v = \delta_1 \cdot \delta_3$, $w = \delta_1 \cdot e$, $s = \delta_2 \cdot f$, $t = \delta_3 \cdot f$,
where $(w^2 - 4)(s^2 - 4) = 4u^2$, $(w^2 - 4)(t^2 - 4) = 4v^2$, $u^2v^2 - 4u^2 - 4v^2 - 2uvw - 3w^2 + 12 = 0$
$(u > 2, v > 2, w > 2, s > 2, t > 2)$.

3°. Classification of the lattices S. We will denote by $\langle D \rangle$ the lattice with matrix D, and by the standard generator of the lattice $\langle D \rangle$ we mean that generator in terms of which it has the matrix D. Besides this, we denote

$$A_1 = \langle -2 \rangle \text{ and } A_2 = \left\langle \begin{matrix} -2 & 1 \\ 1 & -2 \end{matrix} \right\rangle.$$

We note that the lattices $A_1^2 = A_1 \oplus A_1$ and A_2 are the lattices K corresponding to the vertices of $\mathscr{M}$ of type $\mathbf{A}_1^2$ and $\mathbf{A}_2$ respectively, introduced at the beginning of subsection 1°. By the standard generator of the lattice $\langle D_1 \rangle \oplus \langle D_2 \rangle$ we will understand the one in terms of which it has the matrix

$$\left\langle \begin{matrix} D_1 & 0 \\ 0 & D_2 \end{matrix} \right\rangle.$$

If $\langle D \rangle \subseteq S$ is a sublattice of finite index, then elements of S may be expressed by the standard generators of $\langle D \rangle$ with rational coefficients.

THEOREM 1.3. *There are exactly* 6 *even hyperbolic lattices* $S_1, \ldots, S_6$ *of rank* 3 *for which the factor group* $A(S) = O'(S)/W^{(2)}(S)$ *is finite and for which the fundamental polygon* $\mathscr{M}$ *for the group* $W^{(2)}(S)$ *is bounded in the Lobachevsky plane* $\mathscr{L}(S)$. *They are the lattices*

$$S_i = \langle U_i \rangle / \mathrm{Ker}\langle U_i \rangle, \tag{3.1}$$

$i = 1, \ldots, 6$, *where the* U_i *are the matrices of Theorem* 1.2. *Their generators* $\{\delta_1, \ldots, \delta_k\}$, *having the form of standard generators of the lattice* $\langle U_i \rangle$, *give all collections* $P(\mathscr{M}_i)$ *of elements with square* (-2), *orthogonal and "outward pointing" to the sides* r_m *of the fundamental polygon* $\mathscr{M}_i$ *for* $W^{(2)}(S_i)$. *In particular,* U_i *is the Gram matrix of* $\mathscr{M}_i$. *Let* $h \in S_i$ *be a primitive element of the lattice* S_i, *determining a vertex* $v \in \mathscr{M}_i$, *from which radiate sides* r_1 *and* r_2 *of the polygon* $\mathscr{M}$ (*the expression of* h *in terms of* $\delta_1, \ldots, \delta_k$ *is given in Table* 2). *Then* h, δ_1 *and* δ_2 *are standard generators of the sublattice* $\langle V_i \rangle \subseteq S_i$ *of finite index* (*given in Table* 2). *Here* S_i *is the only even over-lattice of the lattice* $\langle V_i \rangle$ *of the given index.*

TABLE 2

Sublattice $\langle V_i \rangle$	Expression of the element h of $\langle V_i \rangle$	Index $[S_i : V_i]$	Discriminant of the lattice S_i
$\langle V_1 \rangle = \langle 6 \rangle \oplus A_1^2$	$2\delta_1 + \delta_3$	1	24
$\langle V_2 \rangle = \langle 36 \rangle \oplus \mathrm{A}_2$	$5\delta_1 + 3\delta_2 + \delta_3$	1	108
$\langle V_3 \rangle = \langle 12 \rangle \oplus A_2$	$3\delta_1 + 2\delta_2 + \delta_3$	1	36
$\langle V_4 \rangle = \langle 60 \rangle \oplus A_2$	$7\delta_1 + 5\delta_2 + 3\delta_3$	3	20
$\langle V_5 \rangle = \langle 4 \rangle \oplus A_2$	$2\delta_1 + \delta_2 + \delta_3$	1	12
$\langle V_6 \rangle = \langle 132 \rangle \oplus A_2$	$10\delta_1 + 5\delta_2 + 3\delta_3$	3	44

PROOF. Since the matrix U_i is integral, the definition (3.1) gives lattices S_i (over $\mathbf{Z}$) which are hyperbolic of rank 3 (see (2.2)). With this, lines r_m in the Lobachevsky plane $\mathscr{L}(S_i)$ orthogonal to the elements δ_m determine polygons $\mathscr{M}_i$ for which the collection $\{\delta_1, \ldots, \delta_k\} = P(\mathscr{M}_i)$ and for which U_i is the Gram matrix $a_{ij} = \delta_i \cdot \delta_j$. The reflections s_{δ_m} of the lattice S_i relative to the elements δ_m with squares (-2) generate in $\mathscr{L}(S_i)$ a group W_i, generated by reflections relative to the sides r_m of the polygon $\mathscr{M}_i$, for which $\mathscr{M}_i$ is fundamental (since all angles of $\mathscr{M}_i$ are equal to $\pi/2$ or $\pi/3$). Since $\mathscr{M}_i$ is bounded, it has in particular a finite area, so then

$$W_i \subset W^{(2)}(S_i) \subset O'(S_i) \tag{3.2}$$

are imbeddings of groups of finite index.

From this it follows that the lattice S_i, $i = 1, \ldots, 6$, meet the condition of the theorem.

We will show that every lattice S meeting the condition of the theorem coincides with some S_i, $i = 1, \ldots, 6$. The fundamental polygon $\mathscr{M}$ for $W^{(2)}(S)$ has as Gram matrix one of the matrices U_i (Theorem 1.2). Therefore S contains a lattice S_i as sublattice of finite index, since S_i is generated by $\delta_1, \ldots, \delta_k$. Even over-lattices $S_i \subseteq S$ of the lattices S_i are described by the subgroups $H = S/S_i \subseteq \mathfrak{A}_{S_i} = S^*/S$ ([7], §1), which are isotropic relative to the discriminant quadratic form q_{S_i}. Using this it is easy to show that the lattices S_1, S_3, S_4, S_5 and S_6 have no nontrivial even over-lattices, and S_2 has as nontrivial even over-lattice only S_5. This proves the claim: S coincides with some S_i, $i = 1, \ldots, 6$.

We show that $W_i = W^{(2)}(S_i)$ (see (3.2)); in particular, $\mathscr{M}_i$ is the fundamental polygon for $W^{(2)}(S_i)$ with Gram matrix U_i. Let $\mathscr{M}'_i$ be a fundamental polygon for $W^{(2)}(S_i)$. By Theorem 1.2, $\mathscr{M}'_i$ has as Gram matrix some matrix U_j. Therefore $S_j \subseteq S_i$, by the definition (3.1) of the lattice S_j and the fact that $W^{(2)}(S_i)$ is a group of 2-reflections of S_i. On the other hand, $W_i \subseteq W^{(2)}(S_i)$; therefore $S_i \subseteq S_j$ by the definition (3.1). Thus, $S_i = S_j$ and $i = j$, for the lattices $S_1, \ldots, S_6$ are different (for example, they have different discriminant). In this way, the Gram matrices U_i and U_j of the polygons $\mathscr{M}_i$ and $\mathscr{M}'_i$ coincide, i.e. the fundamental polygons $\mathscr{M}_i$ and $\mathscr{M}'_i$ are congruent. Hence $W_i = W^{(2)}(S_i)$ for $W_i \subseteq W^{(2)}(S_i)$.

The theorem is proved.

§2. The noncompact case

In this section S is a hyperbolic lattice (even) of rank 3 for which the factor group $A(S) = O(S)/W^{(2)}(S)$ is finite and for which the fundamental polygon $\mathscr{M}$ for the group $W^{(2)}(S)$ in the Lobachevsky space $\mathscr{L}(S)$ has finite volume but is noncompact, i.e. has vertices at infinity. Our goal is to describe all such lattices S.

1°. The maximal lattice S_1. We write $S_1 = U \oplus A_1$, where

$$U = \begin{pmatrix} 0 & 1 \\ 1 & 0 \end{pmatrix}$$

is the standard unimodular hyperbolic lattice of rank 2, and $A_1 = \langle -2 \rangle$.

FUNDAMENTAL THEOREM 2.1. *Every lattice S is a sublattice of the lattice S_1 of finite index.*

PROOF. Since U is an orthogonal component of S_1, the lattice S_1 is single-classed: every lattice of the same genus as S_1 is isomorphic to S_1 (see for every [4], §1). Therefore it is sufficient to check that every lattice S has a sublattice with the same genus as S_1. And for this it is sufficient to check that the local lattices $S \otimes \mathbf{Z}_p$ are sublattices of $S_1 \otimes \mathbf{Z}_p$ for all primes p, where $\mathbf{Z}_p$ is the ring of p-adic integers.

Let c be a primitive zero of S (i.e. $c \in S$, $c^2 = 0$, but $c \neq 0$, and c is irreducible) determining a vertex at infinity $v \in \mathcal{M}$. Let δ_1 and δ_2 be elements of S with square (-2), orthogonal to the sides of $\mathcal{M}$ radiating from v. Then $\delta_1 \cdot \delta_2 = 2$ (by §1, (2.3)) and $\delta_1 \cdot c = \delta_2 \cdot c = 0$. In particular, there is an element $\delta_1 \in S$ for which $\delta_1^2 = -2$ and the lattice $(\delta_1)_S^\perp$ represents zero, i.e. $\operatorname{discr}(\delta_1)_S^\perp = -d^2$, where $d \in \mathbf{Z}$. Then for $p \neq 2$, on account of the invertibility of 2 in $\mathbf{Z}_p$,

$$S \otimes \mathbf{Z}_p = K_{-2}(1) \oplus K_\theta(p^{m_1}) \oplus K_{-\theta}(p^{m_2}),$$

where $m_1 + m_2 \equiv 0 \pmod 2$ and $\theta_1 \cdot \theta_2 = -1$ (we write $K_\alpha(p^k) = \langle \alpha p^k \rangle$, where $\alpha \in \mathbf{Z}_p^*$). From this it easily follows that $K_\theta(p^{m_1}) + K_{-\theta}(p^{m_2}) \subseteq K_\theta(1) + K_{-\theta}(1)$ and $S \otimes \mathbf{Z}_p \subseteq K_{-2}(1) + K_\theta(1) + K_{-\theta}(1)$, where the lattice $K_{-2}(1) + K_\theta(1) + K_{-\theta}(1)$ is unimodular and has discriminant 2. But the lattice $S_1 \otimes \mathbf{Z}_p$ also is unimodular and has discriminant 2; therefore

$$S_1 \otimes \mathbf{Z}_p \simeq K_{-2}(1) \oplus K_\theta(1) \oplus K_{-\theta}(1)$$

(see, for instance, [7], §1). Hence $S \otimes \mathbf{Z}_p \subset S_1 \otimes \mathbf{Z}_p$, which is what is needed.

Let $p = 2$. We consider two cases. Let the form of the lattice S be divisible by 2 (with integral remainder). Then the sublattice $[\delta_1]$ splits off as an orthogonal component of $S \otimes \mathbf{Z}_2$ and,

$$S \otimes \mathbf{Z}_2 = K_{-1}(2) \oplus U(2^m),$$

or

$$S \otimes \mathbf{Z}_2 = K_{-1}(2) \oplus K_\theta(2^{m_1}) \oplus K_{-\theta}(2^{m_2}),$$

where

$$U(2^m) = \begin{pmatrix} 0 & 2^m \\ 2^m & 0 \end{pmatrix} \quad \text{and } m_1 \equiv m_2 \pmod 2.$$

It is easy to see that $U(2^m) \subseteq U(1)$ and

$$K_\theta(2^{m_1}) \oplus K_{-\theta}(2^{m_2}) \subset U(1),$$

from which it follows that

$$S \otimes \mathbf{Z}_2 \subset K_{-1}(2) \oplus U(1) \simeq S_1 \otimes \mathbf{Z}_2.$$

Suppose the form of the lattice S is not divisible by 2. Then

$$S \otimes \mathbf{Z}_2 = U(1) \oplus K_{-1}(2^m),$$

or

$$S \otimes \mathbf{Z}_2 = V(1) \oplus K_3(2^m),$$

where $V(1) = \begin{pmatrix} 2 & 1 \\ 1 & 2 \end{pmatrix}$ and $m \equiv 1 \pmod 2$ (since $\operatorname{discr} S \otimes \mathbf{Q} = 2$). Clearly $K_\theta(2^m) \subseteq K_\theta(2)$, whence in the first case

$$S \otimes \mathbf{Z}_2 = U(1) \oplus K_{-1}(2^m) \subset U(1) \oplus K_{-1}(2) \simeq S_1 \otimes \mathbf{Z}_2$$

and in the second case

$$S \otimes \mathbf{Z}_2 = V(1) \oplus K_3(2^m) \subset V(1) \oplus K_3(2) \simeq U(1) \oplus K_{-1}(2) \simeq S_1 \otimes \mathbf{Z}_2.$$

The theorem is proved.

2°. The topology of decompositions of convex polygons into triangles by nonintersecting diagonals. We consider the decomposition of a convex polygon in the plane (Euclidean or Lobochevsky irrespectively) into triangles by nonintersecting diagonals. By the *order* of a vertex of the polygon we shall mean the number of triangles of the decomposition meeting at that vertex (having it as a common vertex).

THEOREM 2.2. *For every decomposition of the polygon into triangles by nonintersecting diagonals there are two neighboring vertices A and B such that the order of A is* 1 *and the order of B is no greater than* 3.

PROOF. We consider in the Lobachevsky plane a triangle $\tilde{\Delta}$ with zero angles, the group W generated by the reflections through the sides of $\tilde{\Delta}$, and the decomposition $\tilde{\mathfrak{R}}$ of the Lobachevsky plane into triangles congruent to $\tilde{\Delta}$ under the transformations of the group W (Figure 2). Inasmuch as the assertion of the theorem is topological, it is sufficient to prove it for either of two topologically equivalent decompositions of the polygon by diagonals. And any such decomposition is equivalent to a decomposition obtained in the following way: the convex polygon is comprised of triangles of the decomposition $\tilde{\mathfrak{R}}$, and these very triangles determine its decomposition into triangles by nonintersecting diagonals. This is shown by an obvious induction on the number of triangles in the decomposition. Thus, we consider a decomposition of the polygon $\mathcal{M}$ into triangles of the decomposition $\tilde{\mathfrak{R}}$. Let $A_1, \ldots, A_n$ be the vertices of $\mathcal{M}$. According to [10] (geometric Lemma 3.2.1) there are sides A_iA_{i+1} of $\mathcal{M}$ such that the elements δ_{i-1} and δ_{i+1} with squares (-2) and orthogonal (and pointing outward) to the sides $A_{i-1}A_i$ and $A_{i+1}A_{i+2}$ respectively satisfy the inequality

$$\delta_{i-1} \cdot \delta_{i+1} < 14. \tag{2.1}$$

Let the orders of the vertices A_m be k_m and let $\tilde{\Delta}$ be the triangle of the decomposition $\tilde{\mathfrak{R}}$ of $\mathcal{M}$, one side of which coincides with A_iA_{i+1}. Let D be its third vertex and let e_1 and f_1 be elements with square (-2) determining its sides A_iD

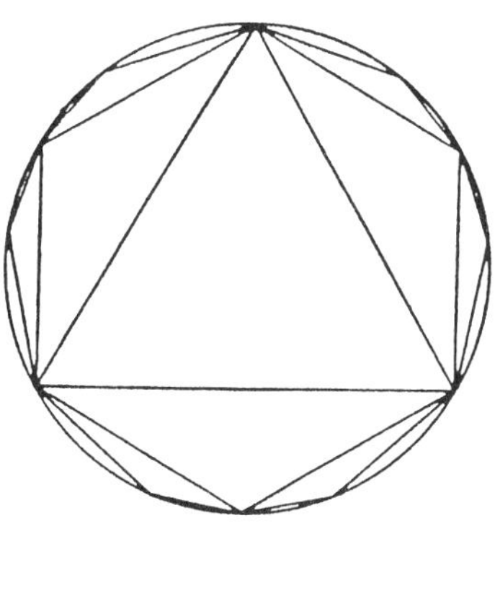

FIG. 2

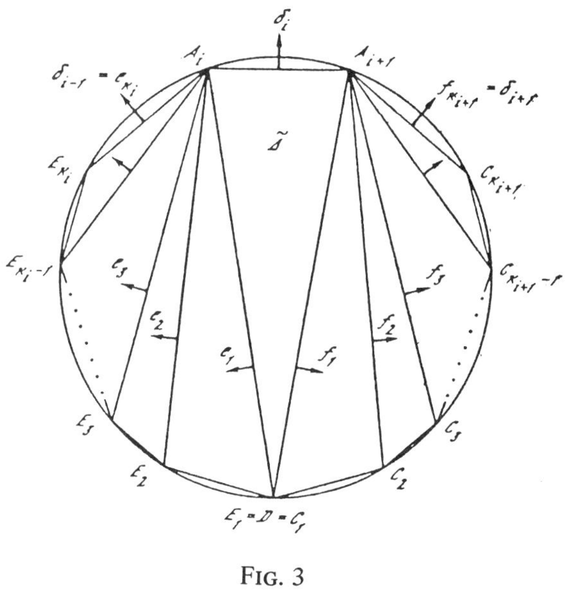

FIG. 3

and $A_{i+1}D$ respectively (Figure 3). In the decomposition $\tilde{\mathfrak{R}}$ of the polygon $\mathscr{M}$ at the vertex A_i there meet, besides $\tilde{\Delta}$, also $k_i - 1$ triangles $A_iE_tE_{t+1}$ $t = 1, \ldots, k_i - 1$, and at the vertex A_{i+1} there meet $k_{i+1} - 1$ triangles $A_{i+1}C_sC_{s+1}$, $s = 1, 2, \ldots, k_{i+1} - 1$, as shown in the figure. Since neighboring triangles of the decomposition $\tilde{\mathfrak{R}}$ are obtained from each other by reflections with respect to the common side, then the elements e_t and e_s with square (-2) orthogonal to the lines A_iE_t and $A_{i+1}C_s$ are equal (see Figure 3):

$$e_2 = s_{e_1}(\delta_i), \qquad e_t = s_{e_{t-1}}(-e_{t-2}), \qquad t = 3, \ldots, k_i;$$
$$f_2 = s_{f_1}(\delta_i), \qquad f_s = s_{f_{s-1}}(-f_{s-2}), \qquad s = 3, \ldots, k_{i+1},$$

where, we recall, $s_\delta(x) = x + (x \cdot \delta)\delta$, if δ is an element with square (-2). Hence.

$$e_2 = \delta_i + 2e_1, \qquad e_t = -e_{t-2} + 2e_{t-1}, \qquad t = 3, \ldots, k_i;$$
$$f_2 = \delta_i + 2f_1, \qquad f_s = -f_{s-2} + 2f_{s-1}, \qquad s = 3, \ldots, k_{i+1},$$

and, as induction shows,

$$e_t = (t-1)\delta_i + te_1, \qquad f_s = (s-1)\delta_i + sf_1.$$

Therefore

$$\delta_{i-1} = e_{k_i} = (k_i - 1)\delta_i + k_ie_1,$$
$$\delta_{i+1} = f_{k_{i+1}} = (k_{i+1} - 1)\delta_i + k_{i+1}f_1.$$

By (2.1) we have

$$\delta_{i-1} \cdot \delta_{i+1} = 2(2k_i \cdot k_{i+1} - 1) < 14,$$

or

$$k_i \cdot k_{i+1} < 4.$$

This proves the theorem.

3°. Classification of the sublattices $s \subset S_1$. According to Theorem 2.1 the lattices of interest to us admit a realization as sublattices $S \subset S_1$ of finite index. Therefore it is natural at first to describe all such sublattices up to automorphisms of the group $W^{(2)}(S_1)$. We shall do so. Let $\mathscr{M}$ be a polygon in the Lobachevsky plane and let D be its Gram matrix $a_{ij} = \delta_i \cdot \delta_j$ $(i, j = 1, \ldots, k)$, where $\{\delta_1, \ldots, \delta_k\} = P(\mathscr{M})$ is the collection of elements with square (-2) orthogonal to the sides of $\mathscr{M}$ and pointing outward. If the matrix D is integral, then the hyperbolic lattice of rank 3

$$S = \langle D \rangle / \mathrm{Ker}\langle D \rangle \tag{3.1}$$

will be called the lattice of the polygon $\mathscr{M}$. Here we identify the Lobachevsky plane with $\mathscr{L}(S)$, and the elements δ_i with the images of the standard generators $\tilde{\delta}_i$, $i = 1, \ldots, k$, of the lattice $\langle D \rangle$.

PROPOSITION 2.3. *S_1 is the lattice of the triangle $\Delta = ABC$ in the Lobachevsky plane with angles $\hat{A} = 0$, $\hat{B} = \pi/3$ and $\hat{C} = \pi/2$ (i.e. a modular figure of the second kind). It has as Gram matrix $P(\Delta) = (a, b, c)$ (see below, Figure* 4a)) *the matrix*

$$\begin{Vmatrix} -2 & 0 & 1 \\ 0 & -2 & 2 \\ 1 & 0 & -2 \end{Vmatrix}.$$

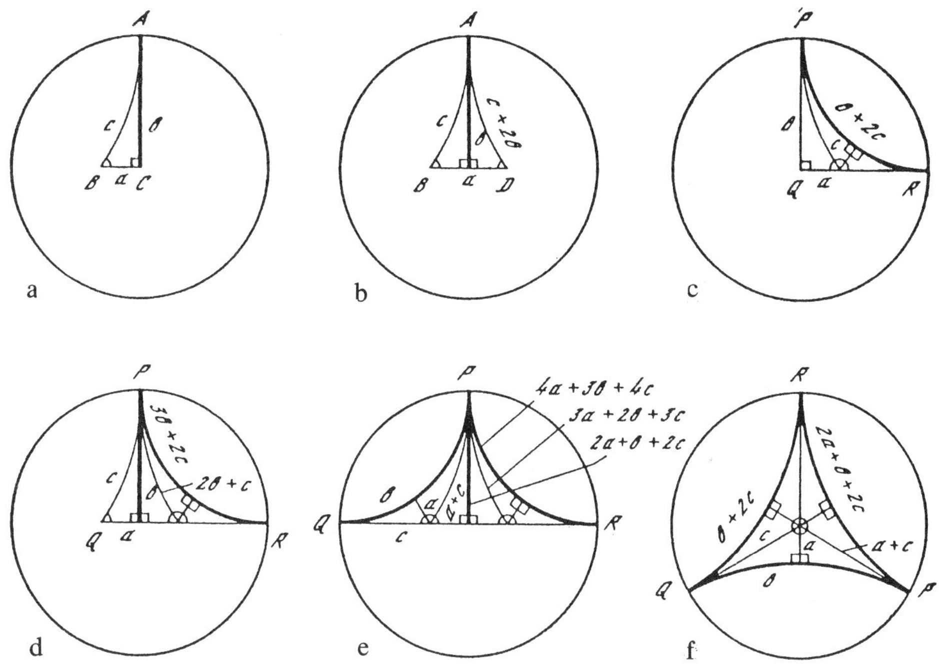

FIG. 4

As standard generators of $S_1 = U \oplus A_1$ *one can take the elements* $b + c$, $a + b + c$ *and* b.

The proof is simple, well known, and left to the reader.

We consider a decomposition $\mathfrak{R}$ of the Lobachevsky plane into triangles $W^{(2)}(S_1) \cdot \Delta$ congruent to Δ (see Figures 4–6). If $S \subset S_1$ is a sublattice of finite index and $\mathscr{M}$ is the fundamental polygon for $W^{(2)}(S)$, then $\mathfrak{R}$ induces a decomposition of the polygon $\mathscr{M}$, which we denote by $\mathfrak{R} \cap \mathscr{M}$. It is an invariant of the imbedding $S \subset S_1$ of lattices. If S satisfies the condition of interest to us (the group $O'(S)/W^{(2)}(S)$ is finite), then $\mathscr{M}$ has finite area; therefore this decomposition is finite. Clearly the converse is also true.

THEOREM 2.4. *Every sublattice* $S \subset S_1$ *for which the factor group* $O'(S)/W^{(2)}(S)$ *is finite is conjugate under the group* $W^{(2)}(S_1)$ *to one and only one of the* 24 *sublattices in Table* 3, *where they are given by the expression of their generators in terms of the generators* a, b *and* c *of the lattice* S (*Proposition* 2.3). *Moreover, in the table is given the number of the figure depicting the decomposition* $\mathfrak{R} \cap \mathscr{M}$ *and the elements* $P(\mathscr{M})$ *for one of the fundamental polygons for* $W^{(2)}(S)$; *the number of vertices of* $\mathscr{M}$; *the ratio of areas* $v(\mathscr{M})/v(\Delta)$, *where* Δ *is a fundamental polygon for* $W^{(2)}(S_1)$; *and* $\sqrt{(\operatorname{discr} S)/2}$.

PROOF. We note that the lines of the decomposition $\mathfrak{R}$ of the Lobachevsky plane may be of two types: lines of type I contain exactly three vertices of triangles of the decomposition $\mathfrak{R}$, and lines of type II contain exactly five vertices (see below, Figures 5 and 6). Another property of these: on lines of type I lie the long legs of the right triangles of $W^{(2)}(S) \cdot \Delta$, and on lines of type II lie the short legs and the

TABLE 3

No. p/p	sublattice $S \subset S_1$	$\mathfrak{R} \cap \mathscr{M}$ and $P(\mathscr{M})$: number of the figure	$\mathfrak{R} \cap \mathscr{M}$ and $P(\mathscr{M})$: $\mathscr{M}$	number of vertices of	$\frac{v(\mathscr{M})}{v(\Delta)}$	$\sqrt{\frac{\operatorname{discr} S}{2}}$
1	$S_1 = [a, b, c]$	4a	$\Delta = ABC$	3	1	1
2	$S_{1,1,2} = [a, b, 2c]$	4b	PQR	3	3	2
3	$S_{1,1,3} = [a, b, 3c]$	5	$PQRS$	4	6	3
4	$S_{1,1,4} = [a, b, 4c]$	5	$PQRT$	4	6	4
5	$S_{1,1,6} = [a, b, 6c]$	5	$PQRU_2U_3U_4$	6	18	6
6	$S_{1,1,8} = [a, b, 8c]$	5	$PQRU_1U_2U_3U_4U_5$	8	24	8
7	$S_{1,2,1} = [a, 2b, c]$	4c	BAD	3	2	2
8	$S_{1,3,1} = [a, 3b, c]$	4d	PQR	3	4	3
9	$S_{1,4,1} = [a, 4b, c]$	6	$PQRS$	4	8	4
10	$S_{1,5,1} = [a, 5b, c]$	6	$PT_2T_4T_5T_6T_9$	6	20	5
11	$S_{1,6,1} = [a, 6b, c]$	6	$PQRU$	4	8	6
12	$S_{1,9,1} = [a, 9b, c]$	6	$PQRT_2T_3T_5T_6T_7T_8$	9	36	9
13	$S_{4,1,1} = [4a, b, c]$	4e	PQR	3	6	4
14	$S_{5,1,1} = [5a, b, c]$	8	$PQRS$	4	12	5
15	$S_{6,1,1} = [6a, b, c]$	8	$PQRT$	4	12	6
16	$S_{7,1,1} = [7a, b, c]$	8	$PQRUTV$	6	24	7
17	$S_{8,1,1} = [8a, b, c]$	8	$PQRW$	4	12	8
18	$S_{10,1,1} = [10a, b, c]$	8	$PQRSTX_1WX_2$	8	36	10
19	$S_{12,1,1} = [12a, b, c]$	8	$PQRTWY$	6	24	12
20	$S_{2,1,2} = [2a, b, 2c]$	4f	PQR	3	6	4
21	$S_{4,1,2} = [4a, b, 2c]$	10	$PQRS$	4	12	8
22	$S_{6,1,2} = [6a, b, 2c]$	10	$PQRT_1ST_2$	6	24	12
23	$S'_{4,1,2} = [2a + c, b, 2c]$	10	$PQRS$	4	12	4
24	$S'_{6,1,2} = [3a + c, b, 2c]$	10	$PQRT_1ST_2$	6	24	6

hypotenuses. We mention that the lines of type I give precisely the decomposition $\tilde{\mathfrak{R}}$ used by us in the proof of Theorem 2.2.

Let $S \subset S_1$ be a sublattice of S_1 of finite index and let $\mathscr{M}$ be a fundamental polygon of finite area for $W^{(2)}(S)$. By the decomposition $\tilde{\mathfrak{R}} \cap \mathscr{M}$ of $\mathscr{M}$ we will understand its decomposition by all lines of type I lying entirely in $\mathscr{M}$.

We claim that $\tilde{\mathfrak{R}} \cap \mathscr{M}$ is a decomposition of $\mathscr{M}$ by nonintersecting diagonals into triangles. The ends of the lines of type I lying entirely in $\mathscr{M}$ are bound to be vertices of $\mathscr{M}$, or else the area of $\mathscr{M}$ would be infinite. Therefore each such line is a diagonal of $\mathscr{M}$. Two different lines of type I can intersect only at their ends; therefore these diagonals are nonintersecting. It remains to be verified that they decompose $\mathscr{M}$ into triangles. This is true for any polygon with acute angles whose sides run along lines of $\mathfrak{R}$. For the proof we carry out an induction on the number of lines of type I lying entirely in the interior of such a polygon $\mathscr{M}$. If there are no such lines, then a sorting out, which is easy to accomplish, glancing at Figure 5 or 6, shows that $\mathscr{M}$ is one of the triangles with decompositions $\mathfrak{R} \cap \mathscr{M}$, shown in Figure 4a–f. If there are such lines, we divide $\mathscr{M}$ into two parts by one of them. Evidently, the inductive assumption is applicable to each of these. This proves the claim.

We apply Theorem 2.2 to the decomposition $\tilde{\mathfrak{R}} \cap \mathcal{M}$: there exist two adjacent (determining a side of $\mathcal{M}$) vertices Q and A_3 of the polygon $\mathcal{M}$, with the order of Q being 1, and the order of A_3 being no greater than 3.

If the order of A_3 is 1, then clearly $\mathcal{M}$ is one of the triangles of Figure 4a–f with the elements $P(\mathcal{M}) \subset S$ indicated on the sketch. Hence it follows that the sublattice $[P(\mathcal{M})] \subset S$, spanned by $P(\mathcal{M})$, in the respective cases a–f is

a) $S_{1,1,1} = [a, b, c]$; b) $S_{1,2,1} = [a, 2b, c]$;
c) $S_{1,1,2} = [a, b, 2c]$; d) $S_{1,3,1} = [a, 3b, c]$;
e) $S_{4,1,1} = [4a, b, c]$; f) $S_{2,1,2} = [2a, b, 2c]$.

Considering all possible intermediate lattices $[P(\mathcal{M})] \subset S \subset S_1$, we satisfy ourselves that in each of these cases $S = [P(\mathcal{M})]$ is the lattice of the triangle $\mathcal{M}$. For example, in case e)

$$[P(\mathcal{M})] = S_{2,1,2} = [2a, b, 2c];$$

therefore an intermediate lattice $[P(\mathcal{M})] \subsetneqq S \subsetneqq S_1 = [a, b, c]$ is one of the intermediate sublattices: $[a, b, 2c]$, $[2a, b, c]$ or $[2a, b, a + c]$. In these cases the fundamental polygon for $W^{(2)}(S)$ will be respectively a triangle $\mathcal{M}' \subsetneqq \mathcal{M}$ with $P(\mathcal{M}') = \{a, b, b + 2c\}$, $\{b, c, 2a + b + 2c\}$ or $\{b, b + 2c, a + c\}$, which is impossible. In this way, we obtain the sublattices $S_{1,1,1}$, $S_{4,1,1}$, $S_{1,2,1}$, $s_{1,3,1}$, $S_{1,1,2}$ and $S_{2,1,2}$ in Table 3.

In the following we will take the order of A_3 to be 2 or 3. Let A_1QA_3 be the unique triangle of the decomposition $\tilde{\mathfrak{R}} \cap \mathcal{M}$, a vertex of which is Q. Then the triangle A_1QA_3 does not contain lines of type I, i.e. it is one of the triangles of Figure 4a–f. Moreover, its side A_1A_3 is a line of type I. Therefore it can be one of the triangles of Figure 4b–e with the vertex Q indicated in this sketch. Accordingly, the vertex A_3 (having order 2 or 3 in $\tilde{\mathfrak{R}} \cap \mathcal{M}$) can be either of the remaining vertices P and R of the triangle, and A_1 is the remaining third vertex. As a result we obtain Cases I–IV, which are investigated below.

Case I. If Figure 5, P, Q and R are consecutive vertices of the polygon $\mathcal{M}$, with either the order of P or the order of R equal to 2 or 3. In Figure 5 is represented the part of the decomposition $\mathfrak{R}$ of the Lobachevsky plane which we need. In it, for each line of $\mathfrak{R}$ is indicated the coordinates in the basis $\{a, b, c\}$ of the vector with square (-2) orthogonal to it, these coordinates being on that side of the line towards which this vector is directed.

Since PQ and QR are sides of $\mathcal{M}$, then $a, b \in S$. Therefore the sublattice $S \subset S_1$ has the form $S_{1,1,k} = [a, b, kc]$, where k is a natural number because $S = [a, b, c]$. Since the order of P or Q is equal to 2 or 3, from Figure 5 it is seen that one of the elements $(m - 1)b + mc$, where $m \leqslant 6$, or $na + (n + 1)b + 2nc$, where $n \leqslant 5$, is bound to be in S. Hence it follows that $S = S_{1,1,k}$, where $k = 2, 3, 4, 5, 6, 8, 10$. From this same figure it is evident that the fundamental polygon $\mathcal{M}$ for $W^{(2)}(S)$ is PQR for $k = 1$, is $PQRS$ for $k = 3$, is $PQRT$ for $k = 4$, is $PQRU_2U_3U_4$ for $k = 6$, and is $PQRU_1U_2U_3U_4U_5$ for $k = 8$. These k give us all lattices of the form $S_{1,1,k}$ in Table 3. It remains to consider $k = 5$ or 10. We show that in these cases the fundamental polygon $\mathcal{M}$ has infinite area. Since P, Q and R are its vertices, from Figure 5 one can see that the line orthogonal to $a + c$ and having an endpoint at U_2 lies in $\mathcal{M}$. If the area of $\mathcal{M}$ is finite, U_2 is a vertex of $\mathcal{M}$. But the line of $\mathfrak{R}$ going out from U_2 is orthogonal to the elements with square (-2) of the form $ma + (3m+3)b + (5m+4)c$, where $m \in \mathbf{Z}$, which do not belong to $S_{1,1,5}$ and $S_{1,1,10}$.

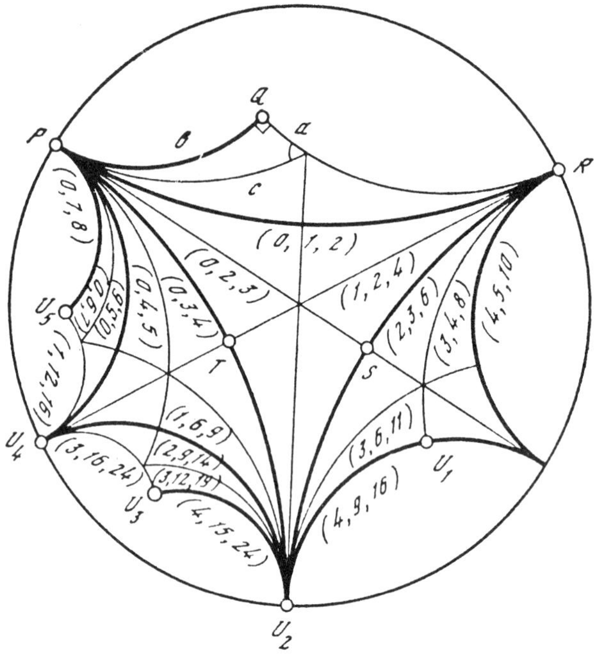

FIG. 5

Case II. In Figure 6, P, Q and R are consecutive vertices or the polygon $\mathscr{M}$, with either the order of P or the order of R being equal to 2 or 3. Just as in the preceding case, it is apparent from Figure 6 that $S = S_{1,k,1} = [a, kb, c]$, where $k = 3, 4, 5, 6, 7, 9, 12, 14$. If $k = 3, 4, 5, 6, 9$, i.e. in the case of lattices of the form $S_{1,k,1}$ from Table 3, from this same figure we find the fundamental polygon of finite area for $W^{(2)}(S_{1,k,1})$ indicated in Table 3.

It remains to check that for the lattices $S_{1,7,1}$, $S_{1,12,1}$ and $S_{1,1,10}$ the fundamental polygon $\mathscr{M}$ has infinite area. For $S_{1,7,1}$ this is verified just as it was for $S_{1,1,5}$ and $S_{1,1,10}$ in Case I, if one considers the line PT_2 and the lines of the decomposition meeting at T_2. For $S_{1,12,1}$ and $S_{1,15,1}$ we note that they are sublattices of $S_{1,3,1}$; therefore their fundamental polygon $\mathscr{M}$ is formed with help of reflections from the fundamental triangle PQR for $W^{(2)}(S_{1,3,1})$ (see Figure 6). Denoting by $\delta = 3b + 2c$ the element with square (-2) orthogonal to its side PR, we obtain that

$$S_{1,12,1} = [a, c, 4\delta] \subset S_{1,3,1}; \qquad S_{1,15,1} = [a, c, 5\delta] \subset S_{1,3,1},$$

where $S_{1,3,1} = [a, c, \delta]$. Using this, it is easy to calculate the series of consecutive vertices P, Q, R, V_1, V_2, V_3 for the fundamental polygon of the group $W^{(2)}(S_{1,12,1})$ and P, Q, R, W_1, W_2, W_3 for $W^{(2)}(S_{1,15,1})$ (see Figure 7, where the topological figure and the coordinates of the vectors orthogonal to the lines and with squares (-2) in the basis $\{a, c, \delta\}$ are shown). The involution of the lattice $S_{1,3,1}$, interchanging a and c and carrying δ into itself, transforms into themselves the sublattices $S_{1,12,1}$ and $S_{1,15,1}$ and gives the symmetry S_{QK} relative to the line QK of the plane

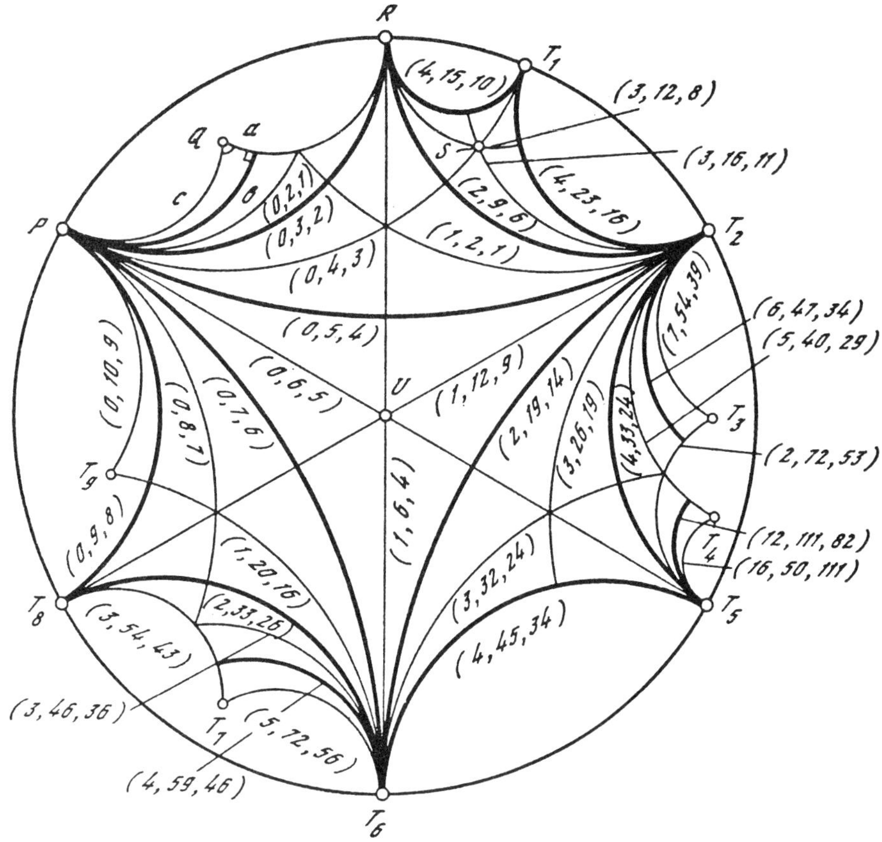

FIG. 6

(Figure 7). Hence the symmetries S_{QK} and S_{V_2L} (see Figure 7) carry into itself the fundamental polygon $\mathscr{M} = \cdots PQRV_1V_2V_3 \cdots$ for $W^{(2)}(S_{1,12,1})$. But the lines QK and V_2L do not intersect, therefore the composition $S_{V_2L} \cdot S_{QK}$ gives a translation of infinite order of the fundamental polygon $\mathscr{M}$. Therefore the area of $\mathscr{M}$ for $S_{1,12,1}$ is infinite. In exactly the same way one shows the infinitude of the area of the fundamental polygon $\mathscr{M} = \cdots PQRW_1W_2W_3 \cdots$ for $W^{(2)}(S_{1,15,1})$, if one uses the symmetries S_{QK} and S_{W_2M} (see Figure 7).

Case III. In Figure 8, P, Q and R are consecutive vertices of the polygon $\mathscr{M}$, with either the order of P or the order of R being equal to 2 or 3. In the same way as in the previous case, from the figure it is apparent, that $S = S_{k,1,1} = [ka, b, c]$, where $k = 4, 5, 6, 7, 8, 10, 12, 17, 20$. If $k = 4, 5, 6, 7, 8, 10, 12$, i.e. in the case of the lattices $S_{k,1,1}$ from Table 3, from this same figure is found the fundamental polygon of finite area for $W^{(2)}(S_{k,1,1})$ indicated in the table.

It remains to verify that for the lattices $S = S_{16,1,1}$ or $S_{20,1,1}$ the area of the fundamental polygon $\mathscr{M}$ for $W^{(2)}(S)$ is infinite. We note that these lattices are sublattices of $S_{4,1,1}$. For $S_{4,1,1}$ the fundamental polygon for $W^{(2)}(S_{4,1,1})$ is the triangle PQR with zero angles. Therefore the fundamental polygon $\mathscr{M}$ for $W^{(2)}(S)$ is formed from the triangles $W^{(2)}(S_{4,1,1}) \cdot (PQR)$ determining a decomposition $\mathfrak{R}_{4,1,1}$ of the Lobachevsky plane into triangles with zero angles. Let A and B be vertices of $\mathscr{M}$ of

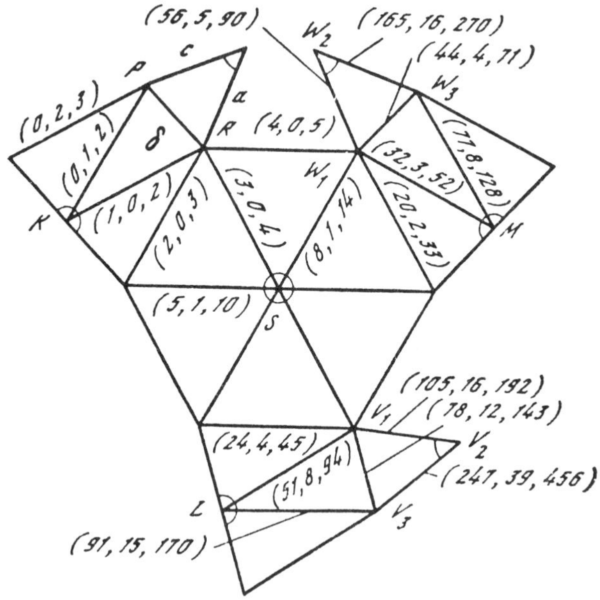

FIG. 7

order q and k, where $k \leqslant 3$, respectively (see Theorem 2.2) and let C, A, B and D be consecutive vertices of $\mathcal{M}$. Let δ_1, δ_2, δ_3 and δ_4 be the vectors with square (-2) orthogonal to the sides CA, AB, BC of the triangle CAB and the side BD of $\mathcal{M}$ respectively, and pointing outward. Then, depending on k, the point $D = D_k$ (see Figure 9, where part of the decomposition $\mathfrak{R}_{4,1,1}$ is presented) and $\delta_4 = \delta_1 + k\delta_3$, where $k = 1, 2, 3$. Taking into account that $S_{4,1,1} = [\delta_1, \delta_2, \delta_3]$ and $[\delta_1, \delta_2, \delta_4] \subseteq S$, we obtain that the index is $[S_{4,1,1}\colon S] = 1$, 2, or 3. For the lattice $S = S_{16,1,1}$ and $S_{20,1,1}$ this is impossible, since $[S_{4,1,1}\colon S_{16,1,1}] = 4$ and $[S_{4,1,1}\colon S_{20,1,1}] = 5$. This contradiction finishes the case.

Case IV. In Figure 10 the order of P is equal to 2 or 3. From the figure we see that $[ka, b, 2c] \subset S$, where $k = 2, 3, 4, 5, 6$. Hence it follows that S is one of the lattices $[ka, b, c]$ or $[ka, b, 2c]$, where $k = 2, 3, 4, 5, 6$, or $[ma + c, b, 2c]$, where $m = 2, 3$. The lattices $S_{k,1,1} = [ka, b, c]$ have already been considered by us in the previous case. If $S = S_{k,1,2} = [ka, b, 2c]$, where $k = 2, 4, 6$, or $S = S_{k,1,2} = [(k/2)a + c, b, 2c]$, where $k = 4, 6$, then we obtain the lattices of Table 3 with the fundamental polygon of finite area indicated in it.

It remains to consider the lattices $S_{3,1,2} = [3a, b, 2c]$ and $S_{5,1,2} = [5a, b, 2c]$. For the lattice $S_{3,1,2}$ the fundamental polygon for $W^{(2)}(S_{3,1,2})$ is the quadrilateral $PQRT_2$ in Figure 10. Therefore the sublattices $S_{3,1,2}$ is equivalent to the sublattice $S_{6,1,1}$ already considered by us in Case III. For the lattice $S_{5,1,2}$ the points $QRUS$ in Figure 10 are consecutive vertices, U having order 1, R having order 3, and the triangle SUR coinciding with the triangle of Figure 4c. Therefore this case has already been investigated by us in Case III.

To finish the proof of the theorem it remains for us to verify that the sublattices of Table 3 are not conjugate under the automorphisms of $W^{(2)}(S_1)$. The invariants (the

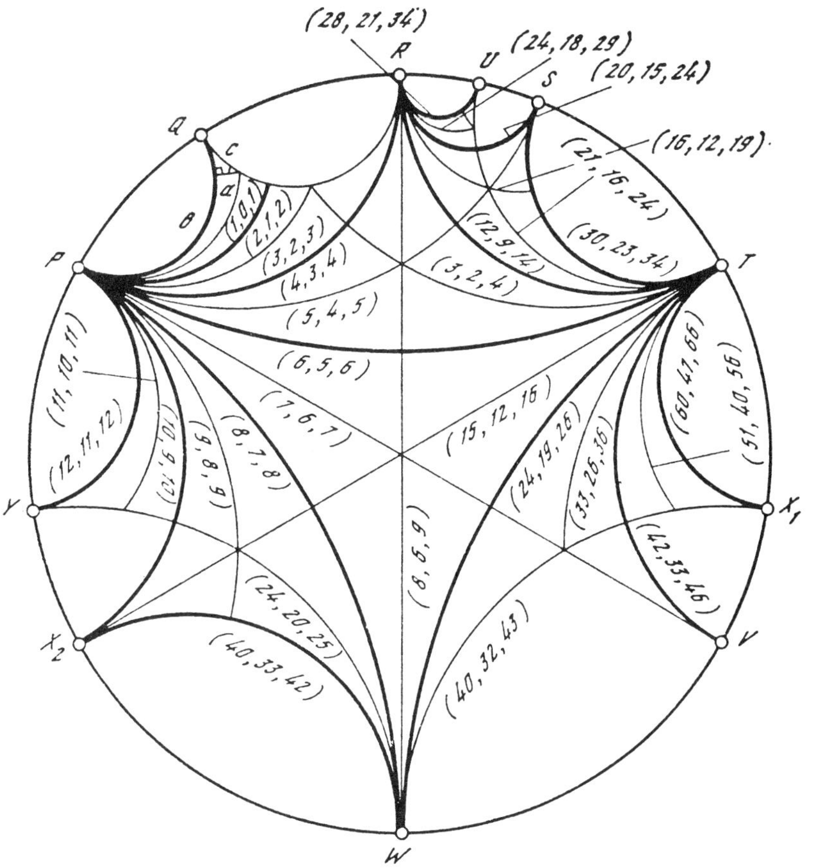

FIG. 8

number of vertices of $\mathscr{M}$, $v(\mathscr{M})/v(\Delta)$, and $\sqrt{(\operatorname{discr} S)/2}$) of the table are invariants of the isomorphism class of S. Therefore it is sufficient to consider only those sublattices S which have these invariants the same. They are the sublattices $S_{4,1,1}$ (No. 13) and $S_{2,1,2}$ (No. 20) with invariants (3, 6, 4); $S_{8,1,1}$ (No. 17) and $S_{4,1,2}$ (No. 21) with invariants (4, 12, 8); and $S_{12,1,1}$ (No. 19) and $S_{6,1,2}$ (No. 22) with invariants (6, 24, 12). But their decompositions $\mathfrak{R} \cap \mathscr{M}$ are clearly different. Therefore they are nonconjugate sublattices under the group $W^{(2)}(S_1)$. We note that nevertheless these pairs of sublattices give isomorphic lattices: $S_{4,1,1} \simeq S_{2,1,2}$, $S_{8,1,1} \simeq S_{4,1,2}$ and $S_{12,1,1} \simeq S_{6,1,2}$, because they are lattices of their fundamental polygons $\mathscr{M}$, and finally they are formed in an identical manner from triangles with zero angles, and so are congruent.

The theorem is proved.

4°. Classification of the lattices S. The following theorem follows at once from Theorems 2.1 and 2.4 and the reasoning at the end of the proof of Theorem 2.4.

THEOREM 2.5. *Every hyperbolic lattice of rank 3 for which the group $A(S) = O'(S)/W^{(2)}(S)$ is finite and for which the fundamental polygon $\mathscr{M}$ of $W^{(2)}(S)$ is noncompact is isomorphic to one and only one of the lattices*

$$S_{1,1,k} = [a, b, kc], \quad \textit{where } k = 1 - 4, 6, 8;$$
$$S_{1,k,1} = [a, kb, c], \quad \textit{where } k = 2 - 6, 9;$$
$$S_{k,1,1} = [ka, b, c], \quad \textit{where } k = 4 - 8, 10, 12;$$

$$S'_{8,1,1} = S'_{4,1,2} = [2a + c, b, 2c];$$
$$S'_{12,1,1} = S'_{6,1,2} = [3a + c, b, 2c],$$

where a, b *and* c *are elements with Gram matrix*

	a	b	c
a	-2	0	1
b	0	-2	2
c	1	2	-2

The isomorphism class of such lattices S *is determined by the invariants*

$$\left(\textit{number of vertices of } \mathcal{M},\ v(\mathcal{M})/v(\Delta),\ \sqrt{(\operatorname{discr} S)/2}\right),$$

where $v(\mathcal{M})$ *is the area of* $\mathcal{M}$, *and* $v(\Delta)$ *is the area of the triangle with angles* 0, $\pi/3$, *and* $\pi/2$.

All of these lattices S, *except the lattices* $S'_{8,1,1} = S'_{4,1,2}$ *and* $S'_{12,1,1} = S'_{6,1,2}$, *are the lattices of their own fundamental polygons* $\mathcal{M}$ (*indicated in Table* 3). *For the lattices* $S'_{8,1,1} = S'_{4,1,2}$ *and* $S'_{12,1,1} = S'_{6,1,2}$ *the sublattices of the fundamental polygons* $\mathcal{M}$ *are* $S_{4,1,2} = [4a, b, 2c]$ *and* $S_{6,1,2} = [6a, b, 2c]$ *respectively, which satisfy* $S_{4,1,2} \simeq S_{8,1,1}$ *and* $S_{6,1,2} \simeq S_{12,1,1}$.

5°. An application: the arithmetic decomposition of a convex polygon. Although we will speak of polygons and their decompositions into smaller polygons in the Lobachevsky plane, the results under discussion are topological. Their translation into topological language we leave to the reader.

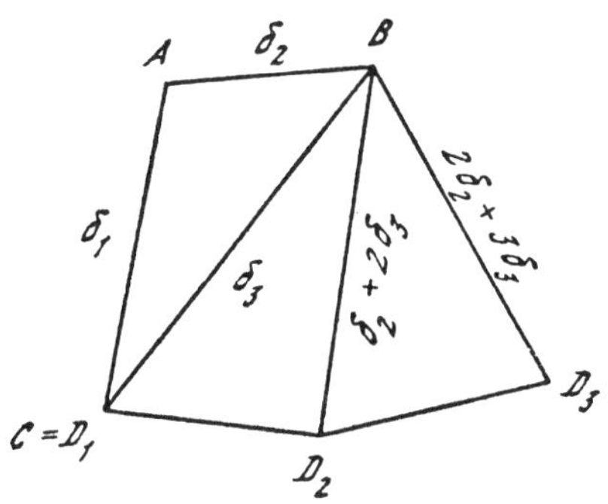

FIG. 9

We consider a decomposition $\mathfrak{R}$ of the Lobachevsky plane into triangles $W^{(2)}(S_1) \cdot \Delta$ with angles 0, $\pi/3$ and $\pi/2$, which we considered in subsection 3 (see Figures 5, 6, 8 and 10). We also consider an arbitrary polygon $\mathcal{N}$ in the Lobachevsky plane with angles 0, $\pi/3$ or $\pi/2$, whose sides run along sides of $\mathfrak{R}$, and its decomposition $\mathfrak{R} \cap \mathcal{N}$ into triangles of $\mathfrak{R}$. Let S be the lattice of this polygon which is, naturally, a sublattice of the lattice S_1 satisfying the conditions of Theorem 2.4 (S is generated in S_1 by the elements $P(\mathcal{N})$ with square (-2), orthogonal to the sides of $\mathcal{N}$).

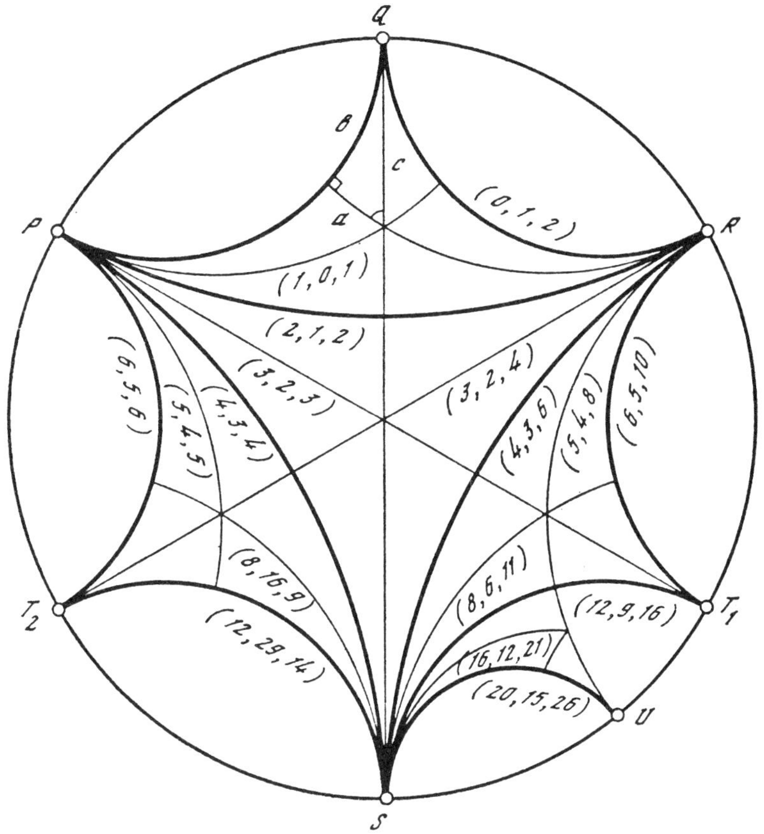

FIG. 10

From this theorem it follows that S is one of the 22 sublattices of Table 3 ($S \neq S'_{4,1,2}$ and $S'_{6,1,2}$), and the polygon $\mathscr{N}$ and its decomposition $\mathfrak{R} \cap \mathscr{N}$ are obtained with the help of reflections from the large polygons $W^{(2)}(S) \cdot \mathscr{M}$ with decompositions $\mathfrak{R} \cap (W^{(2)}(S) \cdot \mathscr{M})$, where $\mathscr{M}$ is a fundamental polygon for $W^{(2)}(S)$. In this way, Theorem 2.4 divides the polygons $\mathscr{N}$ with decomposition $\mathfrak{R} \cap \mathscr{N}$ into 22 types depending on the sublattice S of $\mathscr{N}$. The possibility of "consolidating" the decomposition $\mathfrak{R} \cap \mathscr{N}$ of $\mathscr{N}$ into a decomposition $\mathfrak{R}_S \cap \mathscr{N}$ made up of polygons $W^{(2)}(S) \cdot \mathscr{M}$ with decompositions $\mathfrak{R} \cap (W^{(2)}(S) \cdot \mathscr{M})$, evidently, is determined by the topology of $\mathfrak{R} \cap \mathscr{N}$. Hence, finally, it follows that in fact the sublattice S of $\mathscr{N}$ must have a topological meaning, determined by the combinatorial decomposition $\mathfrak{R} \cap \mathscr{N}$.

It turns out to consist of the fact that the orders of the vertices of $\mathscr{N}$ in $\mathfrak{R} \cap \mathscr{N}$ satisfy a system of congruences of some type. Thus, the solution of this system of congruences proves to be equivalent to the possibility of a "consolidation" of the decomposition $\mathfrak{R} \cap \mathscr{N}$ into a decomposition $\mathfrak{R}_S \cap \mathscr{N}$, and in all there exist 22 types of such equations.

In view of the large number of calculations, we restrict ourselves to only some cases (interested readers can investigate the general case without difficulty on their own). We consider the sublattice $S_{2,1,2}$ of S_1. We restrict ourselves to those polygons $\mathscr{N}$ whose sublattices S lie in $S_{2,1,2}$, i.e. $S \subset S_{2,1,2} \subset S_1$. Since the fundamental polygon $\mathcal{M}$ for $W^{(2)}(S_{2,1,2})$ is a triangle with zero angles, generated by lines of type I (see the beginning of the proof of Theorem 2.4), such polygons $\mathcal{N}$ are exactly those polygons whose sides are lines of type I and which are decomposed into triangles by lines of type I which are diagonals that do not intersect themselves.

We will consider the vertices of the n-gon $\mathcal{N}$ to be A_i, where $i \in \mathbf{Z}/n\mathbf{Z}$, and where A_{i+1} follows A_i. Let k_i be the order of the vertex A_i, i.e. the number of triangles generated by lines of type I and having A_i as a vertex. We denote by e_i, δ_i and f_i the elements with square (-2), orthogonal and pointing outward to the sides D_iA_{i-1}, $A_{i-1}A_i$ and A_iD_i the triangle $\tilde{\Delta}_i \subset \mathcal{M}$ generated by lines of type I, one side of which is $A_{i-1}A_i$. Then the elements e_i, δ_i and f_i are generators of $S_{2,1,2}$, and each sublattice $S \subset S_{2,1,2}$ is given by a system of congruences

$$(T_i):\ a_i^{(m)}\alpha_i + b_i^{(m)}\beta_i + c_i^{(m)}\gamma_i \equiv 0 \pmod{r_i^{(m)}},$$

where $m = 1, 2, \ldots, M_i$; $a_i^{(m)}, b_i^{(m)}, c_i^{(m)} \in \mathbf{Z}$, $r_i^{(m)} \in \mathbf{N}$, if $s = \alpha_i e_i + \beta_i\delta_i + \gamma_i f_i$ is an element of $S_{2,1,2}$, where $\alpha_i, \beta_i, \gamma_i \in \mathbf{Z}$. Since the element δ_i must belong to S, then $b_i^{(m)} \equiv 0 \pmod{r_i^{(m)}}$, i.e. the sublattice S is given by the system of congruences

$$(T_i):\ a_i^{(m)}\alpha_i + c_i^{(m)}\gamma_i \equiv 0 \pmod{r_i^{(m)}}.$$

The systems (T_i) give for different i one and the same sublattice S, i.e. the systems (T_i) for all i are equivalent. We write out the condition for equivalence of (T_i) and (T_{i+1}). The calculations, analogous to those in the proof of Theorem 2.2, show that

$$e_{i+1} = (k_i - 2)\delta_i + (k_i - 1)f_i, \qquad \delta_{i+1} = (k_i - 1)\delta_i + k_i f_i,$$
$$f_{i+1} = e_i + (k_i - 1)(k_i - 2)\delta_i + k_i(k_i - 1)f_i,$$

whence

$$\begin{aligned} s &= \alpha_{i+1}e_{i+1} + \beta_{i+1}\delta_{i+1} + \gamma_{i+1}f_{i+1} \\ &= \gamma_{i+1}e_i + [\alpha_{i+1}(k_i - 2) + \beta_{i+1}(k_i - 1) + \gamma_{i+1}(k_i - 1)(k_i - 2)]\,\delta_i \\ &\quad + [\alpha_{i+1}(k_i - 1) + \beta_{i+1}k_i + \gamma_{i+1}k_i(k_i - 1)]\,f_i \end{aligned}$$

and by the equivalence of (T_i) and (T_{i+1}) the system (T_{i+1}) can be rewritten as

$$\begin{aligned} (T_{i+1}):\ & a_i^{(m)}\gamma_{i+1} + c_i^{(m)}[\alpha_{i+1}(k_i - 1) + \beta_{i+1}k_i + \gamma_{i+1}k_i(k_i - 1)] \\ &\equiv c_i^{(m)}(k_i - 1)\alpha_{i+1} + c_i^{(m)}k_i\beta_{i+1} + [a_i^{(m)} + c_i^{(m)}k_i(k_i - 1)]\,\gamma_{i+1} \\ &\equiv 0 \pmod{r_i^{(m)}}. \end{aligned}$$

Hence

$$M_{i+1} = M_i = M; \quad r_{i+1}^{(m)} = r_i^{(m)} = r^{(m)}, \quad b_{i+1}^{(m)} = c_i^{(m)} \cdot k_i \equiv 0 \pmod{r^{(m)}},$$
$$a_{i+1}^{(m)} \equiv -c_i^{(m)} \pmod{r^{(m)}}, \qquad c_{i+1}^{(m)} \equiv a_i^{(m)} \pmod{r^{(m)}}.$$

Supposing that $a^{(m)} = a_0^{(m)}$ and $c^{(m)} = c_0^{(m)}$, we see that the condition for the elements $P(\mathcal{N}) = \{\delta_i | i \in \mathbf{Z}/n\mathbf{Z}\}$ of the polygon to be elements of some sublattice $S \subset S_{2,1,2}$ of finite index is equivalent to the existence of a tuple $(a^{(m)}, c^{(m)}; r^{(m)})$, $m = 1, \ldots, M$, of integers $a^{(m)}$, $c^{(m)}$ and natural numbers $r^{(m)}$ such that the orders of the vertices of the n-gon $\mathcal{N}$ satisfy the system of congruences

$$\begin{aligned} a^{(m)} \cdot k_i &\equiv 0 \pmod{r^{(m)}}, \quad \textit{if } i = 2l + 1, \\ c^{(m)} \cdot k_i &\equiv 0 \pmod{r^{(m)}}, \quad \textit{if } i = 2l, \end{aligned} \tag{5.1}$$

where $m = 1, \ldots, M$, $i \in \mathbf{Z}/n\mathbf{Z}$ and $l \in \mathbf{Z}$.

The sublattice $S_{2,1,2} \subset S_1$ is clearly invariant to $W^{(2)}(S_1)$, and by Theorem 2.4 up to automorphisms from $O'(S_{2,1,2})$ there are exactly three sublattices $S_{2,1,2} = S_{2,1,2}$, $S_{4,1,2} \subset S_{2,1,2}$ and $S_{6,1,2} \subset S_{2,1,2}$ which might contain the elements $P(\mathscr{N})$. Up to indexing of vertices of $\mathscr{N}$, systems of congruences of the form (5.1) with parameters $M = 1$, $a^{(1)} = 1$, $c^{(1)} = 0$ and $r^{(1)} = 1, 2$, and 3, respectively, correspond to them.

COROLLARY 2.6. *Let $\mathscr{N}$ be a polygon in the Lobachevsky plane, whose sides are lines of type* I *of the decomposition $\mathfrak{R}$.*

Then, up to a permutation of the vertices of $\mathscr{N}$, their orders k_i may satisfy only the following systems of congruences of the form (5.1):

$$k_i \equiv 0 \pmod r \quad \textit{for all } i \in \mathbf{Z}/n\mathbf{Z} \textit{ of the form } i = 2l + 1, \tag{5.2}$$

where $r = 1, 2$, or 3. In this setting the polygon $\mathscr{N}$ satisfies the congruences (5.2) *with a given $r = 1, 2$, or 3 if and only if the decomposition $\mathscr{N} \cap \mathfrak{R}$ is obtained with the help of reflections from $\mathscr{M}_3 \cap \mathfrak{R}$, $\mathscr{M}_4 \cap \mathfrak{R}$ or $\mathscr{M}_6 \cap \mathfrak{R}$ respectively, where $\mathscr{M}_3$, $\mathscr{M}_4$, or $\mathscr{M}_6$ is the fundamental triangle, quadrilateral, or hexagon for the group $W^{(2)}(S)$ of the sublattices $S = S_{2,1,2}$, or $S_{4,1,2}$ or $S_{6,1,2}$ of the lattice S_1 respectively* (*see Table* 3).

BIBLIOGRAPHY

1. A. Hurwitz, *Über algebraische Gebilde mit eindeutigen Transformationen in sich*, Math. Ann. **41** (1893), 403–442.

2. F. Enriques, *Sulle superficie algebriche che ammettono una serie discontinua di transformazioni birazionali*, Atti Reale Accad. Lincei Rend. Cl. Sci. Fis. Mat. Nat. (5) **15** (1906), Sem. 2, 665–669.

3. Lucien Godeaux, *Sur les transformations des surfaces algébriques laissant invariant un système continu de courbes*, Atti Reale Accad. Lincei Rend. Cl. Sci. Fis. Mat. Nat. (5) **21** (1912), Sem. 1, 398–402.

4. Federigo Enriques, *Le superficie algebriche*, Zanichelli, Bologna, 1949.

5. I. R. Shafarevich et al., *Algebraic surfaces*, Trudy Mat. Inst. Steklov. **75** (1965); English transl. Proc. Steklov Inst. Math. **75** (1965).

6. I. I. Pyatetskiĭ-Shapiro [Piatetsky-Shapiro] and I. R. Shafarevich, *A Torelli theorem for algebraic surfaces of type K*3, Izv. Akad. Nauk SSSR Ser. Mat. **35** (1971), 530–572; English transl. in Math. USSR Izv. **5** (1971).

7. V. V. Nikulin, *Integral symmetric bilinear forms and some of their applications*, Izv. Akad. Nauk SSSR Ser. Mat. **43** (1979), 111–177; English transl. in Math. USSR Izv. **14** (1980).

8. ______, *On factor groups of the automorphism groups of hyperbolic forms modulo subgroups generated by 2-reflections*, Dokl. Akad. Nauk SSSR **248** (1979), 1307–1309; English transl. in Soviet Math. Dokl. **20** (1979).

9. ______, *Factor groups of the automorphism groups of hyperbolic forms modulo subgroups generated by 2-reflections: Algebro-geometric applications*, Itogi Nauki: Sovremennye Problemy Mat., vol. 18, VINITI, Moscow, 1981, pp. 3–114; English transl. in J. Soviet Math. **22** (1983), no. 4.

10. ______, *On arithmetic groups generated by reflections in Lobachevsky spaces*, Izv. Akad. Nauk SSSR Ser. Mat. **44** (1980), 637–669; English transl. in Math. USSR Izv. **16** (1981).

11. ______, *On the classification of arithmetic groups generated by reflections in Lobachevsky spaces*, Izv. Akad. Nauk SSSR Ser. Mat. **45** (1981), 113–142; English transl. in Math. USSR Izv. **18** (1982).

12. *Encyklopädie der Mathematischen Wissenschaften*. Vol. III.2.1, Teubner, Leipzig, 1903–1915.

13. N. G. Chebotarev, *Theory of algebraic functions*, OGIZ, Moscow, 1948. (Russian)

14. George Springer, *Introduction to Riemann surfaces*, Addison-Wesley, 1957.

Translated by A. J. BIAGIOLI

Труды Матем. Инст. Стеклов.
165(1984)

Proceedings of the Steklov
Institute of Mathematics
1985, Issue 3

UDC 512.62

LOCAL CLASS FIELD THEORY

A. N. PARSHIN

ABSTRACT. This paper contains a construction of class field theory for local fields of arbitrary dimension and finite characteristic. A structure in a field K consisting of a sequence of complete discrete valuation rings O_i and their quotient fields K_i such that K_{i+1} is the residue field of O_i and $K_0 = K$ is called an *n-dimensional local field*. Taking into account the natural topology in K, one can construct a topological invariant of the higher Milnor K-function $K_n^{\text{top}}(K)$. The main result of the paper is the construction of a reciprocity map φ_k: $K_n^{\text{top}}(K) \to \text{Gal}(K^{\text{ab}}/K)$ having all the properties that are well known in classical class field theory for local fields of dimension 1. The group $K_n^{\text{top}}(K)$ is an everywhere dense subgroup of $\text{Gal}(K^{\text{ab}}/K)$. A complete calculation of it is given in the paper.

Bibliography: 18 titles.

The quotient field of a complete discrete valuation ring is usually called a local field. The theory of such fields is an important part of classical commutative algebra and has many applications in both arithmetic and geometry In [1]–[3] the author introduced the concept of a local field of arbitrary dimension n with respect to which ordinary local fields are local fields of dimension 1. It turns out that this concept is very convenient for the presentation of a number of problems in multidimensional algebraic geometry (see the Introduction in [2]). In addition, for local fields in this sense one can construct an exact analog of class field theory that gives a complete description of abelian extensions. This description is given in terms of higher Milnor K-functors. This aspect of the theory of local fields was discovered independently by K. Kato and developed by him in [4]–[7].

The present paper contains a detailed presentation of a part of the results of local class field theory announced in [3] and pertaining to local fields of finite characteristic. We have tried to give a selfcontained presentation, providing the necessary motivation for the concepts and constructions introduced and using a minimum number of results from algebraic K-theory. In particular, we do not use Quillen's general theory.

The paper consists of four sections. §1 contains general information about local fields. An outline of K-theory and the definition of the groups $K_n^{\text{top}}(K)$ for local fields K and their calculation are presented in §2. §3 is devoted to the construction of Kummer and Artin–Schreier–Witt duality in our situation (and also a transfer map). Class field theory is contained in §4. A number of results, in particular the calculation of the Brauer group of a local field, will be examined in a separate article.

1980 *Mathematics Subject Classification*. Primary 12B25, 12B10; Secondary 12B22.

I am deeply grateful to H. Koch, P. Roquette, J.-P. Serre, and I. R. Shafarevich for their attention to my work.

The notation used in this paper, as a rule, is standard. $a|b$ denotes the divisibility of integers a and b; δ_{ij} is Kronecker's symbol; A/n is the abbreviation of A/nA for an abelian group A. Depending on the context, $\wedge$ is the symbol of a gap or exterior multiplication—convergence in some topology or a mapping; (a, b) is the greatest common divisor of integers a and b or an element of the group K_2; K_i is the ith residue field of a local field or the ith K-functor. The word "symbol" is also used in two senses—to denote an arbitrary generator in a K-functor and to denote specific functions of these generators (for example, the norm residue symbol).

§1. Local fields

Let K and k be fields. We shall introduce a local field structure on K.

DEFINITION 1. By an *n-dimensional local field structure on* K (over a field k) we mean a sequence of complete discrete valuation rings O_i with quotient fields K_i such that the field K_{i+1} is the residue field of O_i, $i = 0, 1, \ldots, n-1$, and $K_0 = K$ and $K_n = k$.

We call K_1 the *first residue field* (and shall denote it by $\bar{K}$) and k the last. We denote by $\mathfrak{p}_i$ ($\mathfrak{p}_0 = \mathfrak{p}$) the maximal ideals of the rings O_i. The ring O_0 (denoted later by O_K) defines a discrete valuation v_K on K. By a *system of parameters* $t_1, \ldots, t_n$ we mean a collection of elements $t_i \in O_0$ such that

$$t_i \bmod \mathfrak{p}_0 \in O_1,$$
$$\cdots\cdots\cdots\cdots$$
$$t_i \bmod \mathfrak{p}_0 \cdots \bmod \mathfrak{p}_{i-2} \in O_{i-1}$$

and the last element is a generator of the ideal $\mathfrak{p}_{i-1}$. A system of parameters determines a valuation of K of rank n, studied in detail in [8] and [9].

An example of a local field is the field $K = \mathbf{F}_q((t_1) \cdots (t_n))$ of power series in n variables over a finite field $\mathbf{F}_q$ of characteristic p. Here

$$O_i = \mathbf{F}_q((t_1) \cdots (t_{n-i-1}))[[t_{n-1}]], \qquad K_i = \mathbf{F}_q((t_1)) \cdots ((t_{n-i})),$$

and $t_1, \ldots, t_n$ is a system of parameters. We note that in this case there are imbeddings of O_i and K_i in K that are compatible with the local structure. This is not the case in general for arbitrary local fields. However, if $\operatorname{char} K = \operatorname{char} k$ and $k = \mathbf{F}_q$, then by the Cohen–Teichmüller structure theorem ([10], Chapter 2, §4) there exists an isomorphism of K with the power series field that preserves the local structure.

The construction of this isomorphism is carried out by induction on the dimension of the field K. Let $\bar{K} = \mathbf{F}_q((\tilde{t}_1)) \cdots ((\tilde{t}_{n-1}))$. Then, to establish an isomorphism $K \cong K((t_n))$, it is necessary to construct an imbedding of $\bar{K}$ in O_K compatible with the projection $O_K/\mathfrak{p} \xrightarrow{\sim} \bar{K}$. By the Cohen–Teichmüller theorem, the required imbedding is uniquely defined if one chooses (arbitrarily) liftings of elements of $\bar{K}$ into O_K that generate a p-basis of $\bar{K}$ over $\bar{K}^p$. Since the variables $\tilde{t}_1, \ldots, \tilde{t}_{n-1}$ are a p-basis, their liftings $t_1, \ldots, t_{n-1}$, together with the generator t_n of the ideal $\mathfrak{p}$, define the required isomorphism $K \cong \mathbf{F}_q((t_1)) \cdots ((t_n))$.

In what follows we shall consider only local fields of this type. Other examples can be found in [3].

If L/K is a finite extension of the local field K, then, setting O_0' equal to the integral closure of O_0 in L, O_1' equal to the integral closure of O_1 in the residue field $L_1 \supset K_1$ of O_1, etc., we obtain a local field structure in L. If there are two finite extensions M/L and L/K, then this construction has a natural transitivity property. Thus, the entire tower of finite extensions of K is naturally endowed with local field structures. If L/K is a Galois extension, then automorphisms of L over K preserve the local structure.

PROPOSITION 1. *Let $K_{\text{sep}}(p)$ and K_{sep} (no p) be maximal separable extensions of K that are respectively p-extensions and extensions of degree prime to p.*

1. *K_{sep} is the compositum of linearly divided extensions $K_{\text{sep}}(p)$ and K_{sep} (no p).*
2. *$K_{\text{sep}}(\text{no } p)$ is the compositum of extensions $(F_q)_{\text{no } p}$ and $E_1, \dots, E_n$, where $E_i = \bigcup_{(m,p)=1} K\sqrt[m]{t_i}$, $i = 1, \dots, n$.*
3. *The field $K^{\text{ab}}_{\text{sep}}(\text{no } p)$ is the compositum of linearly divided extensions $(F_q)_{\text{no } p}$ and $L_1, \dots, L_n$, where*

$$L_i = K\left(t_i^{1/(q-1)}\right) \qquad i = 1, \dots, n.$$

The proof follows immediately from standard properties of extensions of local fields of dimension 1 (see, for example, [10]) and radical extensions. Moreover, property 3 follows from property 2 if we bear in mind that the only roots of unity in K are nonzero elements of $\mathbf{F}_q$.

COROLLARY. *Every Galois extension of a local field K is solvable.*

The following lemma is useful for the study of purely inseparable extensions of K:

LEMMA 1. *Let K be a local field of dimension n.*

1. *$[K^{1/p} : K] = p^n$.*
2. *For any extension $K \subset L \subset K^{1/p}$ of degree p in K there exists a system of parameters $t_1, \dots, t_n$ such that $L = K(t_i^{1/p})$ for some i.*

PROOF. If $K = \mathbf{F}_q((t_1)) \cdots ((t_n))$, then $K^{1/p} = \mathbf{F}_q((T_1)) \cdots ((T_n))$, where $T_i^p = t_i$, $i = 1, \dots, n$. We shall verify the second assertion by induction on n. If $\dim K = 1$, then it obviously follows from the equality $[K^{1/p} : K] = p$. Assuming that it is valid for fields of dimension $n - 1$, we consider a purely inseparable extension L/K of degree p. There are two possibilities:

1. $\bar{L} = \bar{K}$ and $v_L = pv_K$. Let $L = \mathbf{F}_q((T_1)) \cdots ((T_n))$. Then $T_n^p \in K$ and $v_k(T_n^p) = 1$. Setting $t_1 = T_1, \dots, t_{n-1} = T_{n-1}$ and $t_n = T_n^p$, we obtain the required assertion.
2. $[\bar{L} : \bar{K}] = p$ and $v_L = v_K$. In this case if $L = \mathbf{F}_q((T_1)) \cdots ((T_n))$, then $T_n \in K$ and is a generator of the ideal $\mathfrak{p}_K$. Now let $\bar{K} = \mathbf{F}_q((\bar{t}_1)) \cdots ((\bar{t}_{n-1}))$ and for some i let $\bar{L} = \bar{K}(\tilde{T}_i)$ and $\tilde{T}_i^p = \bar{t}_i$, $i \leqslant n - 1$. Lifting $\tilde{T}$ to O_L to some element T_i, we see that $L = K(T_i)$ and that T_i^p occurs in some system of parameters of K.

This proves the lemma.

We shall call an extension L/K *unramified* if it is unramified with respect to the valuation v_K and the residue field extension L_1/K_1 is separable. We denote the maximal extension with these properties by K_{et}. We have the sequence of epimorphisms

$$\begin{gathered}\operatorname{Gal}(K_{\mathrm{sep}}/K) \to \operatorname{Gal}(K_{\mathrm{et}}/K) \xrightarrow{\sim} \operatorname{Gal}(K_{1,\mathrm{sep}}/K_1),\\ \operatorname{Gal}(K_{1,\mathrm{sep}}/K_1) \to \operatorname{Gal}(K_{1,\mathrm{et}}/K_1) \xrightarrow{\sim} \operatorname{Gal}(K_{2,\mathrm{sep}}/K_2)\\ \cdots\cdots\cdots\cdots\cdots\cdots\cdots\cdots\cdots\cdots\cdots\\ \operatorname{Gal}(K_{n-1,\mathrm{sep}}/K_{n-1}) \to \operatorname{Gal}(K_{n-1,\mathrm{et}}/K_{n-1}) \xrightarrow{\sim} \operatorname{Gal}(\mathbf{F}_{q,\mathrm{sep}}/\mathbf{F}_q) = \hat{\mathbf{Z}}.\end{gathered}$$

We now turn to the study of topologies in the field K and its multiplicative group K^*. We shall define them by induction on the degree of the field.

DEFINITION 2. Let $t_1, \ldots, t_n$ be a system of parameters of the field K. If U_m, $m \in \mathbf{Z}$, is a system of neighborhoods of zero in the field $\bar{K} = \mathbf{F}_q((t_1)) \cdots ((t_{n-1}))$, $U_{m+1} \subseteq U_m$, and $U_m = \bar{K}$ for large m, then we shall call subgroups of the form

$$U = \left\{ \sum_m x_m t_n^m,\ x_m \in U_m \right\}$$

neighborhoods of zero in K.

If $n = 0$, then (0) and $\mathbf{F}_q$ are neighborhoods in $\mathbf{F}_q$. We represent K as $\bigcup_m K(m)$, where $x \in K(m)$ if $v_K(x) \geqslant m$. Then the sets U are a fundamental system of neighborhoods of the topology which induces the direct product topology on $K(m) \cong \bar{K}^{\mathbf{Z}}$ (as an additive group). A set $V \subset K$ is open if and only if all intersections $V \cap K(m)$ are open in $K(m)$. For a field of dimension 1 this is the usual locally compact topology. In an arbitrary dimension > 1 this topology is no longer locally compact. Moreover, the sets $K(m)$ are always closed but for $\dim K > 1$ are not open.

PROPOSITION 2. *The topology of the field K does not depend on the choice of the system of parameters $t_1, \ldots, t_n$ and has the following properties*:

1. *If $x_n \to 0$, then all $x_n \in K(m)$ for some m.*
2. *The additive group K_+ of K is a separable topological group. Every fundamental sequence in K_+ converges.*
3. *The mapping x_n: $K \to K$ $(y \to yx)$ is a homeomorphism.*
4. *Every automorphism of the local field K is a homeomorphism.*
5. *If $x_n \to x$ and $y_n \to y$, then $x_n y_n \to xy$ as $n \to \infty$.*

PROOF. If $s_1, \ldots, s_n$ are another system of parameters of K, then

$$s_1^{b_1} \cdots s_n^{b_n} = \sum_{a_n} \cdots \sum_{a_1} a(a_1, \ldots, a_n) t_1^{a_1} \cdots t_n^{a_n}, \tag{1}$$

where if $b_{k+1} = \cdots = b_n = 0$ and $b_k \neq 0$, then $a_{k+1}, \ldots, a_n \geqslant 0$, and if $a_{k+1} = \cdots = a_n = 0$, then $a_k \geqslant b_k$. In addition, there is a nonzero monomial of the form $a t_1^{a_1} \cdots t_{k-1}^{a_{k-1}} t_k^{b_k}$.

We shall prove that the topology is independent of the choice of the system of parameters by induction on the dimension of the field K. Assuming independence for the field $\bar{K}$, we consider the substitution of $s_1, \ldots, s_n$ for the system $t_1, \ldots, t_n$. It is sufficient to consider n cases in turn: $s_1 = t_1, \ldots, s_{n-1} = t_{n-1}$; $\forall i \neq k$, $s_i = t_i, \ldots$; $s_2 = t_2, \ldots, s_n = t_n$. In the first of these the field $\bar{K}$ is imbedded in K in the same way both for $s_1, \ldots, s_n$ and for $t_1, \ldots, t_n$. In the subsequent cases we have two subfields $\overline{K}'$ and $\overline{K}''$ identified as fields (with the induced topology) with $\overline{K}$. It is sufficient to show that the topologies determined by the systems of parameters $s_1, \dots, s_n$ and $t_1, \dots, t_n$ coincide in O_K. Since $K = \overline{K}'((s_n)) = \overline{K}''((t_n))$, each element $x \in K$ has two power series expansions, the passage between which is given by (1). A direct comparison of the coefficients of the series yields the required assertion.

To verify property 1 we set $\alpha_n = v_K(x_n)$ and assume that $\alpha_n \to -\infty$ and $\alpha_{n+1} < \alpha_n$. Then there is a neighborhood of zero $U = (U_k)$, $U_k \subset \overline{K}$, such that the coefficient $y_n \in \overline{K}$ of $t_n^{\alpha_n}$ in x_n does not belong to U_{α_n}. It is obvious that $\forall n\ x_n \notin U$.

Property 2 follows from the fact that $U + U = U$ for any neighborhood U in Definition 2. The convergence of fundamental sequences is obtained by induction on the dimension with the use of 1.

Property 4 is equivalent to the independence of the topology of the system of parameters. To establish 3 we can assume that $v_k(x) = 0$ (for $x = t_n^a$ this is obvious). Then $\forall m\ xK(m) = K(m)$ and the problem reduces to the analogous assertion for O_K. Assuming that U_y is an open set for any neighborhood of zero $U \subset \overline{K}$ and that $y \in \overline{K}$, we find that the same thing is true in O_K. Property 5 is obtained by a similar reduction to O_K (with the help of property 1) and then to a field $\overline{K}$ of lesser dimension.

REMARK 1. Multiplication in K is not continuous in this topology. For if $\dim K > 1$, then for any neighborhoods of zero U and V we have $UV = K$. This does not contradict property 5 since the set of neighborhoods of zero in K is uncountable.

REMARK 2. For any element $x \in K$ we have the unique representation

$$x = \sum_{a_n \geqslant A_n} \sum_{a_{n-1} \geqslant A_{n-1}(a_n)} \cdots \sum_{a_1 \geqslant A_1(a_2, \dots, a_n)} a(a_1, \dots, a_n) t_1^{a_1} \cdots t_n^{a_n}, \qquad a \in \mathbf{F}_q \tag{2}$$

since the series converge. We shall again drop the indices of the coefficients a.

LEMMA 2. *Let $F(x) = x^p$, $x \in K$. Then each element $x \in K$ is uniquely representable modulo the subgroup $(F - 1)K$ by the series* (2) *in which the monomials in the parameters $t_1, \dots, t_n$ satisfy the following conditions*:

1. *The greatest common divisor of all nonzero numbers $a_1, \dots, a_n$ is prime to p.*
2. $a_n \leqslant 0$; $a_{n-1} \leqslant 0$ *if* $a_n = 0; \dots;$ $a_1 \leqslant 0$ *if* $a_2 = \cdots = a_n = 0$.
3. *The coefficients a run through an additive basis of $\mathbf{F}_q$ over $\mathbf{F}_p$, and if $a_1 = \cdots = a_n = 0$ then $\mathrm{Tr}_{\mathbf{F}_q/\mathbf{F}_p}(a) = 0$.*

PROOF. In (2) we consider all monomials for which the last nonzero power a_i of t_i is positive. If their sum is equal to y, then

$$y = y - y^p + \left(y^p - y^{p^2}\right) + \cdots$$

and since $y^{p^n} \to 0$ as $n \to \infty$, we find that $y \in (F - 1)K$. The remaining monomials either satisfy the conditions of the lemma or are pth powers and reduce to the preceding by means of elements in $(F - 1)K$.

To establish that the expansion is unique, we assume that

$$\sum_{a_1, \dots, a_n} a(a_1, \dots, a_n) t_1^{a_1} \cdots t_n^{a_n} = x^p - x, \qquad x = \sum b(b_1, \dots, b_n) t_1^{b_1} \cdots t_n^{b_n},$$

where $a_1, \dots, a_n$ satisfy the conditions of the lemma.

Comparing the coefficients of identical powers of $t_1, \dots, t_n$ on both sides of the equality, we obtain

$$-b(b_1 p, \dots, b_n p) + b(b_1, \dots, b_n)^p = 0.$$

Repeating this, we find that for any $s \geqslant 1$

$$b(b_1 p^s, \dots, b_n p^s) = \pm b(b_1, \dots, b_n)^{p^s}.$$

The indices $b_1, \dots, b_n$ must satisfy condition 2, and so for large s the left side is equal to zero. This yields the required assertion.

LEMMA 3. *Let L/K be a finite extension of local fields. Then K is closed in L and its topology coincides with the topology induced by L.*

In [9] (Chapter 1, Proposition 1.2) there is a stronger assertion. Namely, for any basis $e_1, \dots, e_m$ of L over K the mapping $K^m \to L$ $(x_1, \dots, x_m \to x_1 e_1 + \cdots + x_m e_m)$ is homeomorphic.

The multiplicative group K^* of K has the following representation:

$$K^* = \{t_1\} \times \cdots \times \{t_n\} \mathbf{F}_q^* \mathscr{E}_K, \tag{3}$$

where $\{t_i\} \cong \mathbf{Z}$ and $x = \mathscr{E}_k$ if and only if $x \in O_0$, $x \bmod \mathfrak{p}_0 \in O_2, \dots, x \bmod \mathfrak{p}_0 \cdots \bmod \mathfrak{p}_{n-1} = 1$.

DEFINITION 3. In the group K^* we introduce a topology as the product of the discrete topology in $\{t_1\} \times \cdots \times \{t_n\} \mathbf{F}_q^*$ and the topology in $\mathscr{E}_K$ induced by K.

In this topology K^* is a topological group only for $\dim K \leqslant 2$ (see Remark 1). This fact was noticed by Kato [7].

PROPOSITION 3. *The topology of the group K^* does not depend on the choice of the system of parameters $t_1, \dots, t_n$ and has the following properties*:

1. *The mapping x: $K^* \to K^*$ $(y \to yx)$ is a homeomorphism.*
2. *Every automorphism of the local field K is a homeomorphism in K^*.*
3. *If $x_n \to x$ and $y_n \to y$, then $x_n y_n \to xy$ as $n \to \infty$. If $x_n \to x$, then $x_n^{-1} \to x^{-1}$.*
4. *If $x_n \in \mathfrak{p}^n$, then $\prod(1 + x_n)$ converges in K^*.*
5. *If $x_n \to 0$ in $\overline{K}$ and t is a generator of the ideal $\mathfrak{p}$, then $\prod(1 + x_n t^a)$ converges in K^*.*
6. *For any sequence $y_n \in \mathscr{E}_K$ the sequence $y_n^{p^n}$ converges to* 1.
7. *If $K \subset L$ is a finite extension of local fields, then the group K^* is closed in L^*, and its topology coincides with the topology induced by L^*.*

PROOF. Properties 1–3 correspond to properties 3–5 of Proposition 2 (the transition to x_n^{-1} is found below). Property 7 follows from Lemma 3. Property 4 follows immediately from the definition of a topology.

Let us verify 5. Expanding the product $\prod_{n \geqslant m}(1 + x_n t^a)$, we have

$$1 + \Big(\sum_{n \leqslant m} x_n\Big) t^a + \Big(\sum_{n < k \leqslant m} x_n x_k\Big) t^{2a} + \cdots,$$

and it is necessary to show that the series next to the powers of t converge. Since neighborhoods of zero in K are subgroups (Definition 2), the sequence $\sum_{n \leqslant m} x_n$ is fundamental and therefore converges (Proposition 2.2). Furthermore, we have

$$\sum_{n < k \leqslant m+1} x_n x_k - \sum_{n < k \leqslant m} x_n x_k = \Big(\sum_{n \leqslant m} x_n\Big) x_{m+1} \to 0$$

by what we have proved and Proposition 2.5. Applying the foregoing argument, we find that $\Sigma x_n x_k$ converges. The next series is analyzed in a similar manner. Thus, the convergence of x_n^{-1} in property 3 is obtained.

It remains to consider 6. The group $\mathscr{E}_K$ is the product of its subgroups $\mathscr{E}_{\overline{K}}$, $\overline{K} = \mathbf{F}_q((t_1)) \cdots ((t_{n-1}))$ and $1 + \mathfrak{p}$. Consequently, $y_n = u_n v_n$, $u_n \in \mathscr{E}_{\overline{K}}$ and $v_n \in 1 + \mathfrak{p}$. The topology induced in $\mathscr{E}_{\overline{K}}$ coincides with the resident topology of $\overline{K}^*$, and by induction we can assume that $u_n^{p^n} \to 1$. For $v_n^{p^n}$ this follows from Property 4.

REMARK 3. It follows from 4 that the topology introduced by us is weaker than the topology in K^* determined by the valuation v_K. The latter, of course, is compatible with the structure of K^* but does not completely take into account the topology of the residue field $\overline{K}$. Passing to the valuation of rank n mentioned at the beginning of the section does not change the situation.

PROPOSITION 4. *Let $x \in \mathscr{E}_K$. Then x can be represented in the form of convergent products*

$$x = \prod_{a_n \geqslant A_n} \prod_{a_{n-1} \geqslant A_{n-1}(a_n)} \cdots \prod_{a_1 \geqslant A_1(a_2, \ldots, a_n)} (1 + a t_1^{a_1} \cdots t_n^{a_n}), \tag{4}$$

where $A_n \geqslant 0$; $A_{n-1}(0) \geqslant 0$ if $A_n = 0$; $\cdots$; $A_1(0, \ldots, 0) \geqslant 0$ if $A_2(0, \ldots, 0) = 0$ and runs through an additive basis of the field $\mathbf{F}_q$.

The proof uses induction on the dimension of the field. We represent x in the form yz, where $y \in \mathscr{E}_{\overline{K}}$ and $z \in 1 + \mathfrak{p}$. The expansion of y in $\overline{K}$ yields the part of expansion (4) corresponding to $A_n = 0$. If $z = 1 + (\Sigma_i z_i t_{n-1}^i) t_n + \cdots$, then by Proposition 3.5

$$z = \prod_i (1 + z_i t_{n-1}^i t_n)(1 + z' t_n^2 + \cdots),$$

where $z_i \in \mathbf{F}_q((t_1)) \cdots ((t_{n-2}))$ and $z' \in \overline{K}$. Applying this argument a sufficient number of times to the element $1 + z_i t_{n-1}^i t_n$ and expansion (2) for z_i in the field $\mathbf{F}_q((t_1)) \cdots ((t_{n-2}))$, we obtain the part of (4) corresponding to $A_n = 1$. The subsequent operations proceed in a similar manner.

COROLLARY. *Let $x \in \mathscr{E}_K$. Then x can be (uniquely) represented modulo the subgroup $\mathscr{E}_K^p$ in the form of a product* (4) *with indices $a_1, \ldots, a_n$ whose greatest common divisor is prime to p.*

§2. *K*-theory

We shall give here a brief outline of the concepts and results of algebraic K-theory. The proofs and motivations can be found in [11]–[13].

With each ring A we can associate abelian groups $K_i(A)$, $i = 0, 1, \ldots$. The correspondence $A \to K_i(A)$ is a covariant functor from the category of rings into the category of abelian groups.

If $A = K$ is a field, then $K_0(K) = Z$, $K_1(K) = K^*$ (a multiplicative group), and $K_2(K)$ is the abelian group generated by

$$(x, y), \quad x, y \in K^*,$$

which satisfy the relations

$$(x_1 x_2, y) = (x_1, y)(x_2, y),$$
$$(x, y_1 y_2) = (x, y_1)(x, y_2),$$
$$(x, 1 - x) = 1, \qquad x \neq 0 \text{ or } 1.$$

These generators are called *symbols*. It is not hard to obtain

$$(x, y) = (y, x)^{-1} \tag{1}$$

and

$$(x, x) = (x, -1). \tag{2}$$

For any local field K of dimension 1 over $\overline{K}$ there is defined a boundary homomorphism

$$\partial \colon K_i(K) \to K_{i-1}(\overline{K}).$$

If $i = 2$, then it coincides with the unramified norm residue symbol (see (1) in §3).

Let $\mathfrak{p} \subset O_K$ be the maximal ideal. We set

$$K_2(O_K, \mathfrak{p}^n) = \operatorname{Ker}(K_2(O_K) \to K_2(O_K/\mathfrak{p}^n)).$$

The following canonical exact sequences hold:

$$1 \to K_2(O_K) \to K_2(K) \xrightarrow{\partial} K_1(\overline{K}) \to 1, \tag{3}$$

$$1 \to K_2(O_K, \mathfrak{p}^n) \to K_2(O_K) \to K_2(O_K/\mathfrak{p}^n) \to 1. \tag{4}$$

If $t \in \mathfrak{p} - \mathfrak{p}^2$ and the subgroup $\overline{K}'^*$ is isomorphic mod $\mathfrak{p}$ to the group $\overline{K}^*$, then the subgroup consisting of the symbols (x, t), $x \in \overline{K}'^*$, splits the first exact sequence, and the subgroup consisting of (x, y), $x, y \in \overline{K}'^*$, splits the second sequence for $n = 1$.

For any i and j the multiplication

$$K_i(A) \times K_j(A) \to K_{i+j}(A),$$

which is a bilinear map, is defined. Both the multiplication and the homomorphism ∂ are functorial in the natural sense.

Aside from the usual functorial nature of the ring morphisms $f \colon A \to B$ (which consists of passing to the mapping $f_* \colon K_i(A) \to K_i(B)$), in some cases there is defined a transfer homomorphism

$$N \colon K_i(B) \to K_i(A).$$

It can be constructed if the rings A and B are commutative and B is a projective A-module of finite type. If $i = 1$, then it coincides with the usual norm.

The mapping N has the transitivity property for triples $A \to B \to C$. If K is a local field of dimension 1 over $\overline{K}$, L/K is a finite extension, and L is endowed with the natural local field structure with residue field $\overline{L}$ (§1), then there exist commutative diagrams

$$\begin{array}{ccc} K_i(L) & \xrightarrow{\partial} & K_{i-1}(\overline{L}) \\ N\downarrow & & N\downarrow \\ K_i(K) & \xrightarrow{\partial} & K_{i-1}(\overline{K}) \end{array} \qquad \begin{array}{ccc} K_i(L) & \xrightarrow{\partial} & K_{i-1}(\overline{L}) \\ j_*\uparrow & & e_{L|K}j_*\uparrow \\ K_i(K) & \to & K_{i-1}(\overline{K}) \end{array}, \tag{5}$$

where j corresponds to the imbedding $K \hookrightarrow L$ and $e_{L/K}$ is the ramification index of L over K ([11], p. 373).

The transfer is connected with multiplication by the following *projection formula*. If $x \in K_i(A)$ and $y \in K_j(B)$, then

$$N(f_*(x) \cdot y) = x \cdot N(y)$$

in the group $K_{i+j}(A)$.

If $A \to B$ is a K-algebra homomorphism for which a transfer is defined and $K \subset K'$, then the natural diagram

$$\begin{array}{ccc} K_i(B) & \to & K_i(B \otimes_K K') \\ N \downarrow & & N \downarrow \\ K_i(A) & \to & K_i(A \otimes_K K') \end{array} \tag{6}$$

is commutative; i.e., transfer commutes with change of basis.

LEMMA 1 (BASS?). *Let K be a field, M/K a finite, separable, and normal extension, and $M \supset L \supset K$. If $\sigma \in \mathrm{Hom}_K(L, M)$, then let $\sigma' \in \mathrm{Gal}(M/K)$ be any continuation of σ to M. Then for any $y \in K_i(L)$*

$$\sum_{\sigma \in \mathrm{Hom}_K(L, M)} \beta(y)^{\sigma'} = \beta \circ \alpha(N(y)), \tag{7}$$

where α: $K_i(K) \to K_i(L)$ and β: $K_i(L) \to K_i(M)$ are the natural morphisms.

PROOF. We apply diagram (6) to $A = K$, $B = L$, and $K' = M$ and take into account that $L \otimes M = \prod M$ (product over all $\sigma \in \mathrm{Hom}_K(L, M)$), that the mapping $L \to L \otimes M$ is $\prod \sigma$, and that the mapping $M \to L \otimes M$ coincides with the diagonal imbedding Δ of M into $\prod M$. Our assertion then amounts to the following fact. Let $[L : K] = n$ and let p_i: $\prod M \to M$ be projections. Then

$$\Delta^* = p_{1,*} + \cdots + p_{n,*}.$$

We denote by s_i: $M \to \prod M$ the imbedding of the ith factor. By definition we have

$$\Delta^* = s_1^* + \cdots + s_n^*$$

and

$$1 = s_{1,*} \circ p_{1,*} + \cdots + s_{n,*} \circ p_{n,*},$$

and it is sufficient to show that $s_i^* \circ s_{j,*} = \delta_{ij}$. Let $e \in K_0(M)$, the class of the module M. From the projection formula and the fact that $K_*(\prod M)$ is decomposable into a direct sum, we obtain $(i \neq j)$

$$s_i^* \circ s_{j,*}(x) = e \cdot s_i^*(s_{j,*}(x)) = s_i^*(s_{i,*}(e) s_{j,*}(x)) = 0,$$

and for $i = j$

$$s_i^* \circ s_{i,*}(x) = s_i^*(s_{i,*}(e) \cdot s_{i,*}(x)) = s_i^*(s_{i,*}(e)) \cdot x = x.$$

This proves the lemma.

All these facts are valid for the K-functor $K_i(A)$ constructed by Quillen. In what follows, however, we shall use Milnor's K-functor $K_n^M(K)$, $n \geqslant 0$, defined for any field K [12].

DEFINITION 1. Let K be a field. $K_n^M(K)$ is an abelian group with generators $(x_1, \ldots, x_n)$, $x_j \in K^*$, and relations

$$(x_1, \ldots, x_i' x_i'', \ldots, x_n) = (x_1, \ldots, x_i', \ldots, x_n)(x_1, \ldots, x_i'', \ldots, x_n),$$
$$(x_1, \ldots, x_i, 1 - x_i, x_{i+2}, \ldots, x_n) = 1, \qquad i = 1, \ldots, n.$$

The generators $(x_1, \dots, x_n)$ are called *symbols*. From (1) we immediately find that for any permutation $i_1, \dots, i_n$ of the indices $1, \dots, n$ of parity τ we have

$$(x_{i_1}, \dots, x_{i_n}) = (x_1, \dots, x_n)^{\tau}.$$

Multiplication defines a natural morphism $K_n^M(K) \to K_n(K)$ that is not, in general, an isomorphism for $n > 2$. For example, for $K = \mathbf{F}_q$ all $K_n^M(K) = 0$, $n \geqslant 2$, and the Quillen functors are highly nontrivial (for odd n). Therefore, the concepts and constructions introduced above cannot be immediately transferred to the groups K_n^m. We shall define them for the fields of interest to us without using Quillen's general theory.

There exists a canonical mapping Ψ: $K^* \times \cdots \times K^* \to K_n^M(K)$, and any n-linear function f: $K^* \times \cdots \times K^* \to A$ that carries the trivial symbols from Definition 1 to zero can be uniquely represented in the form f: $f_0 \circ \Psi$, where f_0: $K_n^M(K) \to A$ is a homomorphism. Applying this to the mapping

$$\delta(x_1, \dots, x_n) = x_n^{-1}\, dx_1 \wedge \cdots \wedge x_n^{-1}\, dx_n \in \Omega^n_{K/Z},$$

we obtain the Tate mapping

$$\delta\colon K_n^M(K) \to \Omega^n_{K/\mathbf{Z}}. \tag{8}$$

In the case when the multiplicative group K^* has a topology, it is natural to consider symbols that satisfy the condition of continuity and to change the groups $K_n^M(K)$ in the appropriate way.

We shall consider topologies τ in $K_n^M(K)$ that satisfy the following conditions:

1. Ψ is continuous in each argument with respect to τ and the topology in K^*.
2. If $x_n \to x$ and $y_n \to y$ in the topology τ, then also $x_n y_n \to xy$ and $x_n^{-1} \to x^{-1}$.

The set of these topologies is nonempty (it contains the weakest topology). Let us show that the upper bound of all topologies from this set again beongs to it. For condition 1 this is obvious. To verify condition 2 it is sufficient to apply the following assertion, which is easily obtainable from the definitions (see [14], Chapter 1, §2):

LEMMA 2. *Let τ_α, $\alpha \in I$, be sets of topologies in a topological space X and τ the upper bound of all topologies τ_α, $\alpha \in I$. The sequence $x_n \in X$ converges to $x \in X$ in the topology τ if and only if it converges to x in all topologies τ_α, $\alpha \in I$.*

It follows from condition 2 that the intersection of all neighborhoods of unity is a subgroup.

DEFINITION 2. Suppose that the group K^* is endowed with a topology. We then endow $K_n^M(K)$ with the strongest topology that satisfies conditions 1 and 2, and we set

$$K_n^{\text{top}}(K) = K_n^M(K)/\Lambda,$$

where Λ is the intersection of all neighborhoods of unity.

Of course, this definition is meaningful if the original topology of K^* satisfies condition 2. In this case $K_1^{\text{top}}(K) = K^*$ (always $K_0^{\text{top}}(K) = \mathbf{Z}$). By Proposition 3.3

in §1 this is true for any local field K.

If $\dim K = 0$, then

$$K_m^{\mathrm{top}}(K) = \begin{cases} \mathbf{Z}, & m = 0, \\ \mathbf{F}_q^*, & m = 1, \\ (1), & m > 1. \end{cases}$$

If $\dim K = 1$, then

$$K_m^{\mathrm{top}}(K) = \begin{cases} \mathbf{Z}, & m = 0, \\ K^*, & m = 1, \\ \mathbf{F}_q^*, & m = 2, \\ (1), & m > 2. \end{cases}$$

These are well-known facts from algebraic K-theory [13]. We shall obtain a complete description of the groups $K_m^{\mathrm{top}}(K)$ for local fields.

PROPOSITION 1. *Let K be a local field of dimension n, $t_1, \dots, t_n$ a system of parameters, and $x \in K_{m+1}^{\mathrm{top}}(K)$, $m \geqslant 0$. Then x is a product of symbols of the form*

1) $(t_{i_1}, \dots, t_{i_{m+1}})$, $i_1 < \cdots < i_{m+1}$;
2) $(a, t_{i_1}, \dots, t_{i_m})$, $a \in \mathbf{F}_q^*$, $i_1 < \cdots < i_m$;
3)

$$\prod_{a_n \geqslant A_n} \cdots \prod_{a_1 \geqslant A_1(a_2, \dots, a_n)} \left(1 + a t_1^{a_1} \cdots t_n^{a_n}, t_{i_1}, \dots, t_{i_m}\right), \qquad i_1 < \cdots < i_m, \tag{9}$$

where the indices satisfy the following conditions:

1) $A_n \geqslant 0$; $A_{n-1} \geqslant 0$ *if* $A_n = 0$; $\dots$; $A_1(0, \dots, 0) \geqslant 0$ *if* $A_2(0, \dots, 0) = 0$;

2) *if* $k = k(a_1, \dots, a_n)$ *is such that* $a_k \bmod p \neq 0$ *and* $a_{k+1} = \cdots = a_n = 0 \bmod p$, *then* $k \notin \{i_1, \dots, i_m\}$.

3) *a runs through an additive basis of the field* $\mathbf{F}_q$.

Henceforth the symbols $(x_1, \dots, x_m) \in K_m^M(K)$ are identified with their images in $K_m^{\mathrm{top}}(K)$.

We shall divide the proof into several steps.

Step 1. If $a \in F^*$, $\varepsilon \in \mathscr{E}_K$, and $x_1, \dots, x_{m-1} \in K^*$, then

$$(a, \varepsilon, x_1, \dots, x_{m-1}) = 1. \tag{10}$$

For if $a \in F_q$, then $\forall n \geqslant 1$ $a = a^{q^n}$ and $(a, \varepsilon, x_1, \dots, x_{m-1}) = (a, \varepsilon^{q^n}, x_1, \dots, x_{m-1}) \to 1$ by Proposition 3.6 in §1.

Step 2. In the condition on the indices $i_1, \dots, i_m$ in (9) we can drop the assumption that $k \notin \{i_1, \dots, i_m\}$. We have

$$\begin{aligned} 1 &= \left(1 + a t_1^{a_1} \cdots t_n^{a_n}, -a t_1^{a_1} \cdots t_n^{a_n}, x_1, \dots, x_{m-1}\right) \\ &= \prod_{i=1}^{n} \left(1 + a t_1^{a_1} \cdots t_n^{a_n}, t_i, x_1, \dots, x_{m-1}\right)^{a_i} \end{aligned}$$

(the latter equality by virtue of (10)). We rewrite the factor with $i = k$ as

$$\left((1 + a t_1^{a_1} \cdots t_n^{a_n})^{a_k}, t_k, \dots\right) = \left(1 + a_k^{-1} a t_1^{a_1} \cdots t_n^{a_n}, t_k, \dots\right)(1 + z, t_k, \dots),$$

where $v_K(z) > a_n$ if $a_n > 0$. Induction on a_n then yields the required assertion. If $a_n = 0$, then it is necessary to consider a_{n-1}, etc.

Step 3. Expansion (3) of §1 shows that it is sufficient to consider the elements $x = (x_1, \dots, x_m)$, where $x_i \in \mathscr{E}_K$ or $x_i \in \mathbf{F}_q^*$, or else x_i is the variable t_j. Since $K_2(\mathbf{F}_q) = (1)$, each such symbol contains at most one $x_i \in \mathbf{F}_q^*$. From (2) it is apparent that the indices of the variables t_i can be made distinct. We assume that everything has been proved for $m = 2$. Then any symbol of the form $(x_1, x_2, \dots)$,

$x_{1,2} \in \mathscr{E}_K$, can be represented as the product of $(x_1', t_i, \dots)$ and the symbols $(a, t_j, \dots)$ and $(t_k, t_l, \dots)$. Proceeding in this manner, we see that it is sufficient to consider the case $m = 2$. Moreover, it is only necessary to consider elements $x \in K_2^{\mathrm{top}}(K)$ of the form (x_1, x_2), $x_{1,2} \in \mathscr{E}_K$.

Step 4. MAIN LEMMA. *Let* $\varepsilon_{1,2} \in \mathscr{E}_K$ *and* $\varepsilon_2 \in 1 + \mathfrak{p}^l$, $l > 0$. *Then*

$$(\varepsilon_1, \varepsilon_2) = (\varepsilon_1', \varepsilon_2) \prod_{i=1}^{n} (\eta_i, t_i), \qquad \varepsilon_1', \eta_1, \dots, \eta_n \in \mathscr{E}_K,$$

where $\varepsilon_1' \in 1 + \mathfrak{p}^{k+l}$ *and* $\eta_i \in 1 + \mathfrak{p}^k$ *if* $\varepsilon_1 \in 1 + \mathfrak{p}^k$.

PROOF. We first consider the case when $\varepsilon_1 = 1 + a t_1^{a_1} \cdots t_n^{a_n}$. Setting $\varepsilon_2 = 1 + y t_n^l$, $y \in 0_K$, and $a t_1^{a_1} \cdots t_n^{a_n} = x t_n^k$, we have

$$\begin{aligned}(1 + x t_n^k, \varepsilon_2) &= \left((1 + x t_n^k \varepsilon_2)(1 + x t_n^k)^{-1}, \varepsilon_2\right)^{-1} (1 + x t_n^k \varepsilon_2, \varepsilon_2) \\ &= \left((1 + x t_n^k \varepsilon_2)(1 + x t_n^k)^{-1}, \varepsilon_2\right)(1 + x t_n^k \varepsilon_2, -x)(1 + x t_n^k \varepsilon_2, t_n)^k.\end{aligned}$$

This yields the necessary expansion

$$\begin{aligned}(\varepsilon_1, \varepsilon_2) &= \left(1 + xy(1 + x_n^t)^{-1} t_n^{k+l}, \varepsilon_2\right) \\ &\times \sum_{i=1}^{n-1} (1 + x t_n^k + xy t_n^{k+l}, t_i)\left((1 + x t_n + xy t_n^{k+l})^k, t_n\right).\end{aligned} \tag{11}$$

Now let $\varepsilon_1 \in \mathscr{E}_K$. To obtain the lemma, we represent ε_i in the form of the product (4) in §1. The lemma is valid for each factor and, multiplying the expressions for $(1 + a t_1^{a_1} \cdots t_n^{a_n}, \varepsilon_2)$, we obtain the required expression for $(\varepsilon_1, \varepsilon_2)$ if the products that occur in it converge. So as not to encumber the presentation, we shall prove convergence for $\dim K = 2$. Let $K = \mathbf{F}_q((t_1))((t_2))$ and

$$\varepsilon_1 = \prod_{k \geqslant A_2} \prod_{i \geqslant A_1(k)} (1 + a t_1^i t_2^k), \qquad A_2 \geqslant 0.$$

In this case (11) has the form ($\varepsilon_2 = 1 + y t_2^l$, $y \in O_K$)

$$\begin{aligned}(1 + a t_1^i t_2^k, 1 + y t_2^l) &= \left(1 + a t_1^i (1 + a t_1^i t_2^k)^{-1} y_2^{k+l}, \varepsilon_2\right) \\ &\times (1 + a t_1^i t_2^k + a t_1^i y_2^{k+l}, t_1)^i (1 + a t_1^i t_2^k + a t_1^i y_2^{k+l}, t_2)^k.\end{aligned} \tag{12}$$

The first argument of each of the three symbols can be written in the form

$$1 + \sum_m A_{im}(t_1) t_2^m, \qquad v_{\overline{K}}(A_{im}) \geqslant C_m \cdot i,$$

where the summation over m extends from $k + l$ in the first symbol and from k in the rest (we must bear in mind that necessarily $i > 0$ for $k = 0$). Multiplying (12) over all i (and for fixed k), we see that the product converges ($\forall m\ A_{im} \to 0$ as $i \to \infty$), and the limit of (12) satisfies the condition of the lemma. In particular, it has the form

$$(1 + x_k t_2^{k+l}, \varepsilon_2)(1 + y_k t_2^k, t_1)(1 + w_k t_2^k, t_2),$$

where $x_k, y_k, w_k \in O_K$. Multiplying over $k \geqslant A_2$, we obtain a convergent product that, obviously, satisfies the condition of the lemma. The case of arbitrary dimension is investigated in exactly the same way.

Step 5. By Step 3 we must represent the symbols $(\varepsilon_1, \varepsilon_2)$ in the form of a product $(\eta_1, t_1) \cdots (\eta_n, t_n)$, $\eta_i \in \mathscr{E}_K$. If $\varepsilon_i \in \mathscr{E}_{\overline{K}}$, then we can assume by induction that this is true. Now let $\varepsilon_2 \in 1 + \mathfrak{p}^l$, $l > 0$. We apply the main lemma to the pair $(\varepsilon_1, \varepsilon_2)$. Repeating this process, we obtain the required representation (convergence holds by Proposition 3.4 in §1).

This proves Proposition 1.

REMARK 1. The representation obtained in Step 5 is uniquely defined by the previous constructions and the uniqueness of expansion (4) of §1 (in §3 we shall also show that it is unique). Let $(\varepsilon_1, \varepsilon_2) = \prod_i (\eta_i, t_i)$ be this representation, which we shall call *canonical*. If ε_1 or ε_2 forms a sequence $\varepsilon_{1,m}$ $(\varepsilon_{2,m})$, then there is a corresponding sequence $\eta_{i,m}$. The arguments by which we verify convergence in the proof of the main lemma show that the convergence of $\varepsilon_{1,m}$ $(\varepsilon_{2,m})$ as $m \to \infty$ implies the convergence of $\eta_{i,m}$ as $m \to \infty$ to some $\eta_i \in \mathscr{E}_K$, and the product $\prod_i (\eta_i, t_i)$ is the canonical representation for $(\lim \varepsilon_{1,m}, \varepsilon_2)$ (or $\varepsilon_1, \lim \varepsilon_{2,m})$).

COROLLARY 1. *Let K be a local field of dimension n. If $m > n + 1$, then $K_m^{\mathrm{top}}(K) = (1)$. If $m = n + 1$, then*

$$K_m^{\mathrm{top}}(K) = \left\{(a, t_1, \ldots, t_n), a \in \mathbf{F}_q^*\right\} \simeq \mathbf{F}_q^*.$$

PROOF. In the first case the variables $t_{i_1}, \ldots, t_{i_m}$ in symbols 1–3 cannot have indices that are all distinct. In the second case there is one possibility: $(a, t_1, \ldots, t_n)$, $a \in \mathbf{F}_q^*$, since condition 2 cannot hold for the symbols $(\varepsilon, t_1, \ldots, t_n)$. The isomorphism with $\mathbf{F}_q^*$ follows from Remark 3 in §3.1.

COROLLARY 2. *Let K be a local field of dimension n. Then $x \in K_m^{\mathrm{top}}(K)$ can be represented modulo the subgroup $(K_m^{\mathrm{top}}(K))^p$ as the product of symbols* 1 *and* 3 *in Proposition* 1 *with an additional condition: the greatest common divisor of the numbers $a_1, \ldots, a_n$ is prime to p.*

PROOF. It is sufficient to recall the corollary of Proposition 4 in §1 and to bear in mind that the group $\mathbf{F}_q^*$ is p-divisible.

COROLLARY 3. *Let K be a local field of dimension n and l an integer, $(l, p) = 1$. Then $x \in K_m^{\mathrm{top}}(K)$ can be represented modulo the subgroup $(K_m^{\mathrm{top}}(K))^l$ as the product of symbols* 1 *and* 2 *in Proposition* 1.

PROOF. By general properties of complete discrete valuation rings the group $\mathscr{E}_K$ is l-divisible.

The groups $K_m^{\mathrm{top}}(K)$ have a number of unusual subgroups. In particular, analogs of the subgroups $K_m(O_k)$ and $K_m(O_K, \mathfrak{p}^k)$ can be defined (see the beginning of this section). We shall do this for the case when $m = n = \dim K$ and $k = 1$.

DEFINITION 3. Let $K_n^{\mathrm{top}}(O_K, \mathfrak{p})$ be the subgroup of $K_n^{\mathrm{top}}(K)$ generated by the symbols $(\varepsilon, t_1, \ldots, \hat{t}_k, \ldots, t_n)$, $\varepsilon \in 1 + \mathfrak{p}$, and $K_n^{\mathrm{top}}(O_K)$ the subgroup generated by $K_n^{\mathrm{top}}(O_K, \mathfrak{p})$ and the symbols $(a, t_1, \ldots, t_{n-1})$, $a \in \mathbf{F}_q^*$.

PROPOSITION 2. *There exists a unique continuous homomorphism*

$$\partial\colon K_n^{\mathrm{top}}(K) \to K_{n-1}^{\mathrm{top}}(\bar{K}),$$

such that the following conditions are satisfied.

1. $\partial(x_1, \ldots, x_{n-1}, t_n) = (x_1 \bmod \mathfrak{p}, \ldots, x_{n-1} \bmod \mathfrak{p})$ *if* $v_K(x_1) = \cdots = v_K(x_{n-1}) = 0$.

2. $\partial(x_1, \ldots, x_n) = 1$ *if* $v_K(x_1) = \cdots = v_K(x_n) = 0$.

3. *The sequence*

$$1 \to K_n^{\mathrm{top}}(O_K) \to K_n^{\mathrm{top}}(K) \xrightarrow{\partial} K_{n-1}^{\mathrm{top}}(\bar{K}) \to 1 \tag{12}$$

is exact.

PROOF. The existence and uniqueness of the mapping ∂ with properties 1 and 2 in our situation can be obtained by repeating verbatim Milnor's arguments in §2 of [12]. We note that it uses only the structure of a local field of dimension 1 in K (with residue field $\bar{K}$). Property 3 can be obtained without difficulty from Proposition 1.

REMARK 2. The following analog of sequence (4) also holds:

$$1 \to K_n^{\mathrm{top}}(O_K, \mathfrak{p}) \to K_n^{\mathrm{top}}(O_K) \to K_n^{\mathrm{top}}(\bar{K}) \to 1. \tag{13}$$

This follows from Corollary 1 of Proposition 1 and Proposition 3 below. One can define even finer filtrations in the group $K_n^{\mathrm{top}}(K)$, using the ideals of the valuation ring of rank n connected with a system of parameters (see [8] and [9]).

LEMMA 3. *Let* $U \subset K_{m+1}^{\mathrm{top}}(K)$ *be the subgroup generated by symbols of the form* $(\varepsilon, t_{i_1}, \ldots, t_{i_m})$, $\varepsilon \in \mathscr{E}_K$. *Then* $\bigcap_{k \geqslant 0} U^{p^k} = (1)$.

PROOF. It is sufficient to apply Proposition 3.6.

We again assume that $m = n = \dim K$, and we denote by U_K the subgroup $U \subset K_n^{\mathrm{top}}(K)$. Attaching the products of the symbols $(a, t_1, \ldots, \hat{t}_i, \ldots, t_n)$ to U_K, we obtain the subgroup V_K.

We introduce further the subgroups $\mathscr{E}_{i,K} \subset \mathscr{E}_K$ consisting of those elements x whose expansion (4) in §1 contains only the powers $a_1, \ldots, a_n$ of the variables $t_1, \ldots, t_n$ with the condition $k(a_1, \ldots, a_n) = i$ (see condition 2 of Proposition 1). We have the mapping

$$\Phi_K\colon \prod_{i=1}^{n} \mathscr{E}_{i,K} \to U_K,$$

which associates with a collection $\varepsilon_1, \ldots, \varepsilon_n$ the product

$$\prod_i (\varepsilon_i, t_1, \ldots, \hat{t}_i, \ldots, t_n).$$

PROPOSITION 3. *Let K be a local field of dimension n. Then the filtration* $U_K \subset V_K \subset K_n^{\mathrm{top}}(K)$ *does not depend on the choice of the system of parameters* $t_1, \ldots, t_n$. *Moreover,*

1. $K_n^{\mathrm{top}}(K) = \{(t_1, \ldots, t_n)\} V_K$, $K_n^{\mathrm{top}}(K)/V_K \cong \mathbf{Z}$.

2. $V_K = \{(a, t_1, \ldots, \hat{t}_i, \ldots, t_n),\ a \in \mathbf{F}_q^*,\ i = 1, \ldots, n\}\ U_K$, $V_K/U_K \cong (\mathbf{F}_q^*)^n$.

3. *The mapping Φ_K is an isomorphism that carries the product topology in* $\prod_i \mathscr{E}_{i,K}$ *into the topology of the group* U_K.

PROOF. The independence of V_K on the choice of the system of parameters and Property 1 follow from the properties of the symbol c_K (see Remark 3 in §3.1).

The subgroup U_K is characterized by the fact that $x^{p^k} \to 1$ for $x \in U_K$ (Lemma 3). property 2 is a consequence of Kummer duality (Corollary 2 of Proposition 3 in §3.1).

It follows from Pıoposition 1 that the homomorphism Φ_K is surjective, and the corollary of Proposition 5 in §3 shows that Φ_K is one-to-one. By the definition of the topology in U_K, this mapping is continuous. We now consider the composition

$$\Phi_K^{-1} \circ \Psi \colon \mathscr{E}_K \times \overset{n}{\cdots} \times \mathscr{E}_K \to U_K \to \prod_i \mathscr{E}_{i,K}. \tag{14}$$

From the properties of canonical representations indicated in Remark 1 it follows that this mapping is continuous in each argument.

We now denote by τ the topology in U_K induced by the topology of $K_n^{\mathrm{top}}(K)$, and by τ' the topology obtained by transferring the product topology in $\prod \mathscr{E}_i$ to U_K by means of Φ_K. We have $\tau' \geqslant \tau$. The topology τ can be defined by using the same conditions 1 and 2 as for the topology of $K_n^{\mathrm{top}}(K)$ (replacing K^* in condition 1 by $\mathscr{E}_K$). The continuity of the mapping (14) means that condition 1 holds for τ'. Condition 2 also holds. Consequently, $\tau' = \tau$ and Φ_K is a homeomorphism.

This proves the proposition.

COROLLARY. *The group $K_n^{\mathrm{top}}(K)$ has no p-torsion.*

Indeed, the multiplicative group K^* has none.

REMARK 3. Thus, we have obtained an explicit description of the group $K_n^{\mathrm{top}}(K)$. The approach used by us, as usual, splits into two parts in K-theory: the "upper" bound produced in this section, and a "lower" bound consisting of constructing nontrivial symbols in the group $K_n^{\mathrm{top}}(K)$ (§3). The constructions presented above do not contradict, for example, the equality $K_n^{\mathrm{top}}(K) = 1$.

§3. Symbols and duality

1. Kummer duality. If K is a local field of dimension 1 with residue field k, then we can define a tamely ramified norm residue symbol $(\cdot, \cdot)_K$. Namely, if v_K is a valuation of K and $f \bmod \mathfrak{p}$ is the image of an element f from the ring of integers of K in the group k^*, then

$$(f, g)_K = (-1)^{mn} f^n g^{-m} \bmod \mathfrak{p}, \tag{1}$$

where $v_K(f) = m$ and $v_K(g) = n$. This is a bilinear skew-symmetric form, whose properties are well known (see [10] and [13]).

In this section we shall introduce an analogous symbol in the $(n+1)$st argument for fields of dimension n and show that the usual properties of the symbol $(\cdot, \cdot)_K$ are preserved for it (in an appropriate form). We shall assume for a while that the last residue field k of the local field K is arbitrary.

DEFINITION 1. Let K be a local field of dimension n over k and $x_1, \dots, x_{n+1} \in K^*$. We set

$$(x_1, \dots, x_{n+1})_{K/k} = \pm \prod_{i=1}^{n} x^{(-1)^{i-1} v_n(x_1, \dots, \hat{x}_i, \dots, x_{n+1})_{K/K_{n-1}}} \bmod \mathfrak{p}_0 \cdots \bmod \mathfrak{p}_{n-1} \in k^*,$$

where the symbol in the n arguments occurring in the exponent refers to the field K regarded as a local field of dimension $n-1$ over the next-to-the-last residue field K_{n-1}, and $v_n\colon K_{n-1}^* \to \mathbf{Z}$ is a discrete valuation of K_{n-1}.

If $\dim K = 1$, then this is the symbol (1) up to a sign. As regards the sign in the general case, see below. We shall also abbreviate the index K/h to K if this does not cause confusion.

PROPOSITION 1. *The symbol* $(x_1, \dots, x_{n+1})_K$ *has the following properties*:

1. $(x_1, \dots, x_i' x_i'', \dots, x_{n+1})_K = (x_1, \dots, x_i', \dots, x_{n+1})_K (x_1, \dots, x_i'', \dots, x_{n+1})_K$.
2. $(x_1, \dots, x_i, 1 - x_i, \dots, x_{n+1})_K = 1$ *up to a sign.*
3. $(x_1, \dots, x_{n+1})_K = 1$ *up to a sign if* $\exists i\; x_i \in \mathscr{E}_K$.

PROOF. Properties 1 and 3 are obvious. To obtain 2, we note first that $(x_1, \dots, x_{n+1})_K$ is skew-symmetric in each pair of arguments. Let us now calculate the integers occurring in the exponent. To do so, we choose a system of parameters $t_1, \dots, t_n$ of K and associate with each $x \in K^*$ integers $a_1, \dots, a_n$:

$$
\begin{aligned}
a_n &= v_K(x), \\
a_{n-1} &= v_{K_1}(x t_n^{-a_n} \bmod \mathfrak{p}_0), \\
&\dots\dots\dots\dots\dots\dots\dots\dots \\
a_1 &= v_{K_{n-1}}(x t_n^{-a_n} \bmod \mathfrak{p}_0 t_{n-1}^{-a_{n-1}} \bmod \mathfrak{p}_1 \cdots).
\end{aligned}
$$

This is the valuation of rank n in K mentioned in §1.

Now it is not hard to show that $v_n(x_1, \dots, \hat{x}_i, \dots, x_n)$ is equal to the determinant of order n composed of the integers $a_1, \dots, a_n$ for the elements $x_1, \dots, x_{i-1}, x_{i+1}, \dots, x_{n+1}$.

Returning to property 2, we note that it follows immediately from this calculation that $(x_1, \dots, x_i, -x_i, \dots, x_{n+1}) = \pm 1$ and $(x_1, \dots, x_{n+1}) = \pm 1$ if $x_i,\ x_{i+1} \in k^*$. Furthermore, we have

$$
\begin{aligned}
(\cdots, x_i, 1 - x_i, \dots) &= (\cdots, x_i, -x_i, \dots)(\dots, x_i, 1 - x_i^{-1} \cdots) \\
&= \pm(\dots, x_i^{-1}, 1 - x_i^{-1}, \dots).^{-1}
\end{aligned}
$$

If $a_1, \dots, a_n$ are the integers associated with x_i, then by virtue of this identity we can assume that $(a_1, \dots, a_n) \geqslant (0, \dots, 0)$ (lexicographically). But then either $x_i \in k^*$ or $1 - x_i \in \mathscr{E}_K$, and the required property holds by what we proved here.

REMARK 1. Of course, properties 2 and 3 do not seem very natural. If $\dim K = 2$, then Definition 1 can be supplemented by explicitly indicating the sign. Namely, we set

$$
\begin{aligned}
(x_1, x_2, x_3)_K = {} & (-1)^{a_{12}a_{22}a_{31} + a_{12}a_{21}a_{32} + a_{12}a_{21}a_{31} + a_{11}a_{22}a_{32} + a_{11}a_{22}a_{31} + a_{11}a_{21}a_{31}} \\
& \times x_1^{a_{22}a_{32} - a_{32}a_{21}} x_2^{a_{32}a_{11} - a_{12}a_{31}} x_3^{a_{12}a_{21} - a_{22}a_{11}} \bmod \mathfrak{p}_1 \bmod \mathfrak{p}_1. \qquad (2)
\end{aligned}
$$

Property 2 then holds in the form $(x_1, 1 - x_1, x_3) = 1$.

To obtain property 2 in the general case, it is necessary to use the following proposition:

PROPOSITION 2. *Let K be a local field of dimension n. Up to a sign,*

$$
(x_1, \dots, x_{n+1})_{\overline{K}} = \partial_n \circ \cdots \circ \partial_1(x_1, \dots, x_{n+1}).
$$

The mappings $\partial = \partial_n\colon K_n^{\text{top}}(K) \to K_{n-1}^{\text{top}}(\overline{K})$ were introduced in §2.

The proof of this proposition can easily be obtained by a direct calculation on the generators (it is necessary to use formula (3) in §1, Proposition 2 in §2, and Proposition 1.3).

We shall assume that the sign in Definition 1 is chosen so that property 2 with $+1$ holds.

Question. What is the explicit formula for the sign in Definition 1 *that generalizes formulas* (1) *and* (2)?

REMARK 2. When $n = 2$, a symbol similar to the one introduced by us was defined independently by Y. Pourchet in 1970 (unpublished). As Serre has shown, this symbol is a trilinear skew-symmetric form (a, b, c) defined for elements a, b, c of an arbitrary local field K (of dimension 1) with residue field k and taking the values in the group $\operatorname{Br}(k)_n$ (the nth order elements of the Brauer group). It is uniquely defined by the following properties:

1. $(a, -a, c) = 0$.
2. If $v_K(a) = 1$ and $b, c \in \mathcal{E}$ (the group of units), then $(a, b, c) = (\bar{b}, \bar{c})_K$ ($\bar{b}$ and $\bar{c}$ are the images in the residue field).
3. If $a, b, c \in \mathcal{E}$, then $(a, b, c) = 0$.

To construct this symbol, we must take the product of the elements $a, b, c \in K^*$ in the group $K_3(K)$, apply $\partial\colon K_3(K) \to K_2(k)$, and then the symbol $(\bar{b}, \bar{c})_k\colon K_2(k) \to \operatorname{Br}(k)_n$ introduced in [13] (§15). Proposition 2 shows the connection between our symbol and Pourchet's symbol. The latter arose in connection with the following applications to the theory of quadratic forms. Namely, let $(1, -a)$ be a form of the type $x_1^2 - ax_2^2$. If the form

$$(1, -a) \otimes (1, -b) \otimes (1, -c)$$

represents zero, then $(a, b, c) = 0$ and $n = 2$. The converse is true if every form in five variables over k represents zero (for example, $k = \mathbf{Q}_p$).

Now let K again be a local field of dimension n with a finite residue field $\mathbf{F}_q$. We denote by K' its next-to-last residue field (it is a local field of dimension 1 over $\mathbf{F}_q$). By Proposition 2 the following definition makes sense:

DEFINITION 2. We introduce the mappings

$$a_K\colon K_{n+1}^{\text{top}}(K) \to \mathbf{F}_q^* \tag{3}$$

and

$$c_K\colon K_n^{\text{top}}(K) \to \mathbf{Z}, \tag{4}$$

by setting

$$a_K(x_1, \ldots, x_{n+1}) = (x_1, \ldots, x_{n+1})_{K/\mathbf{F}_q}$$

and

$$c_K(x_1, \ldots, x_n) = v_{K'}(x_1, \ldots, x_n)_{K/K'}.$$

We note that for the definition of c_K the question of the choice of sign is unimportant.

REMARK 3. The mappings (3) and (4) are surjective since $a_K(a, t_1, \ldots, t_n) = a$ and $c_K(t_1, \ldots, t_n) = 1$.

As we know, Kummer duality for local fields of dimension 1 can be constructed with the help of symbol (1). We shall show that the same is true in an arbitrary dimension.

If l is an integer, $(l, p) = 1$, and ζ is a primitive lth root, then $l \in K$ if and only if $l | q - 1$, and then $\zeta \in \mathbf{F}_q^*$. We set

$$\lambda(x) = x^{(q-1)/l}, \qquad x \in \mathbf{F}_q^*.$$

PROPOSITION 3. *Let $m \geqslant 0$. The mapping $\lambda \circ a_K \colon K_{n+1}^{\text{top}}(K) \to \mathbf{Z}/l$ defines a continuous and nondegenerate pairing*

$$K_m^{\text{top}}(K)/l \times K_{n+1-m}^{\text{top}}(K)/l \to \mathbf{Z}/l.$$

PROOF. The case $m = 0$ was analyzed above (Corollary 1 of Proposition 1 in §2 and Remark 3). By Proposition 1 in §2 and the l-divisibility of $\mathcal{E}_K$, the group $K_m^{\text{top}}(K)$ is generated by the elements $(t_{i_1}, \ldots, t_{i_m})$ and $(\zeta, t_{i_1}, \ldots, t_{i_{m-1}})$, $i_1 < \cdots < i_m$. Let us show that they are linearly independent. If $m = 1$, then this is the case (see (4) in §1). The generators of the group $K_n^{\text{top}}(K)/l$ can be associated with the generators of the group K^*/l. Namely,

$$\begin{array}{cccc} (t_1, \ldots, t_n) & (\zeta, t_1, \ldots, t_{n-1}) & \cdots & (\zeta, t_2, \ldots, t_n). \\ \zeta & t_n & \cdots & t_1 \end{array}$$

Applying our pairing (see Remark 3), we find that the generators in $K_n^{\text{top}}(K)$ are also independent, and the pairing is nondegenerate.

The investigation of subsequent groups $K_m^{\text{top}}(K)$ proceeds by induction downward on m. If the assertion has been proved for m, then it is necessary to turn to the pairing $K^* \times K_{m-1}^{\text{top}}(K) \to K_m^{\text{top}}(K)$. This yields the required assertion.

COROLLARY 1 (KUMMER DUALITY). *Let K be a local field of dimension n. The mapping $\lambda \circ a_K \colon K_{n+1}^{\text{top}}(K)/l \xrightarrow{\sim} \mathbf{Z}/l$ defines a continuous and nondegenerate pairing*

$$(\cdot, \cdot)_K \colon K_n^{\text{top}}(K)/l \times K^*/l \to \mathbf{Z}/l.$$

From the proof of the proposition we also have the following corollary:

COROLLARY 2. $\operatorname{rk} K_m^{\text{top}}(K)/l = C_n^m + C_n^{m-1}$.

This was used in §2 (Proposition 3).

2. Artin-Schreier duality. If we again turn to local fields of dimension 1, then in the case $K = k((t))$ there is a symbol

$$(x, y]_K = \operatorname{res}_K(yx^{-1}\,dx) \in k, \qquad x \in K^*,\ y \in K, \tag{5}$$

which plays an essential role in describing the p-extensions of K ([10], Chapter XIV, and [15] and [16]). We now give the appropriate definitions for arbitrary local fields.

DEFINITION 3. Let K be a local field of dimension n, $\omega \in \Omega^n_{K/\mathbf{F}_q}$, and $t_1, \ldots, t_n$ a system of parameters. We set

$$\operatorname{res}_K(\omega) = a(-1, \ldots, -1),$$

if

$$\omega = \Sigma a(a_1, \ldots, a_n) t_1^{a_1} \cdots t_n^{a_n}\, dt_1 \wedge \cdots \wedge dt_n.$$

This is the usual expansion ((1) in §1). As shown in [2] and [9], the residue $\operatorname{res}_K(\omega)$ does not depend on the choice of the system of parameters.

DEFINITION 4. Let $x_1, \dots, x_n \in K^*$ and $y \in K$. We set

$$(x_1, \dots, x_n | y]_K = \operatorname{Tr}_{\mathbf{F}_q/\mathbf{F}_p} \circ \operatorname{res}_K\left(y x_1^{-1} dx_1 \wedge \cdots \wedge x_n^{-1} dx_n\right).$$

When considering a single field we shall again drop the index K.

PROPOSITION 4. The symbol $(x_1, \dots, x_n | y]_K$ has the following properties:

1. $(x_1, \dots, x_i' x_i'', \dots, x_n | y]_K = (x_1, \dots, x_i', \dots, x_n | y]_K + (x_1, \dots, x_i'', \dots, x_n | y]_K$.
2. $(x_1, \dots, x_n | y_1 + y_2]_K = (x_1, \dots, x_n | y_1]_K + (x_1, \dots, x_n | y_2]_K$.
3. $(x_1, \dots, x_i, 1 - x_i, \dots, x_n | y]_K = 0$.
4. $(x_1, \dots, x_n | y]_K = 0$ *if* $x_i \in (K^*)^p$ *for some* i.
5. $(x_i, \dots, x_n | y^p]_K = (x_1, \dots, x_n | y]_K^p$.
6. *Let*

$$x_i = \prod_{k=1}^{n} t_k^{m_{ik}} a_i \prod_{a_1, \dots, a_n} (1 + a(a_1, \dots, a_n) t_1^{a_1} \cdots t_n^{a_n})$$

be (1) *and* (4) *in* §1, *and let* $y = \Sigma b(b_1, \dots, b_n) t_1^{b_1} \cdots t_n^{b_n}$. *Then for fixed numbers* m_{ik} *the symbol* $(x_1, \dots, x_n | y]_K$ *is a polynomial in* $a_1^{-1}, \dots, a_n^{-1}$, $a(a_1, \dots, a_n)$, *and* $b(b_1, \dots, b_n)$ *with coefficients in* $\mathbf{Z}$.

PROOF. Properties 1–4 are obvious. Property 5 is obtained for $n = 2$ in [2] (Proposition 1.7 in §1). The general case is investigated in a similar manner. The last property follows immediately from the definition of residue. This yields the following corollary:

COROLLARY. *The symbol* $(x_1, \dots, x_n | y]_K$ *is continuous in each argument in the topology of the group* K^* *and the field* K.

PROPOSITION 5 (Artin-Schreier duality). *Let* K *be a local field of dimension* n. *The symbol* $(x_1, \dots, x_n | y]_K$ *defines a continuous and nondegenerate pairing*

$$(\cdot | \cdot]_K \colon K_n^{\mathrm{top}}(K)/p \times K/(F-1)K \to \mathbf{Z}/p,$$

where $F(x) = x^p$, $x \in K$.

PROOF. By what we said above it is only necessary to verify that the form $(\cdot | \cdot]_K$ is nondegenerate. Let us find its values on the generators of the groups $K_n^{\mathrm{top}}(K)/p$ and $K(F-1)K$. To do so, we assume that the bases of $\mathbf{F}_q$ over $\mathbf{F}_p$ that appear in expansion (4) and Lemma 2 in §1 are dual to each other with respect to the bilinear form $\operatorname{Tr}_{\mathbf{F}_q/\mathbf{F}_p}(ab)$ on F_q. Then the following symbols are "generators" in $K_n^{\mathrm{top}}(K)/p$:

$$(t_1, \dots, t_n),\ e(a_1, \dots, a_n) = (1 + a t_1^{a_1} \cdots t_n^{a_n}, t_1, \dots, \hat{t}_k, \dots, t_n),$$

where the indices $a_1, \dots, a_n$ satisfy the conditions of Proposition 1 in §2 and its Corollary 2. In $K/(F-1)K$ there are elements $b_0 \in \mathbf{F}_q$, $\operatorname{Tr}_{\mathbf{F}_q/\mathbf{F}_p}(b_0) = 1$, and $b t_1^{b_1} \cdots t_n^{b_n}$ (Lemma 2 in §1). Calculation of the residue shows that

$$(t_1, \dots, t_n | b_0]_K = 1, \tag{6}$$

$$(t_1, \dots, t_n | b t_1^{b_1} \cdots t_n^{b_n}]_K = 0, \tag{7}$$

$$(e(a_1, \dots, a_n) | b_0]_K = 0, \tag{8}$$

$$(e(a_1, \dots, a_n) | b t_1^{b_1} \cdots t_n^{b_n}]_K = 0 \text{ if } (a_1, \dots, a_n) \neq (-b_1, \dots, -b_n). \tag{9}$$

But if $(a_1, \ldots, a_n) = (-b_1, \ldots, -b_n)$, then

$$\mathrm{Tr}_{\mathbf{F}_p/\mathbf{F}_p} \circ \mathrm{res}_K \left(bt_1^{-a_1} \cdots t_n^{-a_n} \frac{d(at_1^{a_1} \cdots t_n^{a_n})}{1 + at_1^{a_1} \cdots t_n^{a_n}} \wedge \frac{dt_1}{t_1} \wedge \cdots \wedge \widehat{\frac{dt_k}{t_k}} \wedge \cdots \wedge \frac{dt_n}{t_n} \right)$$

$$= \mathrm{Tr}_{\mathbf{F}_q/\mathbf{F}_p} \circ \mathrm{res}_K \left(abt_1^{-a_1} \cdots t_n^{-a_n} \left(\sum_{i=1}^{n} a_i t_1^{a_1} \cdots t_i^{a_i} \cdots t_n^{a_n} \frac{dt_i}{t_i} \right) \right.$$

$$\left. \wedge \frac{dt_1}{t_1} \wedge \cdots \wedge \widehat{\frac{dt_k}{t_k}} \wedge \cdots \wedge \frac{dt_n}{t_n} \right)$$

$$= \mathrm{Tr}_{\mathbf{F}_q/\mathbf{F}_p}(aba_k) = \delta_{ab} \cdot a_k. \tag{10}$$

This is equal to zero if $a \neq b$ and is nonzero (by the definition of the index $k = k(a_1, \ldots, a_n)$; see Proposition 1 in §2) for $a = b$. Since each element of $K_n^{\mathrm{top}}(K)/p$ and $K(F-1)K$ has an expansion with respect to the generators considered by us (and in $K/(F-1)K$ it is unique), we find that the pairing $(\cdot \mid \cdot]_K$ is nondegenerate. At the same time, we have the following corollary:

COROLLARY. *Each element $x \in K_n^{\mathrm{top}}(K)$ has only one representation in the form of product* 3 *in* (9) *of Proposition* 1 *in* §2.

REMARK 4. Artin-Schreier duality can be generalized to the groups $K_m^{\mathrm{top}}(K)$ for any m. Namely, there is a pairing

$$(\cdot \mid \cdot]_K\colon K_m^{\mathrm{top}}(K)/p \times \Omega_K^{n-m}/(C-1)\Omega_K^{n-m} \to \mathbf{Z}/p,$$

where $C\colon \Omega_K^{n-m} \to \Omega_K^{n-m}$ is Cartier's operator, and

$$(x_1, \ldots, x_m \mid \omega]_K = \mathrm{Tr}_{\mathbf{F}_q/\mathbf{F}_p} \circ \mathrm{res}_K (\omega \wedge x_1^{-1} dx_1 \wedge \cdots \wedge x_m^{-1} dx_m).$$

It is continuous and nondegenerate.

3. Witt duality. For fields of dimension 1 it is developed by Witt in [16], and its application to the construction of class field theory is given in [15]. Our construction is a direct generalization of the constructions in those papers. Since there is essentially no difference between the general case and the case of dimension 2, for simplicity of exposition we shall confine ourselves to the latter.

Let A be an arbitrary field of characteristic 0 and $\tilde{K} = A((t_1)) \cdots ((t_n))$. We consider the Witt vector ring $W(\tilde{K})$ over $\tilde{K}$. Each element of $W(\tilde{K})$ has the form $\mathbf{x} = (\mathbf{x}_0, \mathbf{x}_1, \ldots), \mathbf{x}_0, \mathbf{x}_1, \ldots, \in \tilde{K}$.

If $\tilde{K}$ has a topology, then we have a natural topology in $W_m(\tilde{K})$.

We introduce the auxiliary coordinates

$$\mathbf{x}(m) = \mathbf{x}_0^{p^m} + p\mathbf{x}_1^{p^{m-1}} + \cdots + p^m \mathbf{x}_m, \qquad m = 0, 1, \ldots.$$

We then have

$$\mathbf{x}_m = P_m(\mathbf{x}(0), \mathbf{x}(1), \ldots, \mathbf{x}(m)), \qquad m = 0, 1, \ldots, \tag{11}$$

where the P_m are polynomials with coefficients in $\mathbf{Z}[p^{-1}]$. Addition and multiplication in the ring $W(\tilde{K})$ are defined by universal polynomials in the variables $\mathbf{x}_0$, $\mathbf{x}_1, \ldots$ so that $(\mathbf{x} + \mathbf{y})(m) = \mathbf{x}(m) + \mathbf{y}(m)$ and $(\mathbf{xy})(m) = \mathbf{x}(m)\mathbf{y}(m)$. Moreover, the mth component of the sum and product depends only on the components of the vectors $\mathbf{x}$ and $\mathbf{y}$ with indices $\leqslant m$. In addition,

$$W(\tilde{K}) = \varprojlim W_m(\tilde{K}),$$

where $W_m(\tilde{K})$ is the ring consisting of the vectors of length m. For more details, see [10] and [16].

Now let $x, y \in \tilde{K}^*$ and $\mathbf{z} \in W(\tilde{K})$. We set ($n = 2$)

$$\mathbf{w}_m = P_m\left(\operatorname{res}_{\tilde{K}}\left(\mathbf{z}(0)\frac{dx}{x} \wedge \frac{dy}{y}\right), \quad \operatorname{res}_{\tilde{K}}\left(\mathbf{z}(1)\frac{dx}{x} \wedge \frac{dy}{y}\right), \dots\right).$$

Then $\mathbf{w} = (\mathbf{w}_0, \mathbf{w}_1, \dots) \in W(A)$.

LEMMA 1. *If*

$$x = t_2^m t_1^p(a_0 + a_1 t_2 + \cdots), \qquad a_i = a_{i0} + a_{i1}t_1 + \cdots, \; a_{00} \neq 0,$$

$$y = t_2^k t_1^q(b_0 + b_1 t_2 + \cdots), \qquad b_j = b_{j0} + b_{j1}t_1 + \cdots, \; b_{00} \neq 0,$$

$$\mathbf{z}_m = \sum_{i,j} a_{mij} t_1^i t_2^j$$

are expansions in the field $\tilde{K}$, then for fixed m, p, k, and q the components $\mathbf{w}_m$ of the vector $\mathbf{w}$ are polynomials in a_{00}^{-1}, b_{00}^{-1}, a_{ik}, b_{jl}, and c_{mij} with coefficients in $\mathbf{Z}$ that do not depend on the field A.

PROOF. By construction, the vector $\mathbf{w}$ depends linearly on $\mathbf{z}$ and multiplicatively on x and y. Let us introduce two operations Ω_{t_1} and Ω_{t_2} on Witt vectors. Namely, we set

$$\mathbf{z}'_m = \sum_{j>0} c_{mij} t_1^i t_2^j, \qquad \mathbf{z}''_m = \sum_{j<0} c_{mij} t_1^i t_2^j,$$

$$\Omega_{t_2}(\mathbf{z}) = \mathbf{z} - \mathbf{z}' - \mathbf{z}''$$

and similarly for Ω_{t_1}. If the components $\mathbf{z}_0, \dots, \mathbf{z}_{m-1}$ do not depend on t_2, then the same is true for the components $(\Omega_{t_2}\mathbf{z})_0, \dots, (\Omega_{t_2}\mathbf{z})_m$. Similarly, if $\mathbf{z}_0, \dots, \mathbf{z}_{m-1}$ do not depend on t_1 and t_2, then $(\Omega_{t_1}\mathbf{z})_0, \dots, (\Omega_{t_1}\mathbf{z})_m$ also do not depend on t_1 and t_2.

Let us now prove the lemma for $x = t_1$ and $y = t_2$. In this situation if $\mathbf{z}$ contains only positive powers of t_2 or only negative ones, then $\mathbf{w} = 0$ by the definition of residue. The same is true with respect to t_1. We write $\mathbf{w}$ as $(x, y, \mathbf{z})$. Then, by what we have said,

$$(t_1, t_2, \mathbf{z}) = (t_1, t_2, \mathbf{z}) - (t_1, t_2, \mathbf{z}') - (t_1, t_2, \mathbf{z}'') = (t_1, t_2, \Omega_{t_2}\mathbf{z}).$$

Repeating this process, we obtain

$$(t_1, t_2, \mathbf{z}) = \left(t_1, t_2, \Omega_{t_2}^N \mathbf{Z}\right),$$

where the coordinates $(\Omega_{t_2}^N\mathbf{z})_m$ do not depend on t_2 for $m < N$. Applying Ω_{t_1} next, we find that

$$(t_1, t_2, \mathbf{z}) = \left(t_1, t_2, \Omega_{t_1}^M \Omega_{t_2}^N \mathbf{z}\right),$$

where the coordinates $(\Omega_{t_1}^M\Omega_{t_2}^N\mathbf{z})_m$ do not depend on t_1 and t_2. Consequently,

$$(t_1, t_2, \mathbf{z})_m = \left(\Omega_{t_1}^M \Omega_{t_2}^N \mathbf{z}\right)_m \text{ for } m < N, \; m < M,$$

which yields the lemma.

Since a residue is invariant under changes of parameters t_1 and t_2 in $\tilde{K}$, the lemma is also true for any pairs of parameters t_1' and t_2' in $\tilde{K}$. But we can always write x as $t_2^{m-1}t_1^p t_2'$ and y as $t_2^k t_1^{q-1} t_1'$ and, using the multiplicativity, we can reduce everything to the case already examined (the expressions $(t_1, t_1', \mathbf{z})$ and $(t_2, t_2', \mathbf{z})$ are also encountered on this path, but they are equal to $(t_1, t_2 t_2'^{-1}, \mathbf{z})$ and $(t_2, t_1 t_1'^{-1}, \mathbf{z})$, respectively).

This completes the proof. We now turn to the field $K = \mathbf{F}_q((t_1)) \cdots ((t_n))$ and denote by A the quotient field of the ring $W(\mathbf{F}_q)$ and by $\tilde{f} \in \tilde{K}$ an element of the ring $W(\mathbf{F}_q)((t_1) \cdots (t_n))$ such that $\tilde{f} \bmod p = f \in K$.

DEFINITION 5. Let $x_0, \ldots, x_n \in K^*$ and $\mathbf{y}_0, \ldots, \mathbf{y}_{m-1} \in K$. Then

$$(x_1, \ldots, x_n | \mathbf{y}_0, \ldots, \mathbf{y}_{m-1}]_K = (\mathbf{w}_0, \mathbf{w}_1, \ldots, \mathbf{w}_{m-1}) \in W_m(\mathbf{F}_q),$$

where

$$\mathbf{w}_i = P_i\big(\operatorname{res}_{\tilde{K}}(\tilde{\mathbf{y}}(0) x_1^{-1} dx_1 \wedge \cdots \wedge x_n^{-1} dx_n), \\ \ldots, \operatorname{res}_{\tilde{K}}(\tilde{\mathbf{y}}(m-1) x_1^{-1} dx_1 \wedge \cdots \wedge x_n^{-1} dx_n)\big) \mod p$$

and $\operatorname{res}_{\tilde{K}}$ is the residue in the local field $\tilde{K} = A((t_1) \cdots (t_n))$.

The fact that the values of the polynomials P_i belong to the subring $W(\mathbf{F}_q)$ of the field A follows from Lemma 1, so that the transition to mod p makes sense.

For brevity we shall again denote the arguments $x_1, \ldots, x_n$ by x and $\mathbf{y}_0, \ldots, \mathbf{y}_{m-1}$ by $\mathbf{y}$.

PROPOSITION 6. *The symbol* $(x_1, \ldots, x_n | \mathbf{y}_0, \ldots, \mathbf{y}_{m-1}]_K$ *depends only on the structure of the local field K and has the following properties*:

1. $(x_1, \ldots, x_i' x_i'', \ldots, x_n | \mathbf{y}]_K = (x_1, \ldots, x_i', \ldots, x_n | \mathbf{y}]_K + (x_1, \ldots, x_i'', \ldots, x_n | \mathbf{y}]_K$.
2. $(x | (\mathbf{y} + \mathbf{z})_0, \ldots, (\mathbf{y} + \mathbf{z})_{m-1}]_K = (x | \mathbf{y}_0, \ldots, \mathbf{y}_{m-1}]_K + (x | \mathbf{z}_0, \ldots, \mathbf{z}_{m-1}]_K$.
3. $(x_1, \ldots, x_i, 1 - x_i, \ldots, x_n | \mathbf{y}]_K = 0$.
4. $(x_1, \ldots, x_n | \mathbf{y}]_K = 0$, *if* $x_i \in (K^*)^{p^m}$ *for some i*.
5. $(x | \mathbf{y}_0^p, \ldots, \mathbf{y}_{m-1}^p]_K = (\mathbf{w}_0^p, \ldots, \mathbf{w}_{m-1}^p)$, *if* $(x | \mathbf{y}_0, \ldots, \mathbf{y}_{m-1}]_K = (\mathbf{w}_0, \ldots, \mathbf{w}_{m-1})$.
6. $(x | \mathbf{y}]_K$ *is continuous in each argument in the topology of the groups* K^* *and* $W_m(K)$.
7. $(x | \mathbf{0}, \mathbf{y}_1, \ldots, \mathbf{y}_{m-1}]_K = (\mathbf{0}, (x | \mathbf{y}_1, \ldots, \mathbf{y}_{m-1}]_K)$.
8. $(x | \mathbf{y}_0, \ldots, \mathbf{y}_{m-2}]_K = (\mathbf{w}_0, \ldots, \mathbf{w}_{m-2})$, *if* $(x | \mathbf{y}_0, \ldots, \mathbf{y}_{m-1}]_K = (\mathbf{w}_0, \ldots, \mathbf{w}_{m-1})$.

PROOF. If $B = W(\mathbf{F}_q)((t_1)) \cdots ((t_n))$, then every parameter change in K can be lifted to a parameter change in B, and the usual proof that the residue is invariant yields the fact that the symbol $(\cdot | \cdot]_K$ is independent of the choice of $t_1, \ldots, t_n$.

Properties 1–3 are obvious. To obtain 4, we note that in $\tilde{K}$

$$\operatorname{res}_{\tilde{K}}\left(\tilde{\mathbf{y}}(k) \frac{d\tilde{x}_1^{p^m}}{\tilde{x}_1^{p^m}} \wedge \frac{d\tilde{x}_2}{\tilde{x}_2} \wedge \cdots \wedge \frac{d\tilde{x}_n}{\tilde{x}_n}\right) = p^m \operatorname{res}_{\tilde{K}}\left(\tilde{\mathbf{y}}^{(k)} \frac{d\tilde{x}_1}{\tilde{x}_1} \wedge \cdots \wedge \frac{d\tilde{x}_n}{\tilde{x}_n}\right),$$

and so in the transition from the auxiliary variables $w(m)$ to $\mathbf{w}_m$ we obtain a vector $p^m \mathbf{w}$ equal to 0 in $W_m(K)$.

Properties 7 and 8 can also be obtained from the transition formulas (11). In particular, from 7 we have

$$(x | \mathbf{0}, \ldots, \mathbf{0}, \mathbf{y}_{m-1}] = (\mathbf{0}, \ldots, \mathbf{0}, (x | \mathbf{y}_{m-1}]),$$

and since

$$(\mathbf{y}_0, \dots, \mathbf{y}_{m-1}) = (\mathbf{y}_0, \dots, \mathbf{y}_{m-2}, \mathbf{0}) + (\mathbf{0}, \dots, \mathbf{0}, \mathbf{y}_{m-1}),$$

we deduce 5 from 8 and Proposition 4.5 by induction on m.

Continuity (property 6) is an obvious consequence of Lemma 1.

PROPOSITION 7 (Witt duality). *Let K be a local field of dimension n. The symbol $(x_1, \dots, x_n | \mathbf{y}_0, \dots, \mathbf{y}_{m-1}]_K$ defines a continuous and nondegenerate pairing*

$$(\cdot | \cdot]_K\colon K_n^{\mathrm{top}}(K)/p^m \times W_m(K)/(F-1)W_m(K) \to \mathbf{Z}/p^m,$$

where $F(\mathbf{y}_0, \dots, \mathbf{y}_{m-1}) = (\mathbf{y}_0^p, \dots, \mathbf{y}_{m-1}^p)$.

PROOF. As before, it is only necessary to verify that the pairing is nondegenerate. Let

$$(x | \mathbf{y}_0, \dots, \mathbf{y}_{m-1}]_K = 0$$

for any $(\mathbf{y}_0, \dots, \mathbf{y}_{m-1}) \in W_m(K)$. From Proposition 6 (properties 4 and 8) we obtain $x = g^p$. Then

$$0 = (g^p | \mathbf{y}_0, \dots, \mathbf{y}_{m-1}] = p(g | \mathbf{y}_0, \dots, \mathbf{y}_{m-1}] = (\mathbf{0}, \mathbf{w}_0^p, \dots, \mathbf{w}_{m-2}^p),$$

if $(g | \mathbf{y}_0, \dots, \mathbf{y}_{m-1}] = (\mathbf{w}_0, \dots, \mathbf{w}_{m-1})$ (this is the well-known relation $p = VF$ in the Witt vector ring [10], [16]). Thus, induction on m can be carried out.

Nondegeneracy with respect to the second argument can also be proved by induction by using (3) and Proposition 6.8.

This proves the proposition, and we see that the symbols constructed by us have precisely the same properties as the analogous symbols in the one-dimensional case. We can consider, therefore, the inductive limit $\mathfrak{M}(K)$ of the groups $W_m(K)/(F-1)W_m(K)$ with respect to the mappings that carry $(\mathbf{y}_0, \dots, \mathbf{y}_{m-1})$ to $(\mathbf{0}, \mathbf{y}_0, \dots, \mathbf{y}_{m-1})$.

Standard arguments ([15], §2, p. 373) give us the pairing

$$K_n^{\mathrm{top}}(K) \times \mathfrak{M}(K) \to \mathbf{Q}/\mathbf{Z},$$

which is nondegenerate with respect to the second argument. Its kernel with respect to the first argument is equal to $\bigcap_{m \geqslant 1} K_n^{\mathrm{top}}(K)^{p^m}$.

It is clear that the mapping c_K (see (4)) annihilates this kernel, and if we use Proposition 3 and Lemma 3 in §2, we find that products of symbols $(a, t_1, \dots, \hat{t}_i, \dots, t_n)$, $a \in \mathbf{F}_q^*$, are the only elements of $K_n^{\mathrm{top}}(K)$ that belong to the kernel. Thus, the kernel of the pairing consists of the torsion elements prime to p. Since there is no p-torsion in $K_n^{\mathrm{top}}(K)$, we obtain the following corollary:

COROLLARY. *The symbols $(\cdot | \cdot]_K$ define a continuous pairing*

$$K_n^{\mathrm{top}}(K)/K_n^{\mathrm{top}}(K)_{\mathrm{tors}} \times \mathfrak{M}(K) \to \mathbf{Q}/\mathbf{Z}.$$

4. Transfer. As we saw at the beginning of §2, for any finite extension of fields L/K a transfer map $N\colon K_2(L) \to K_2(K)$ is defined which satisfies, in particular, the projection formula in §2. This definition does not immediately carry over to Milnor's functor $K_n^M(K)$ for $n > 2$. In addition, we need a transfer on the groups $K_n^{\mathrm{top}}(K)$ that are factors of the groups $K_n^M(K)$ (Definition 2 in §2). To overcome these difficulties, we shall consider the following roundabout route. We shall confine ourselves to a Galois extension L/K.

LEMMA 2. *Let L/K be a cyclic extension of prime degree of a local field K of dimension n. Then there exist in L and K systems of parameters $s_1, \dots, s_n$ and $t_1, \dots, t_n$ such that all $s_1, \dots, s_n$, except one, are equal to $t_1, \dots, t_n$.*

PROOF. By general properties of local fields (of dimension 1 over $\overline{K}$) ([10], Chapter 1, §7) there are three possibilities: 1) The extension L/K is totally ramified, i.e., $\overline{L} = \overline{K}$. 2) The extension L/K is unramified, and so there is a compatible imbedding of residue fields $\overline{L} \supset \overline{K}$ in L/K, and the parameter t_n in K occurs in the system of parameters for L. 3) The extension L/K has ramification index 1 and $\overline{L}/\overline{K}$ is purely inseparable of degree p. In the first two cases the required property is obvious. In the third it is necessary to note that the parameter t_n is preserved, as in case 2, and to apply Lemma 1 in §1.

From Proposition 1 in §2 and this lemma it follows that in a cyclic extension L/K the elements of $K_n^{\mathrm{top}}(L)$ are products of elements of the form

$$(x, y, x_1, \dots, x_{n-2}), \qquad (x, y) \in K_n^{\mathrm{top}}(L),\ x_1, \dots, x_{n-2} \in K^*. \tag{12}$$

DEFINITION 6. Let L/K be a finite Galois extension and $L = L_0 \supset L_1 \supset \cdots \supset L_m = K$ a tower of cyclic extensions of prime degree. If L/K is a cyclic extension and $x \in K_n^{\mathrm{top}}(L)$ has the form (12), then we set

$$N(x) = \big(N(x, y), x_1, \dots, x_{n-2}\big) \in K_n^{\mathrm{top}}(K).$$

For an arbitrary L/K we define $N\colon K_n^{\mathrm{top}}(L) \to K_n^{\mathrm{top}}(K)$ as the composition of transfer maps for cyclic extensions in the tower $L_0 \supset L_1 \supset \cdots \supset L_m$.

We note that such a tower always exists (corollary of Proposition 1 in §1).

PROPOSITION 8. *The transfer map $N\colon K_n^{\mathrm{top}}(L) \to K_n^{\mathrm{top}}(K)$ is well-defined. It does not depend on the choice of tower and the representations* (12). *For any finite Galois extension L/K the following relations are valid*:

1. $(N(x), y)_K = (x, y)_L$, $x \in K_n^{\mathrm{top}}(L)$, $y \in K^*$.
2. $(x, N(y))_K = (x, y)_L$, $x \in K_n^{\mathrm{top}}(K)$, $y \in L^*$.
3. $(N(x)|\mathbf{y}]_K = (x|\mathbf{y}]_L$, $x \in K_n^{\mathrm{top}}(L)$, $\mathbf{y} \in W_m(K)$.
4. $(x|\mathrm{Tr}(\mathbf{y})]_K = (x|\mathbf{y}]_L$, $x \in K_n^{\mathrm{top}}(K)$, $\mathbf{y} \in W_m(L)$.

Here $\mathrm{Tr}\colon W_m(L) \to W_m(K)$ is a generalization of the trace (see [15], §2). Relations 2 and 4, of course, are in no way connected with the transfer and will be examined in the course of the proof.

PROOF. Let B and A be abelian groups; we wish to construct a mapping $f\colon B \to A$. Moreover, there also exist abelian groups B' and A' and nondegenerate pairings

$$(\cdot,\cdot)_A\colon A \times A' \to \mathbf{Q}/\mathbf{Z}, \qquad (\cdot,\cdot)_B\colon B \times B' \to \mathbf{Q}/\mathbf{Z}$$

and a mapping $f'\colon A' \to B'$. It is required that the mapping $f\colon B \to A$ satisfy the relation $(f(b), a')_A = (b, f'(a'))_B$, i.e., be conjugate to f'. It is clear that at most one f satisfies these conditions. In addition, if $M \subset B$ is a subset that generates the group B and

$$\forall b \in M\ \exists a \in A\ \forall a' \in A' \qquad \big(b, f'(a')\big)_B = (a, a')_A,$$

then the mapping f exists.

Let us apply these arguments in the situation when $B = K_n^{\text{top}}(L)$, $A = K_n^{\text{top}}(K)$, L/K is a cyclic extension of prime degree, and $f = N$ is the transfer that we wish to construct. We take as M the set of symbols of the form (12), and as B' and A' the corresponding groups in the theories of Kummer and Artin-Schreier (depending on the degree $[L: K]$). From the corollary of Proposition 3 and Proposition 5 we find that it is sufficient to prove 1 and 3 for the symbols (12).

Relations 1 (and also 2) can be verified by elementary means on the generators of the groups $K_n^{\text{top}}(L)/l$ and $K_n^{\text{top}}(K)/l$ for Kummer extensions (see the symbols indicated in the proof of Proposition 3). The transfer can be calculated with the help of Lemma 1 in §2. We can argue differently and use the first diagram (5) in §2 and Proposition 2 in §3.

Formulas (3) and (4) can be obtained from the case $m = 1$ by induction on m by applying Proposition 6 (properties 7 and 8). If $m = 1$, then the transfer occurring in (3) can be calculated explicitly with the help of the following result:

LEMMA 3. *Let L/K be a finite separable extension of K. The diagram*

$$\begin{array}{ccc} K_2(L) & \xrightarrow{\delta} & \Omega^2_{L/K} \\ N\downarrow & & \mathrm{Tr}\downarrow \\ K_2(K) & \xrightarrow{\delta} & \Omega^2_{K/h} \end{array},$$

where N is a transfer, δ is the Tate mapping (*see* (8) *in* §2), *and* Tr *is the trace, is commutative.*

The required assertion then follows from the fact that $\mathrm{Tr}(y\omega) = y\,\mathrm{Tr}(\omega)$, $y \in K$, $\omega \in \Omega^n_{L/K}$, and the following general property of residues:

LEMMA 4. *Let L/K be a finite separable extension of a local field K of dimension n, l and k the last residue fields of L and K, and $\omega \in \Omega^n_{L/l}$. Then*

$$\mathrm{Tr}_{l/k}(\mathrm{res}_L(\omega)) = \mathrm{res}_K(\mathrm{Tr}\,\omega),$$

where $\mathrm{Tr}\colon \Omega^n_{L/l} \to \Omega^n_{K/k}$ *is the trace.*

This completes the construction of a transfer for cyclic extensions. The independence of the choice of tower also follows from the duality.

PROOF OF LEMMA 3. We use Lemma 1 in §2. For any $y \in K_2L$ we have

$$\delta\Big(\sum_\sigma \beta(y)^{\sigma'}\Big) = \sum_\sigma \delta\circ\beta(y)^{\sigma'} \sum_\sigma j\circ\delta(y)^{\sigma'} = j\circ i(\mathrm{Tr}(\delta(y))),$$

where $i\colon \Omega^2_K \to \Omega_L$ and $j\colon \Omega^2_L \to \Omega^2_M$ are the natural imbeddings and

$$\delta\circ(\beta\circ\alpha(Ny)) = j\circ\delta\circ\alpha(Ny) = j\circ i\circ\delta(N(y)).$$

This yields the required assertion, since $j \circ i$ is an imbedding.

PROOF OF LEMMA 4 (see [9] and [18]). We have retained here the primitive proof from the first version of this paper, since it contains useful examples and calculations.

As above, the lemma can be reduced to the case of a cyclic extension L/K of prime degree. Furthermore, the three possibilities encountered in the proof of Lemma 2 must be examined. In the first two there exists an imbedding of the residue field $\bar{L}$ in L compatible with the imbedding of $\bar{K}$ in K, and we can argue by induction on $\dim K$ using Lemma 5 from Chapter 2 in [17]. That lemma is just our lemma for fields of dimension 1. The last case—when the extension L/K is purely inseparable—requires direct calculations. We shall carry them out for the two dimensional case.

By Artin-Schreier theory and Lemma 2 in §1, $L = K(x)$, where

$$x^p - x = \lambda = at_1^i t_2^{-k} + \cdots \in K, \qquad k > 0.$$

Since L/K is unramified with respect to the valuation v_K, t_2 is also a parameter for the field L, and if $v_L(x) = -m$, then $k = mp$. Setting $y = t_2^m x$ and $\mu = t_2^k \lambda$, we obtain $L = K(y)$ and

$$y^p - t_2^{m(p-1)} y = \mu = bt_1^i + \cdots . \tag{13}$$

Since $p|k$, by Lemma 2 in §1 we have $(i, p) = 1$, and by a change of parameters in K we can obtain $\mu = bt_1^i$. Now let $K' = \mathbf{F}_q((\mu))((t_2))$ and let L'/K' be the extension defined by (13), i.e., $L' = K'(y)$. We have the diagram

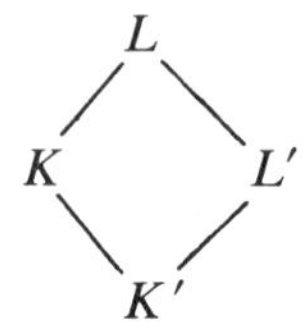

in which the lemma is true for the extensions L/L' and K/K'.

We deduce from this that it is sufficient to consider the extension L'/K', i.e., the case when $b = 1$ and $i = 1$ in (13).

We note that by construction $y \in O_L$ and $L = \mathbf{F}_q((y))((t_2))$. Let $\omega = y^i t_2^j \, dt_1 \wedge dt_2$. Then

$$\operatorname{res}_L(\omega) = \operatorname{res}_L\left(-y^i t_2^{j+m(p-1)} \, dy \wedge dt_2\right) = \begin{cases} -1, & i = 1,\ j = -m(p-1) - 1, \\ 0 & \text{in the remaining cases.} \end{cases}$$

On the other hand,

$$\operatorname{res}_K(\operatorname{Tr} \omega) = \operatorname{res}_K\left(\operatorname{Tr}(y^i) t_2^j \, dt_1 \wedge dt_2\right).$$

This equals 0 for $i \geqslant 0$ (since $\operatorname{Tr}(y^i) \in O_K$), equals -1 for $i = -1$ and $j = -m(p-1) - 1$ ($\operatorname{Tr}(y^{-1}) = -t_1^{-1} t_2^{m(p-1)}$), and equals 0 for $i = -1$ and $j \neq -m(p-1) - 1$. But if $i < -1$, then from (13) we have the recurrence relation

$$\operatorname{Tr}(y^{-i}) = t_1^{-1} \operatorname{Tr}(y^{p-i}) - t_1^{-1} t_2^{m(p-1)} \operatorname{Tr}(y^{-i+1}).$$

In addition, $\operatorname{Tr}(y^k) = \cdots = \operatorname{Tr}(y^{p-2}) = 0$, $1 \leqslant k \leqslant p - 2$ (by virtue of Newton's formulas, which express power sums in terms of elementary symmetric functions). Hence by induction on i we easily find that $\operatorname{Tr}(y^{-i})$, $i > 1$, equals

$$P_0(t_1) + P_1(t_1) t_2 + \cdots ,$$

where the series $P_k(t_1)$ contains only the powers $t_1^{-2}, t_1^{-3}, \ldots,$ so that the residue of the form $\operatorname{Tr}(\omega)$ equals zero for $i < -1$.

This proves the lemma.

Let $x \in W_m(K)$ and $(F - 1)y = x$, $y \in W_m(K^{\mathrm{ab}})$. Then

$$\forall \sigma \in \operatorname{Gal}\left(K^{\mathrm{ab}}/K\right)(F - 1)(\sigma(y) - y) = 0,$$

and so $\sigma(y) - y \in W_m(F_p)$. This yields a character of the group $\mathrm{Gal}(K^{\mathrm{ab}}/K)$ with values in $\mathbf{Z}/p^m$. Passing to the dual group with the help of Proposition 7, for any Galois extension of degree p^m we obtain a mapping $\varphi_K\colon W_m(K) \to \mathrm{Gal}(L/K)$.

LEMMA 5. *For any cyclic extension L/K of degree p^m the sequence*

$$K_n^{\mathrm{top}}(L) \overset{N}{\to} K_n^{\mathrm{top}}(K) \overset{\varphi_K}{\to} \mathrm{Gal}(L/K) \to 1$$

is exact.

The proof will be given in a separate paper.

§4. Class field theory

We now have everything necessary to prove the main result of this paper.

THEOREM 1. *Let K be a local field of dimension n. There exists a canonical mapping*

$$\varphi_K\colon K_n^{\mathrm{top}}(K) \to \mathrm{Gal}(K^{\mathrm{ab}}/K),$$

such that the following conditions hold:

1. $\operatorname{Ker}\varphi_K = (1)$ *and* $\operatorname{Im}\varphi_K$ *is a dense subgroup in* $\mathrm{Gal}(K^{\mathrm{ab}}/K)$.
2. *For any abelian extension L/K the sequence*

$$K_n^{\mathrm{top}}(L) \overset{N}{\to} K_n^{\mathrm{top}}(K) \overset{\varphi_K}{\to} \mathrm{Gal}(L/K) \to 1$$

is exact.

3. *For any finite separable extension L/K the natural diagrams*

$$\begin{array}{ccc} K_n^{\mathrm{top}}(L) & \xrightarrow{\varphi_L} & \mathrm{Gal}(L^{\mathrm{ab}}/L) \\ \uparrow & & v\uparrow \\ K_n^{\mathrm{top}}(K) & \xrightarrow{\varphi_K} & \mathrm{Gal}(K^{\mathrm{ab}}/K) \end{array} \qquad \begin{array}{ccc} K_n^{\mathrm{top}}(L) & \xrightarrow{\varphi_L} & \mathrm{Gal}(L^{\mathrm{ab}}/L) \\ N\downarrow & & \downarrow \\ K_n^{\mathrm{top}}(K) & \xrightarrow{\varphi_K} & \mathrm{Gal}(K^{\mathrm{ab}}/K), \end{array}$$

where V is a group-theoretic transfer map, are commutative.

4. *The diagram*

$$\begin{array}{ccc} K_n^{\mathrm{top}}(K) & \overset{\varphi_K}{\to} & \mathrm{Gal}(K^{\mathrm{ab}}/K) \\ & & \downarrow \\ \partial\Big\downarrow & & \mathrm{Gal}(K_{\mathrm{et}}^{\mathrm{ab}}/K) \\ & & \| \\ K_{n-1}^{\mathrm{top}}(\overline{K}) & \overset{\varphi_{\overline{K}}}{\to} & \mathrm{Gal}(\overline{K}^{\mathrm{ab}}/\overline{K}) \end{array}$$

is commutative.

PROOF. By Proposition 1 in §1 the field K^{ab}/K contains the following subfields:

$L_0 = \mathbf{K}\overline{\mathbf{F}}_q$, where $\mathbf{F}_q$ is the algebraic closure of $\mathbf{F}_q$.

$L_1 = K(t_1^{1/(q-1)}, \dots, t_n^{1/(q-1)})$,

L_2 is the maximal abelian p-extension; L_1 is linearly divided from L_0 and L_2, and the fields L_0 and L_2 intersect in the obvious way. Let G_0, G_1, and G_2 be the corresponding Galois groups.

The groups $G_0 \cong \hat{\mathbf{Z}}$ contains a canonical generator—the Frobenius automorphism Fr. We define

$$\varphi_K\colon K_n^{\mathrm{top}}(K) \to G_0,$$

by setting $\varphi_K(x_1, \dots, x_n) = (\mathrm{Fr})^{c_K(x_1, \dots, x_n)}$, where c_k is the symbol (4) in §3.

The extension L_1/K is Kummer, and by Kummer theory the group G_1 is dual to $K^*/q-1$. Let

$$\varphi_K\colon K_n^{\mathrm{top}}(K) \to G_1,$$

be the mapping that arises from the Kummer duality (Corollary 1 of proposition 3 in §3).

Artin-Schreier-Witt theory is applicable to the extension L_2/K. Therefore, the group G_2 is dual to the discrete group $\mathfrak{M}(K)$ from §3.3. Witt duality (corollary of Proposition 7 in §3) yields the mapping

$$\varphi_K\colon K_n^{\mathrm{top}}(K) \to G_2.$$

The mappings into the groups G_0 and G_2 are compatible with each other (this follows from their explicit calculation on the generator $(t_1, \dots, t_n)$ of the group $K_n^{\mathrm{top}}(K)$ with the help of (6)–(10)). By the intrinsic nature of the dualities constructed in §3, the mappings depend only on the local structure in K, and they can be pasted together into a single mapping $\varphi_K\colon K_n^{\mathrm{top}}(K) \to \mathrm{Gal}(K^{\mathrm{ab}}/K)$. The nondegeneracy of the pairings in §3 yields property 1 of the theorem.

Property 2 can be verified separately in Kummer extensions and Artin-Schreier-Witt extensions. In these cases the exactness of the sequences follows from Proposition 8 in §3.4 and standard exact sequences. Namely, in the Kummer case it is sufficient to consider the commutative diagram

$$\begin{array}{ccccc} H^1(\mathrm{Gal}(L_{\mathrm{sep}}/L), \mathbf{Z}/l) & \longleftarrow & H^1(\mathrm{Gal}(K_{\mathrm{sep}}/K), \mathbf{Z}/l) & \longleftarrow & H^1(\mathrm{Gal}(L/K), \mathbf{Z}/l). \\ \uparrow \cong & & \uparrow \cong & & \uparrow \\ L^*/l & \xleftarrow{\quad i \quad} & K^*/l & \longleftarrow & \mathrm{Ker}\, i \end{array}$$

The exactness of the upper row yields the exactness in the lower and, passing to the groups $K_n^{\mathrm{top}}(K)$ and $K_n^{\mathrm{top}}(L)$ (Proposition 8 in §3.4), we obtain the required assertion (the exactness in property 2 follows from Proposition 3 in §3).

Artin-Schreier extensions can be handled in a similar manner (Lemma 5 in §3). The same arguments yield property 3.

To prove the last property, it is necessary to consider an abelian unramified extension L/K and verify the commutativity of the diagram

$$\begin{array}{ccc} K_n^{\mathrm{top}}(K) & \xrightarrow{\varphi_K} & \mathrm{Gal}(L/K) \\ \partial \downarrow & & \downarrow \cong \\ K_{n-1}^{\mathrm{top}}(\overline{K}) & \xrightarrow{\varphi_{\overline{K}}} & \mathrm{Gal}(\overline{L}/\overline{K}). \end{array}$$

This can be done if we but again turn to extensions of two types, using Proposition 2 in §2. Regarding $\overline{K}$ as a subfield of K and choosing a system of parameters compatibly, we see that Kummer extensions L/K are generated by $\sqrt[h]{a}$, $a \in K^*$, and that Artin-Schreier extensions are generated by an element x with $x^p - x = b \in \overline{K}$. The diagram can easily be calculated if we use the explicit form of the duality for extensions L_α/K such that a or b is a "generator" of the group $\overline{K}^*/l$ or $\overline{K}/(F-1)\overline{K}$, respectively (see Proposition 4 and Lemma 2 in §1). Since

$$\bigcap_\alpha \mathrm{Ker}\left[\mathrm{Gal}(L/K) \to \mathrm{Gal}(L_\alpha/K)\right] = (1),$$

if L is the maximal abelian extension, we obtain the required property.

This proves the theorem.

REMARK 1. Since Proposition 3 in §2 completely describes the group $K_n^{\mathrm{top}}(K)$, Theorem 1 not only contains class field theory for K, but also gives a complete calculation of the Galois group of the maximal extension. It is the profinite completion of the discrete group $K_n^{\mathrm{top}}(K)$. Thus, the compact group $\mathrm{Gal}(K^{\mathrm{ab}}/K)$ contains a subgroup that depends only on the structure of the local field K. If $\dim K = 0$, then

$$K_0^{\mathrm{top}}(K) = Z \hookrightarrow \hat{Z} = \mathrm{Gal}(K^{\mathrm{ab}}/K).$$

REMARK 2. In the case $\dim K = 1$ we can supplement Theorem 1 with information about the behavior of the function φ_K with respect to the filtration $\{1 + \mathfrak{p}^k\}$ ([10], Chapter XV). The analogous question for $\dim K > 1$ remains open. A partial answer for $\dim K = 2$ was obtained in [8] and [9].

REMARK 3. Extensions of a local field K that have no higher ramification (with respect to v_K) can be described with the help of the group $K_n^{\mathrm{top}}(K)/K_n^{\mathrm{top}}(O_K, \mathfrak{p})$ (see (13) and Corollary 1 of Proposition 1 in §2).

BIBLIOGRAPHY

1. A. N. Parshin, *Class fields and algebraic K-theory*, Uspekhi Mat. Nauk **30** (1975), no. 1(181), 253–254. (Russian)

2. ______, *On the arithmetic of two-dimensional schemes*, I, Izv. Akad. Nauk. SSSR Ser. Mat. **40** (1976), 736–773; English transl. in Math. USSR Izv. **10** (1976).

3. ______, *Abelian coverings of arithmetic schemes*, Dokl. Akad. Nauk SSSR **243** (1978), 855–858; English transl. in Soviet Math. Dokl. **19** (1978).

4. Kazuya Kato, *A generalization of local class field theory by using K-groups*. I, Proc. Japan Acad. Ser. A Math. Sci. **53** (1977), 140–143.

5. ______, *A generalization of local class field theory by using K-groups*. II, Proc. Japan Acad. Ser. A Math. Sci. **54** (1978), 250–255.*

6. ______, *A generalization of local class field theory by using K-groups*. I, J. Fac. Sci. Univ. Tokyo Sect. I A Math. **26** (1979), 303–376.

7. ______, *The existence theorem for higher local class field theory*, Preprint, Bures-sur-Yvette, 1980.

8. V. G. Lomadze, *On the ramification theory of two-dimensional local fields*, Mat. Sb. **109(151)** (1979), 378–394; English transl. in Math. USSR Sb. **37** (1980).

9. ______, *Multidimensional local fields and their applications*, Candidate's Dissertation, Steklov Inst. Math. Acad. Sci. USSR, Moscow, 1981. (Russian)

10. Jean-Pierre Serre, *Corps locaux*, 2nd ed., Hermann, Paris, 1968.

11. H. Bass, editor, *Higher K-theory*. II (Proc. Conf., Seattle, 1972), Lecture Notes in Math., vol. 342, Springer-Verlag, 1973.

12. John Milnor, *Algebraic K-theory and quadratic forms*, Invent. Math. **9** (1969/70), 318–344.

13. ______, *Introduction to algebraic K-theory*, Princeton Univ. Press, Princeton, N.J., and Univ. of Tokyo Press, Tokyo, 1971.

14. N. Bourbaki, *Topologie générale*, Chaps. 1–3, 2nd ed., Actualités Sci. Indust., nos. 1142, 1143, Hermann, Paris, 1951; English transl. of 4th ed., Hermann, Paris, and Addison-Wesley, Reading, Mass., 1966.

15. Yukiyosi Kawada and Ichiro Satake, *Class formations*. II, J. Fac. Sci. Univ. Tokyo Sect. I **7** (1956), 353–389.

16. Ernst Witt, *Zyklische Körper und Algebren der Charakteristik p vom Grad p^n*, J. Reine Angew. Math. **176** (1936/37), 126–140.

17. Jean-Pierre Serre, *Groupes algébriques et corps de classes*, Actualités Sci. Indust., no. 1264, Hermann, Paris, 1959.

18. V. G. Lomadze, *On residues in algebraic geometry*, Izv. Akad. Nauk SSSR Ser. Mat. **45** (1981), 1258–1287; English transl. in Math. USSR Izv. **19** (1982).

Translated by R. LENET

Editor's note.* For the full version, see J. Fac. Sci. Univ. Tokyo Sect. IA Math. **27 (1980), 603–683.

Труды Матем. Инст. Стеклов.
165(1984)

Proceedings of the Steklov
Institute of Mathematics
1985, Issue 3

THE MULTIPLICATIVE STRUCTURE OF DIVISION RINGS OVER NUMBER FIELDS AND THE HASSE NORM PRINCIPLE

UDC 512.7 + 511

V. P. PLATONOV AND A. S. RAPINCHUK

ABSTRACT. Let D be a finite-dimensional central division ring of index n over an algebraic number field K, and let

$$\mathrm{SL}(1, D) = \{ x \in D^* \,|\, \mathrm{Nrd}_{D/K}(x) = 1 \},$$

where $\mathrm{Nrd}_{D/K}(x)$ is the reduced norm of x. In Soviet Math. Dokl. **20** (1979), 693–697, the authors developed the multiplicative arithmetic of quaternion divison rings and described the commutator of the group $\mathrm{SL}(1, D)$ in the case when D is a quaternion division ring.

In the present article, the results and methods of that paper are generalized to division rings of arbitrary index. The multiplicative arithmetic of division rings of arbitrary index is developed, and is then applied to prove the following theorem describing the commutator of the group $D^{(1)} = \mathrm{SL}(1, D)$:

THEOREM. *Let D be a division ring of index n over an algebraic number field K, and let $(n, 2)$ be the greatest common divisor of n and 2. If $v(n, 2) = 0$ for all $v \in T$, then the commutator satisfies*

$$[D^{(1)}, D^{(1)}] = D^1 \cap \prod_{v \in T} [D_v^{(1)}, D_v^{(1)}];$$

in particular, if $T = \varnothing$, then $D^{(1)} = [D^{(1)}, D^{(1)}]$.

In proving this theorem, we make essential use of a generalization of the Hasse norm principle to certain extensions of K which are not normal.

Bibliography: 22 items. titles.

§1. Introduction and statement of basic results

Let D be a finite-dimensional central division ring of index n over the field K. The multiplicative group D^* is a classical algebraic object, the study of which is of interest from various points of view. The study of D^* essentially reduces to that of the group $\mathrm{SL}(1, D) = \{x \in D^* | \mathrm{Nrd}_{D/K}(x) = 1\}$, where $\mathrm{Nrd}_{D/K}(x)$ is the reduced norm of x. Unlike the groups $\mathrm{SL}(n, D)$, $n > 1$, which have been investigated with considerable success (see [1]–[3]), the groups $\mathrm{SL}(1, D)$ do not seem to offer any realistic hope of yielding definitive results in the general case. The situation for division rings over algebraic number fields is more promising, since here it is conjecturally possible to pass from the local investigation of $\mathrm{SL}(1, D)$ to the global case. This conjecture was stated by the first author in [4] in the general context of

1980 *Mathematics Subject Classification*. Primary 12B35, 12A80, 12E15, 16A39; Secondary 12A65, 12A25.

algebraic groups (see also [5], §9). In [6] this conjecture was somewhat generalized, so that, in the situation under consideration, it can be stated as follows: Let T be the set of all nonarchimedean valuations v of K for which $D_v = D \otimes_K K_v$ is a division ring; then for any noncentral normal divisor $N \subset \mathrm{SL}(1, D)$ there exists an open normal divisor $W \subset \prod_{v \in T} \mathrm{SL}(1, D_v)$ such that $N = \mathrm{SL}(1, D) \cap W$. In particular, if $T = \varnothing$, then the group $\mathrm{SL}(1, D)$ is projectively simple (i.e., the quotient by the center is simple), and we obtain a new type of local-global principle: the group $\mathrm{SL}(1,D)$ is projectively simple if and only if the groups $\mathrm{SL}(1, D_v)$ are projectively simple for all nonarchimedean v.

In 1956 Kneser [7] stated this conjecture for a quaternion division ring D with $T = \varnothing$, but even in this minimal case a long time passed before the conjecture was proved. It was only in 1979 [8] that the multiplicative arithmetic was developed for quaternion division rings, and was then used first to study the structure of the commutator of $\mathrm{SL}(1, D)$, and then in [9] to complete the proof for quaternion division rings.

In the present article the results and methods of [8] are generalized to division rings of arbitrary index. Namely, the multiplicative arithmetic of division rings of arbitrary index is developed and then applied to prove the following fundamental theorem. In what follows we write $D^{(1)}$ for $\mathrm{SL}(1, D)$, and we denote by (a, b) the greatest common divisor of a and b.

THEOREM 1. *Let D be a division ring of index n over an algebraic number field K, and let T be the set of all nonarchimedian valuations v of K for which $D_v = D \otimes_K K_v$ is a division ring. Suppose that*

$$v((n,2)) = 0 \quad \textit{for all } v \in T. \tag{$*$}$$

Then the following equality holds for the commutator:

$$[D^{(1)}, D^{(1)}] = D^{(1)} \cap \prod_{v \in T} [D_v^{(1)}, D_v^{(1)}].$$

In particular, if $T = \varnothing$, then $D^{(1)} = [D^{(1)}, D^{(1)}]$.

It is possible that the multiplicative arithmetic method used to prove Theorem 1 would enable one to prove the above conjecture in complete generality. This method can also be used in other problems, for example, in the study of the congruence-problem.

In what follows we shall adhere to the following standard notation: V is the set of all pairwise inequivalent valuations of the algebraic number field K; V_∞ and V_f are, respectively, the sets of archimedean and nonarchimedean valuations; if $v \in V_f$, we let U_v (and $\mathfrak{p}_v$ denote the group of units and the valuation ideal in the v-adic completion K_v of the field K; for $a \in K^*$ we set $V(a) = \{v \in V_f | a \notin U_v\}$; $T = \{v \in V_f | D_v \text{ is a division ring}\}$. Every valuation $v \in T$ can be uniquely extended to D_v by means of the formula $\tilde{v}(x) = \frac{1}{n} v(\mathrm{Nrd}_{D_v/K_v}(x))$, $x \in D_v$; let $\mathfrak{P}_v = \{x \in D_v | \mathrm{Nrd}_{D_v/K_v}(x) \in \mathfrak{p}_v\}$ be the corresponding valuation ideal.

It is well known that over the global field K the division ring D is a cyclic algebra, i.e., it can be represented as the cross product of a maximal cyclic subfield L and the Galois group $\mathrm{Gal}(L/K)$. Let $L^{(1)} = L \cap D^{(1)}$. The following result plays an important role in the proof of Theorem 1, and is also of independent interest.

THEOREM 2. *For L a maximal cyclic subfield of the division ring D, $D^{(1)} = L^{(1)}[D^{(1)}, D^{(1)}]$ if and only if for any $v \in T$ the field L_v is a maximal unramified subfield of the division ring D_v.*

If condition $(*)$ holds, then there always exists such a field L.

We note that the existence of a maximal cyclic subfield L with unramified completions $L_{\bar{v}}/K_v$, $\bar{v}/v$, of degree n for $v \in T$ under the condition $(*)$ (which means either that $D_{\bar{2}} = D \otimes_K K_{\bar{2}}$ is not a division ring or else the index of D is odd), can be derived from the Grunwald-Wang theorem [10]. In the general case our condition is necessary and cannot be removed. A minimal counterexample is obtained as follows. Let D be a division ring of index 8 over the rational number field K, where $D \otimes_K K_2$ is a division ring, for example, we have Hasse invariant $i_2(D) = 1/8$. Then it can be verified that $L_{\bar{2}}$ is ramified over K_2 for any maximal cyclic subfield $L \subset D$; consequently, by Theorem 2, $D^{(1)} \neq L^{(1)}[D^{(1)}, D^{(1)}]$.

THEOREM 3. *Suppose that the division ring D satisfies the conditions of Theorem 1, and L is a maximal cyclic submanifold for which $D^{(1)} = L^{(1)}[D^{(1)}, D^{(1)}]$. Then*

$$L^{(1)} \cap \prod_{v \in T} \left[D_v^{(1)}, D_v^{(1)}\right] \subset \left[D^{(1)}, D^{(1)}\right].$$

Theorem 1 follows immediately from Theorem 2 and 3. The method of proof of Theorems 2 and 3 is a rather natural extension of the method in [8], although it requires some new technical devices and more complicated constructions. From a technical point of view, the most complicated part of the proof is concentrated in the following theorem.

THEOREM 4. *Let $x, y \in D^*$, where $y \in 1 + \mathfrak{P}_v$ for all $v \in T$. Then*

$$[x, y] = x^{-1}y^{-1}xy \in \left[D^{(1)}, D^{(1)}\right].$$

The proof of Theorem 4 is based on several assertions, of which the most important are the following two facts:

PROPOSITION 4. *Set $U = \{x \in D^* \mid \mathrm{Nrd}_{D/K}(x) \in U_v\ \forall v \in T\}$. Suppose that $x \in U$, $y \in D^*$, and for any $v \in T \cap V(\mathrm{Nrd}_{D/K}(y))$ suppose that $x \in U_v[D_v^{(1)}, D_v^{(1)}]$. Then $[x, y] \in [D^{(1)}, D^{(1)}]$.*

PROPOSITION 5. *Let $x, y \in D^*$ and $x \in 1 + \mathfrak{P}_v$ for all $v \in T$. Then $[x, y] \in [D^*, D^{(1)}]$.*

We note that, if the division ring D satisfies not only $(*)$ but also the stronger condition $v(n) = 0$ for all $v \in T$, then Theorem 4 in this case becomes equivalent to Proposition 4.

In the proof of Proposition 4 an essential role is played by a generalization of the Hasse norm principle to certain extensions of K which are not normal; and this is of independent interest. We recall that a finite extension F/K satisfies the local-global condition for the norm (or the "Hasse principle") if

$$K^* \cap N_{F/K}(I_F) = N_{F/K}(F^*), \tag{1}$$

where I_F is the idele group of F and $N_{F/K}$ is the norm map.

A classical theorem of Hasse [11] says that the local-global principle holds for cyclic extensions F/K. Later, Tate obtained a cohomological criterion in any Galois extension for the Hasse principle to hold (see [12], Chapter 7): if $G = \mathrm{Gal}(F/K)$ and if G^v is the decomposition group of the valuation v in the extension F for any $v \in V$, then the equality (1) is equivalent to injectivity of the canonical mapping $H^3(G, \mathbf{Z}) \to \prod_v H^3(G^v, \mathbf{Z})$. Since almost all of the decomposition groups G^v are cyclic, it follows that $H^3(G^v, \mathbf{Z}) = 0$ for almost all v. This means that it is possible to characterize the Hasse principle for extensions with a fixed group by a finite number of "local conditions". For our purposes what is important is the converse problem: given a finite set $S \subset V$, we would like to find a class of extensions of given degree, not necessarily normal, for which the Hasse principle is given by the local behavior at the places in S. Here the smaller $\mathrm{Card}(S)$ the better. In this direction we obtain the following result.

THEOREM 5. *Let $S = \{v_1, v_2\}$ be an arbitrary two-element subset of V_f. Then for any integer $n \geqslant 1$ there exists an extension $L = L(S)$ of degree n over K with the property that if $M \supset K$ and $M \otimes_K K_{v_i} \simeq L \otimes_K K_{v_i}$ as K_{v_i}-algebras ($i = 1, 2$), then the Hasse principle holds for M/K.*

The proof of Theorem 5 makes use of the properties of the Schur multiplier in the classical situation (see [13]). Theorem 5 can be reformulated in *algebro-geometric* terms, and this is of independent interest. One should recall that for the norm torus $R^{(1)}_{L/K}$ corresponding to the extension L/K, triviality of the Shafarevich-Tate group $Ш(R^{(1)}_{L/K})$ is equivalent to the validity of the Hasse principle for L/K. Now suppose that $G = \mathrm{SL}(n, \mathbf{C})$, and $\mathscr{T}$ is the variety of maximal tori of the group G. It then follows from Theorem 5 and the construction of the fields $L(S)$ that for any two-element subset $S \subset V_f$ there exists an open subset

$$W \subset \mathscr{T}_S = \prod_{v \in S} \mathscr{T}_{K_v}$$

such that the Shafarevich-Tate group $Ш(R)$ equals zero for any torus $R \in \mathscr{T}_K \cap W$. It is not known whether the analogous assertion holds for an arbitrary semisimple group G, even if we restrict ourselves to the weaker formulation: there exist a finite $S \subset V_f$ and an open subset $W \in \mathscr{T}_S$ such that when $R \in \mathscr{T}_K \cap W$ we have $Ш(R) = 0$.

The proofs of the key assertions of this paper are based on class field theory; all of the necessary facts can be found in the lectures by Serre and Tate in [12]. A brief presentation of the fundamental results is contained in [22].

The first author would like to acknowledge the stimulating influence of I. R. Shafarevich on the formation of his arithmetic tastes.

§2. The Hasse principle

In this section we prove

THEOREM 5. *Let $S = \{v_1, v_2\}$ be an arbitrary two-element subset of V_f. Then for any integer $n \geqslant 1$ there exists an extension $L = L(S)$ of degree n over K with the property that if $M \supset K$ and $M \otimes_K K_{v_i} \simeq L \otimes_K K_{v_i}$ as K_{v_i}-algebras ($i = 1, 2$), then the Hasse principle holds for M/K.*

We first note that for $n \leqslant 3$ an extension L/K of degree n always satisfies the Hasse principle. Moreover, every extension of prime degree satisfies the Hasse principle (see [5], Proposition 10.11). For completeness, we give a proof of this fact, following [5].

PROPOSITION 1. *Every extension L/K of prime degree p satisfies the Hasse principle.*

PROOF. We let F denote the normal closure of L over K, and set $G = \mathrm{Gal}(F/K)$. Since $[G]|p!$, it follows that the p-Sylow subgroup G_p has order p. We let F^p denote the subfield of F corresponding to G_p. It is obvious that $F = LF^p$. Let $a \in K^* \cap N_{L/K}(I_L)$. Then $a \in N_{L_v/K_v}(L_v^*)$ if $v \in V$ is such that $[L_v : K_v] = p$. But then for all $\bar{v}/v$ we have $a \in N_{F_{\bar{v}}/F^p_{\bar{v}}}(F_{\bar{v}}^*)$. Since the extension F/F^p is cyclic, it follows by the Hasse norm theorem that $a \in N_{F/F^p}(F^*)$. Thus, $a = N_{F/F^p}(b)$ and $N_{L/K}(N_{F/L}(b)) = a^{[F^p:K]}$. On the other hand, $a^p \in N_{L/K}(L^*)$. But $(p, [F^p : K]) = 1$, so that $a \in N_{L/K}(L^*)$, as was to be proved.

Now suppose that $n \geqslant 4$. Our conditions on L have the following form:

1) $L \otimes_K K_{v_1} \simeq P_{v_1} \oplus K_{v_1}$, where P_{v_1} is an unramified extension of degree $n - 1$ over the field K_{v_1}.

2) $L \otimes_K K_{v_2} \simeq P_{v_2}^{(1)} \oplus P_{v_2}^{(2)} \oplus K_{v_2}^{(n-4)}$, where $P_{v_2}^{(1)}$ and $P_{v_2}^{(2)}$ are, respectively, an unramified and a totally ramified quadratic extension of K_{v_2}.

The existence of such a field L can be deduced from the weak approximation theorem for K and the following immediate generalization of Krasner's lemma (see [14]).

LEMMA 1. *Suppose that $v \in V_f$, $g(x) = x^n + a_{n-1}x^{n-1} + \cdots + a_0 \in K_v[x]$ is a polynomial with no multiple roots, and $g(x) = g_1(x) \cdots g_r(x)$ is the decomposition of $g(x)$ into irreducible factors. There exists $\varepsilon > 0$ with the following property: if $h(x) = x^n + b_{n-1}x^{n-1} + \cdots + b_0 \in K_v|x|$ and $\|g - h\| = \max_i |a_i - b_i|_v < \varepsilon$, then the irreducible factorization $h(x) = h_1(x) \cdots h_s(x)$ has length r, i.e., $s = r$, and the factors can be ordered in such a way that for each $i = 1, \ldots, r$ the polynomials g_i and h_i determine isomorphic extensions of K_v.*

To prove the existence of L it is now difficult to take a nonarchimedean valuation $v_3 \neq v_1, v_2$, polynomials $f_i \in K_{v_i}[x]$ $(i = 1, 2, 3)$ of degree n such that $K_{v_1}[x]/(f_1) \simeq P_{v_1} \oplus K_{v_1}$, $K_{v_2}[x]/(f_2) \simeq P_{v_2}^{(1)} \oplus P_{v_2}^{(2)} \oplus K_{v_2}^{(n-4)}$ and $K_{v_2}[x]/(f_3)$ is a field, and then use the weak approximation theorem to construct a polynomial $f(x)$ which is close enough to $f_i(x)$, $i = 1, 2, 3$, and which then will give L.

Now let $M = K(\alpha)$ be an extension of K of degree n, let $f(x)$ be the minimal polynomial of α over K, and let

$$M \otimes_K K_{v_i} \simeq L \otimes_K K_{v_i}, \quad i = 1, 2.$$

We let R denote the normal closure of M over K.

LEMMA 2. *The Galois group $G = \mathrm{Gal}(R/K)$ is isomorphic to the symmetric group S_n.*

PROOF. The action of $\mathrm{Gal}(R/K)$ on the roots of the equation $f(x) = 0$ gives an injective homomorphism $\theta \colon G \to S_n$, where the image $\theta(G)$ is a transitive subgroup by the irreducibility of f. We have the equality of completions $R_{v_1} = P_{v_1}$, and hence the decomposition group G^{v_1} is cyclic of order $n-1$. Let σ be a generator of G^{v_1}. Then the substitution $\theta(\sigma)$ is a cycle of order $n-1$. In fact, over K_{v_1} we can write $f(x)=\varphi(x)(x-a)$, $a\in K_{v_1}$, where $\varphi(x)$ gives an unramified extensions of K_{v_1} of degree $n-1$, and the powers of σ permute the roots of $\varphi(x)$ transitively. Similarly, it follows from the construction of the field M that $R_{v_2}=P_{v_2}^{(1)}P_{v_2}^{(2)}$ and the Galois group

$$G^{v_2}=\mathrm{Gal}(R_{v_2}/K_{v_2})\simeq \mathrm{Gal}\left(P_{v_2}^{(1)}/K_{v_2}\right)\times \mathrm{Gal}\left(P_{v_2}^{(2)}/K_{v_2}\right)$$

is a product of two cyclic groups of order 2. If τ_1 and τ_2 are generators of $\mathrm{Gal}(P_{v_2}^{(1)}/K_{v_2})$ and $\mathrm{Gal}(P_{v_2}^{(2)}/K_{v_2})$ respectively, then $\theta(\tau_1)$ and $\theta(\tau_2)$ are transpositions. In fact, over K_{v_2} we have $f(x)=\varphi_1(x)\varphi_2(x)\prod_{i=1}^{n-4}(x-a_i)$, $a_i\in K_{v_2}$, φ_1 and φ_2 determine $P_{v_2}^{(1)}$ and $P_{v_2}^{(1)}$ respectively, and τ_i permutes the roots of $\varphi_i(x)$, leaving the other roots fixed.

Thus, $\theta(G)$ is a transitive subgroup of S_n which contains a cycle of length $n-1$ and a transposition; consequently (see [15], §66), $\theta(G)=S_n$.

Lemma 2 is proved.

Obviously, θ induces an isomorphism of the group $H=\mathrm{Gal}(R/M)\subset G$ onto the stabilizer of the element $i\in\{1,2,\dots,n\}$ corresponding to the root α. Since $G\simeq S_n$, it follows that $H\simeq S_{n-1}$. The commutator $[S_n,S_n]=A_n$, and hence $[G,G]\cap H=[N_G(H),H]$. We are now in the situation when the following fact from [16] can be applied.

PROPOSITION 2. *Let $R\supset M\supset K$ be a tower of finite extensions in which R/K is normal with Galois group G and H is the subgroup corresponding to M. If the Hasse principle holds for R/K and if $[N_G(H),H]=H\cap[G,G]$, then the Hasse principle holds for M/K.*

Thus, to complete the proof of Theorem 5 it is sufficient to establish that the Hasse principle holds for R/K. As noted above, this is equivalent to injectivity of the canonical mapping $H^3(G,\mathbf{Z})\to\prod_v H^3(G^v,\mathbf{Z})$. It turns out that in our situation we have injectivity of the map $H^3(G,\mathbf{Z})\to H^3(G^{v_2},\mathbf{Z})$. By construction, $\theta(G^{v_2})$ is a subgroup of S_n generated by two independent transpositions; hence the required conclusion follows from the following fact.

LEMMA 3. *Suppose that $G=S_n$, $n\geqslant 4$, and $F\subset G$ is a subgroup generated by two independent transpositions. Then the restriction mapping $H^3(G,\mathbf{Z})\to H^3(F,\mathbf{Z})$ is injective.*

Before proving Lemma 3, we recall a well-known fact about the connection between $H^3(G,\mathbf{Z})$ and the Schur multiplier $M(G)=H^2(G,\mathbf{C}^*)$ (all cohomology groups are considered with trivial action of G). Because there seems to be no suitable reference, we shall give the proof here.

LEMMA 4. *For any finite group G and integer $q\geqslant 2$ there exists a natural isomorphism $H^{q+1}(G,\mathbf{Z})\simeq H^q(G,\mathbf{C}^*)$.*

PROOF. Let Σ be the group of all roots of unity in $\mathbf{C}^*$. Then the quotient group $D=\mathbf{C}^*/\Sigma$ is a group in which one can uniquely extract roots of any integer degree; hence, $H^q(G,D)=0$ for $q\geqslant 1$. The exact sequence

$$0\to\Sigma\to\mathbf{C}^*\to D\to 0$$

gives us the standard exact cohomology sequence

$$0 = H^{q-1}(G, D) \to H^q(G, \Sigma) \to H^q(G, \mathbf{C}^*) \to H^q(G, D) = 0,$$

which leads to the isomorphism $H^q(G, \Sigma) = H^q(G, \mathbf{C}^*)$.

The group Σ is isomorphic to $\mathbf{Q}/\mathbf{Z}$; hence from the exact sequence $0 \to \mathbf{Z} \to \mathbf{Q} \to \mathbf{Z}/\mathbf{Q} \to 0$ we obtain

$$0 = H^q(G, \mathbf{Q}) \to H^q(G, \mathbf{Q}/\mathbf{Z}) \to H^{q+1}(G, \mathbf{Z}) \to H^{q+1}(G, \mathbf{Q}) = 0,$$

and this gives $H^{q+1}(G, \mathbf{Z}) \simeq H^q(G, \mathbf{Q}/\mathbf{Z}) \simeq H^q(G, \mathbf{C}^*)$.

PROOF OF LEMMA 3. Taking Lemma 4 into account, we see that it suffices to show that the map $H^2(G, \mathbf{C}^*) \to H^2(F, \mathbf{C}^*)$ is injective.

We recall that a representation group of a finite group G is defined to be a finite group G^* of minimal order which satisfies the following two conditions: 1) G^* is a central extension of G; 2) every complex projective representation of G can be linearized by means of G^*. As Schur showed in 1904 [17], any finite group has at least one representation group. The kernel of the natural homomorphism $G^* \to^{\pi} G$ depends only on G and is called the Schur multiplier $M(G)$ of the group G; it is well known that $M(G) \simeq H^2(G, \mathbf{C}^*)$. In the case $G = S_n$, Schur [13] made explicit computations showing that for $n \geqslant 4$ there exist two representation groups G^* and $\bar{G}^*$, which are isomorphic only when $n = 6$ and $[M(G)] = 2$. These representation groups are distinguished by the following property: the preimage of a transposition has order 4 in G^* and order 2 in $\bar{G}^*$. We shall work with the group G^*. If $M(G) = \{1, z\}$, and $x_1 = (i_1 j_1)$ and $x_2 = (i_2 j_2)$ are independent transpositions, i.e., $\{i_1, j_1\} \cap \{i_2, j_2\} = \varnothing$, then the commutator $[\pi^{-1}(x_1), \pi^{-1}(x_2)] = z$. We denote $y_i = \pi^{-1}(x_i)$, $i = 1, 2$.

Since T, it follows that injectivity can be proved by constructing a projective representation $f\colon G \to \mathrm{PGL}_r(\mathbf{C})$ such that the restriction $f|_F$ is not linearizable. There exists an irreducible representation $\varphi\colon G^* \to \mathrm{GL}_r(\mathbf{C})$ such that $\varphi(z) \neq E_r$, where E_r is the identity matrix. Then f can be taken to be the corresponding projective representation. In fact, suppose that the restriction $f|_F$ is linearizable by means of the representation $g\colon F \to \mathrm{GL}_r(\mathbf{C})$. Then $[g(x_1), g(x_2)] = g([x_1, x_2]) = E_r$, where E_r is the identity matrix of order r. On the other hand, $g(x_i) = \varphi(y_i)\lambda_i$, $\lambda_i \in \mathbf{C}^*$, so that

$$[g(x_1), g(x_2)] = [\varphi(y_1), \varphi(y_2)] = \varphi([y_1, y_2]) = \varphi(z) \neq E_r.$$

This is a contradiction.

This proves Lemma 3, and completes the proof of Theorem 5.

§3. Multiplicative and norm properties

Let D be a central division ring of index n over an algebraic number field K, and set $D^{(1)} = \{x \in D^* \mid \mathrm{Nrd}_{D/K}(x) = 1\}$. It is well known (see [1], [21]) that over the global field $D^{(1)}$ coincides with the commutator $[D^*, D^*]$ of the multiplicative group, i.e., it is generated by all possible commutators $[x, y] = x^{-1}y^{-1}xy$, $x, y \in D^*$. Hence, to compute $[D^{(1)}, D^{(1)}]$ it would be desirable to know which commutators $[x, y]$ belong to the group $[D^{(1)}, D^{(1)}]$. Our immediate goal is to give an appropriate sufficient condition (Theorem 4).

Recall that we denote $T = \{v \in V_f | D \otimes_K K_v$ is a division ring$\}$; if $a \in K^*$, then $V(a) = \{v \in V_f | a \notin U_v\}$, where U_v is the group of v-adic units in K_v.

LEMMA 5. *For any finite $S \subset V$ the imbedding*

$$\delta_S\colon [D^{(1)}, D^{(1)}] \to \prod_{v \in S \setminus T} D_v^{(1)} \times \prod_{v \in S \cap T} [D_v^{(1)}, D_v^{(1)}]$$

is dense.

PROOF. By the weak approximation theorem for D^* the group $D^{(1)}$ is densely imbedded in $\prod_{v \in S} D_v^{(1)}$, and hence $\delta_S([D^{(1)}, D^{(1)}])$ is dense in $\prod_{v \in S}[D_v^{(1)}, D_v^{(1)}]$. But for $v \notin T$ we have $[D_v^{(1)}, D_v^{(1)}] = D_v^{(1)}$ (see, for example, [18]), and this proves Lemma 5.

LEMMA 6. *For any $v \in V$ and any maximal semisimple commutative K_v-subalgebra $H \subset D_v$ the set*

$$W_H = \{x \in D_v^* | K_v[x] \simeq H\}$$

is open in D_v^*, and $W_H \cap D_v^{(1)}$ is nonempty and open in $D_v^{(1)}$.

PROOF. The algebra $H \simeq H_1 \oplus \cdots \oplus H_r$, where the H_i are finite field extensions of K_v. Since any extension has a primitive element with norm one (see [19]), it follows that $D_v^{(1)} \cap W_H \neq \varnothing$. It is well known that the set R of elements regular in D_v^* is open in the Zariski topology. From this it is not hard to see that the intersection of R with the set $\Delta = \{xhx^{-1} | h \in H^*, x \in D_v^*\}$ is open in the v-adic topology. This proves Lemma 6.

LEMMA 7. *Suppose that $x \in D^*$, $S \subset V$ is a finite subset, and for every $v \in S$ we are given a maximal semisimple commutative K_v-algebra H_v in D_v. Assume that for any $v \in S$ there exists $z_v \in [D_v^{(1)}, D_v^{(1)}]$ with the property that $xz_v \in H_v$. Then there exists an element $z \in [D^{(1)}, D^{(1)}]$ such that $K_v[xz] \simeq H_v$ for all $v \in S$.*

PROOF. The lemma is equivalent to the assertion that the intersection $[D^{(1)}, D^{(1)}] \cap x^{-1}W$, where $W = \prod_{v \in S} W_{H_v}$, is nonempty (for brevity we omit the symbol δ_s). Since the set of regular elements in D_v^* is open in the Zariski topology, it follows that the regular elements form open subsets of H_v^* and $D_v^{(1)}$. We note that $H_v \cap [D_v^{(1)}, D_v^{(1)}]$ is dense in $H_v^{(1)} = H_v \cap D_v^{(1)}$ with respect to the Zariski topology. In fact, if D_v is not a division ring, then $D_v^{(1)} = [D_v^{(1)}, D_v^{(1)}]$ (see [18]), and otherwise $H_v \cap [D_v^{(1)}, D_v^{(1)}]$ is a subgroup of finite index in the connected group $H_v^{(1)}$ and so is obviously dense. Since the set of regular elements is open, it now easily follows that there exists $h_v \in H_v \cap [D_v^{(1)}, D_v^{(1)}]$ with the property that $K_v[xz_v h_v] = H_v$. In other words, the intersection

$$\prod_{v \in S} [D_v^{(1)}, D_v^{(1)}] \cap x^{-1}W$$

is nonempty and, according to Lemma 6, is open; to complete the proof it remains to apply Lemma 5.

LEMMA 8. *For any $v \in V_f$ and any $x \in D_v^*$ with $\mathrm{Nrd}_{D_v|K_v}(x) \in U_v$ there exists an element $z_v \in [D_v^{(1)}, D_v^{(1)}]$ such that $K_v[xz_v]$ is an unramified extension of K_v of degree n.*

PROOF. It is well known that there exists an unramified extension L_v/K_v of degree n contained in D_v. We shall show that

$$\left\{d \in D_v | \mathrm{Nrd}_{D_v|K_v}(d) \in U_v\right\} \subset L_v^*\left[D_v^{(1)}, D_v^{(1)}\right].$$

In fact, since L_v/K_v is unramified, it follows that $U_v \subset \mathrm{Nrd}_{D_v|K_v}(L_v^*)$, and so it suffices to show that

$$D_v^{(1)} = L_v^{(1)}\left[D_v^{(1)}, D_v^{(1)}\right], \quad \text{where } L_v^{(1)} = L_v \cap D_v^{(1)}.$$

If the algebra D_v is not a division ring, then $D_v^{(1)} = [D_v^{(1)}, D_v^{(1)}]$ and there is nothing to prove. Otherwise, it is well known that $[D_v^{(1)}, D_v^{(1)}]$ coincides with the first congruence-subgroup of $D_v^{(1)}$ (see [20]), so that $L_v^{(1)}$ contains all representatives of the quotient group $D_v^{(1)}/[D_v^{(1)}, D_v^{(1)}]$. Thus, $x = st$, $s \in L_v^*$, $t \in [D_v^{(1)}, D_v^{(1)}]$. As in Lemma 7, the fact that the subgroup $L_v \cap [D_v^{(1)}, D_v^{(1)}]$ is dense in $L_v^{(1)}$ in the Zariski topology and the set of regular elements is open implies the existence of $d \in L_v \cap [D_v^{(1)}, D_v^{(1)}]$ for which the element sd is regular, i.e., $K_v[sd] = L_v$. Then z_v can be taken to be $t^{-1}d$.

Lemma 8 is proved.

LEMMA 9. *Let L_1 and L_2 be two extensions of the field K of degree n which satisfy the Hasse principle, and let $a \in K^*$. Let P_1 and P_2 denote the normal closures of L_1 and L_2, and let $P = P_1P_2$ denote their compositum. Further let $S_0 \subset V$ be a finite subset containing V_∞, and for $i = 1, 2$ let T_i be the set of places where the extension P_i/K ramifies. Suppose that the following conditions are fulfilled*:

(i) $P_1 \cap P_2 = K$.

(ii) $P_2 \subset K_v$ *for any* $v \in T_1 \setminus (V(a) \cup S_0)$.

(iii) $a \in N_{P_{\bar v}/K_v}(P_{\bar v}^*)(\bar v/v)$ *for any* $v \in V(a) \cup S_0$.

Then $a \in N_{L_1/K}(L_1^*)N_{L_2/K}(L_2^*)$.

The proof of Lemma 9 uses the following fact.

LEMMA 10. *Using the notation and assumptions of Lemma 9, suppose that there exist $v_0 \in V_f$ and a uniformizing element $\pi_0 \in K^*$ such that $v(\pi_0) = 0$ for $v \in V_f \setminus \{v_0\}$ and the following conditions are fulfilled*:

(i) $P \subset K_{v_0}$;

(ii) $\pi_0 \in U_v^m$ *for* $v \in T_1 \cup V(a) \cup S_0$, $\pi_0 \in aU_v^m$ *for* $v \in T_2 \setminus (V(a) \cup S_0)$,

where $m = n!$ (for archimedean v we set $U_v = K_v^*$).

Then $a \in N_{L_1/K}(L_1^*)N_{L_2/K}(L_2^*)$.

PROOF. It suffices to show that

$$\pi_0 \in N_{L_1/K}\left(L_1^*\right) \quad \text{and} \quad a\pi_0^{-1} \in N_{L_2/K}\left(L_2^*\right).$$

We prove the first of these claims. Since L_1/K satisfies the Hasse principle, it is enough to show that for any $v \in V$ and any extension $\bar v/v$ we have

$$\pi_0 \in N_{L_{1\bar v}/K_v}\left(L_{\bar v}^*\right).$$

First suppose that $v \notin T_1 \cup V_\infty \cup \{v_0\}$. Then the extension $L_{1\bar v}/K_v$ is unramified, and $\pi_0 \in N_{L_{1\bar v}/K_v}(L_{1\bar v}^*)$. For $v \in T_1 \cup V_\infty$, by condition (ii) we have

$$\pi_0 \in U_v^m \Rightarrow \pi_0 \in N_{L_{1\bar v}/K_v}\left(L_{1\bar v}^*\right),$$

since the degree $[L_{1\bar{v}}:K_v]$ divides $m = n!$. In the remaining case, according to (i) we have $L_{1\bar{v}} = K_v$, and the claimed inclusion is obvious.

The second claim is proved similarly. As before, it is enough to show that $a\pi_0^{-1} \in N_{L_{2\bar{v}}/K_v}(L_{2\bar{v}}^*)$ for all $v \in V(\bar{v}/v)$, where we really need to check only $v \in T_2 \cup V_\infty \cup V(a) \cup \{v_0\}$. If $v \in T_2 \setminus (S_0 \cup V(a))$, then $a\pi_0^{-1} \in U_v^m \subset N_{L_{2\bar{v}}/K_v}(L_{2\bar{v}}^*)$. Otherwise, $v \in V(a) \cup S_0 \cup \{v_0\}$. For $v \in V(a) \cup S_0$ we have

$$a\pi_0^{-1} \in aU_v^m \subset N_{L_{2\bar{v}}/K_v}\left(L_{2\bar{v}}^*\right)$$

since $a \in N_{P_{\bar{v}}/K_v}(P_{\bar{v}}^*) \subset N_{L_{2\bar{v}}/K_v}(L_{2\bar{v}}^*)$ (see part (iii) of Lemma 9). But if $v = v_0$, then $L_{2\bar{v}} = K_v$ and $a\pi_0^{-1} \in K_v^* = N_{L_{2\bar{v}}/K_v}(L_{2\bar{v}}^*)$.

Lemma 10 is proved.

PROOF OF LEMMA 9. We let W denote the following subgroup of the idele group I_k:

$$W = K^*\left(\prod_{v \notin V_0} U_v \times \prod_{v \in V_0} U_v^d\right),$$

where $V_0 = S_0 \cup V(a) \cup S$, S consists of the valuations which ramify at P, and $d = m^2$. According to the existence theorem (see [12], Chapter VIII), the group W corresponds to a unique abelian extension R of K having W as its norm subgroup. Let $\sigma \in \mathrm{Gal}(R|K)$ be the automorphism which corresponds under the Artin map to the class cW, where the idele e has the following local components:

$$c_v = \begin{cases} a, & v \in V_1, \\ 1, & v \notin V_1, \end{cases}$$

$V_1 = S \setminus (T_1 \cup V(a) \cup S_0)$. Let M denote the maximal abelian subextension of P. Since $P_1 \cap P_2 = K$, it follows that $M = M_1 M_2$, where M_i is the maximal abelian subextension of P_i $(i = 1, 2)$, and hence the degree $[M:K] = [M_1:K][M_2:K]$ divides $d = m^2$. From this it easily follows that the norm subgroup N_M for the extension M contains $W = N_R$, so that $M \subset R$ and $P \cap R = M$. We shall show that $\sigma \in \mathrm{Gal}(R|M)$. To do this, it is well known that one must show that $\prod_{v \in V_1} \theta_v(a) = 1$, where the θ_v are the local Artin maps for the extension M/K. We also let θ_{1v} denote the local Artin maps for the extension M_1/K.

By the reciprocity law we have $\prod_{v \in V} \theta_v(a) = 1$, so that it suffices to show that $\prod_{v \in V \setminus V_1} \theta_v(a) = 1$. We have $V \setminus V_1 \subset (V \setminus S) \cup (T_1 \cup V(a) \cup S_0)$. If $v \in (V \setminus S) \cup V(a) \cup S_0$, then $a \in U_v \subset N_{M_{\bar{v}}/K_v}(M_{\bar{v}}^*)$ and hence $\theta_v(a) = 1$. In fact, this follows from (iii) for $v \in V(a) \cup S_0$. But if $v \in V \setminus (S \cup V(a) \cup V_\infty)$, then $a \in U_v \subset N_{M_{\bar{v}}/K_v}(M_{\bar{v}}^*)$, since the extension $M_{\bar{v}}/K_v$ is unramified. We note that $T_1 \subset V \setminus V_1$, and hence it is enough for us to establish the equality $\prod_{v \in S_1} \theta_v(a) = 1$, where $S_1 = T_1 \setminus (V(a) \cup S_0)$. But for $v \in S_0$ we have $\theta_v(a) = \theta_{1v}(a)$, since $M_{\bar{v}} = M_{1\bar{v}}$ from the fact that $M = M_1 M_2$ and $M_2 \subset K_v$ (see (ii)). Using the reciprocity law for M_1, we obtain

$$\prod_{v \in S_1} \theta_v(a) = \prod_{v \in S_1} \theta_{1v}(a) = \left(\prod_{v \in V \setminus S_1} \theta_{1v}(a)\right)^{-1}.$$

Furthermore, $V \setminus S_1 \subset (V \setminus T_1) \cup (V(a) \cup S_0)$ and, as above, one shows that $\theta_{1v}(a) = 1$ for $v \in V \setminus S_1$.

Thus, we have proved the equality $\prod_{v \in V_1} \theta_v(a) = 1$.

We have the isomorphism of Galois groups

$$\mathrm{Gal}(PR/M) \simeq \mathrm{Gal}(P/M) \times \mathrm{Gal}(R/M).$$

Let $\tau \in \mathrm{Gal}(PR/M)$ correspond to the pair $(1, \sigma)$. By the Chebotarev density theorem (see [12], Chapter VII, §2.4), there exists a valuation $v_0 \in V_f \setminus V_0$ such that the Frobenius automorphism $\mathrm{Fr}(PRK_{v_0}/K_{v_0}) = \tau$. Then $P \subset K_{v_0}$, and the idele $\tilde{\pi}_0$ with components

$$(\tilde{\pi}_0)_v = \begin{cases} 1, & v \neq v_0, \\ \pi_{v_0}, & v = v_0, \end{cases}$$

where π_{v_0} is a uniformizing element, belongs to cW. Recalling the definition of W, we find that there exists $\pi_0 \in K^*$ with the property

$$\pi_0^{-1}\tilde{\pi} \in c\Big(\prod_{v \notin V_0} U_v \times \prod_{v \in V_0} U_v^d\Big).$$

From this last inclusion it easily follows that the element π_0 satisfies the conditions of Lemma 10, which can now be applied to give our claim.

Lemma 9 is proved.

LEMMA 11. *Suppose we are given a finite extension P of the field K and a finite subset $S \subset V_f$. Then there exists a finite subset $T(P) \subset V_f \setminus S$ with the property that if $R \supset K$ and $R_{\bar{v}} = K_v$ for all $v \in T(P)$ and all $\bar{v}/v$, then $P \cap R = K$.*

PROOF. Without loss of generality we may assume that P/K is a Galois extension. Let $\sigma_1, \dots, \sigma_r$ be all of the nonidentity elements of the Galois group $\mathrm{Gal}(P/K)$. By the Chebotarev density theorem, for each $i = 1, \dots, r$ there exist a valuation $v_i \in V_f \setminus S$ and an extension $\bar{v}_i$ such that the Frobenius automorphism $\mathrm{Fr}(P_{\bar{v}_i}/K_{v_i}) = \sigma_i$. We set $T(P) = \{v_1, \dots, v_r\}$. Now if $R \supset K$ and $L = P \cap R \neq K$, then there exists σ_i which acts nontrivially on L. Let $\bar{v}$ denote the valuation of R which extends the restriction of $\bar{v}_i$ to L. Then $\bar{v}/v_i$ and $R_{\bar{v}} \neq K_{v_i}$, since the automorphism $\sigma_i \in \mathrm{Gal}(P_{\bar{v}_i}/K_{v_i})$ is not the identity on the subfield $L_{\bar{v}}$; consequently, $L_{\bar{v}} \neq K_{v_i}$.

Lemma 11 is proved.

§4. Commutator properties. Proof of Proposition 4

We shall need some additional notation. We set

$$T_f = \{v \in V_f | D \otimes_K K_v \not\simeq M_n(K_v)\}, \quad T_\infty = \{v \in V_\infty | D \otimes_K K_v \not\simeq M_n(K_v)\}.$$

In addition, we introduce a subgroup U_f analogous to U by setting

$$U_f = \{x \in D^* | \mathrm{Nrd}_{D/K}(x) \in U_v, \forall v \in T_f\}.$$

Obviously, T_f and T_∞ are finite sets with $T \subset T_f$, and hence $U_f \subset U$.

We now prove a proposition which generalizes Lemma 3 of [8] and will play a fundamental role in the proof of Proposition 4.

PROPOSITION 3. *Let x, $y \in U_f$. Then there exist elements z_1, $z_2 \in [D^{(1)}, D^{(1)}]$ and $d_i \in K(yz_i)$ $(i = 1, 2)$ such that* $\mathrm{Nrd}_{D/K}(x) = \mathrm{Nrd}_{D/K}(d_1 d_2)$.

PROOF. We set $a = \mathrm{Nrd}_{D/K}(x)$ and $b = \mathrm{Nrd}_{D/K}(y)$, $S_0 = T_f \cup V_\infty$. The Chebotarev density theorem implies that there exists a two-element subset $S_1 = \{v_1, v_2\} \subset V_f \setminus (V(a) \cup S_0)$ such that $b \in U_{v_i}^m$ for $i = 1, 2$; $m = n!$. Let $L(S_1)$ be the degree n extension of K that was constructed in Theorem 5 for the subset S_1. We choose an element $z_1 \in [D^{(1)}, D^{(1)}]$ in such a way that the following conditions hold:

1) $K_v[yz_1]/K_v$ is an unramified extension of degree n for $v \in T_f$.
2) $K_v[yz_1] \simeq K_v^{(n)}$ for $v \in V(a) \cup (V_\infty \setminus T_\infty)$;
3) $K_v[yz_1] \simeq L(S_1) \otimes_K K_v$ for $v \in S_1$.

The existence of such a z_1 follows from Lemmas 7 and 8. In fact, according to Lemma 7 it is enough to see that for every $v \in T_f \cup V(a) \cup (V_\infty \setminus T_\infty)$ there exists $z_v \in [D_v^{(1)}, D_v^{(1)}]$ such that the corresponding condition holds for $K_v[yz_v]$. The existence of z_v for the first condition follows from Lemma 8. For the second and third condition it follows easily from the fact that for $v \in V(a) \cup (V_\infty \setminus T_\infty) \cup S_1$ we have

$$D_v = D \otimes_K K_v \simeq M_n(K_v), \qquad D_v^{(1)} = \left[D_v^{(1)}, D_v^{(1)}\right];$$

one need only note that $b \in U_v^m \subset \mathrm{Nrd}_{D_v|K_v}(L(S_1) \otimes_K K_v)$ for $v \in S_1$. For example, the argument for 3) is as follows: *$L(S_1) \otimes_K L_v$ is contained in D_v, where for suitable $t_v \in D_v^{(1)}$* we have $K_v[t_v] \simeq L(S_1) \otimes_K K_v$ and hence

$$K_v\left[yy^{-1}t_v\right] = K_v\left[yy^{-1}\alpha t_v\right] = K_v\left[t_v\right], \quad \text{where } \alpha \in K,\ \alpha^n = b;$$

where $\alpha \in K$ and $\alpha^n = b$; then $y^{-1}\alpha t_v = z_v \in D_v^{(1)}$.

Furthermore, let L_1 be the image of $K(yz_1)$ under an imbedding in a fixed algebraic closure $\overline{K}$ of K, let P_1 be the normal closure of L_1, and let T_1 be the set of places of P_1/K which ramify. We further let $T(P_1)$ denote the subset in Lemma 11 disjoint with S_0 corresponding to P_1. Let $S_2 = \{v_3, v_4\}$ be a two-element subset of $V_f \setminus (V(a) \cup S_0 \cup T_1 \cup T(P_1))$ such that $b \in U_{v_i}^m$, $i = 3, 4$, $m = n!$, and let $L(S_2)$ be the degree n extension in Theorem 5 which corresponds to S_2. We make our choice of z_2 subject to the following conditions:

1) $K_v[yz_2]/K_v$ is an unramified extension of degree n for $v \in T_f$.
2) $K_v[yz_2] \simeq K_v^{(n)}$ for $v \in V(a) \cup (V_\infty \setminus T_\infty) \cup T_1 \cup T(P_1)$.
3) $K_v[yz_2] \simeq L(S_2) \otimes_K K_v$ for $v \in S_2$.

As before, the existence of z_2 can be derived from Lemma 7 and 8.

We let L_2 denote the image of $K(yz_2)$ in $\overline{K}$, P_2 denote the normal closure of L_2, and T_2 denote the set of places of P_2/K which ramify; we set $P = P_1P_2$.

To prove the proposition it suffices to show that

$$a \in N_{L_1/K}\left(L_1^*\right) N_{L_2/K}\left(L_2^*\right).$$

Since it follows from our constructions and Theorem 5 that the extensions L_1/K and L_2/K satisfy the Hasse principle, it remains for us to verify conditions (i)–(iii) of Lemma 9. Condition (i) follows immediately if we use the inclusion $P_2 \subset K_v$ for $v \in T(P_1)$ and Lemma 11; condition (ii) follows from the construction of P_2. We verify (iii). If $v \in V(a) \cup (V_\infty \setminus T_\infty)$, for $P = P_1P_2 \subset K$ and the claim is obvious. If $v \in T_f$, then $P_{\bar{v}}/K_v$ is an unramified extension of degree n, so that $a \in U_v \subset N_{P_{\bar{v}}/K_v}(P_{\bar{v}}^*)$. For $v \in T_\infty$ we have $K_\infty = \mathbf{R}$, $a = \mathrm{Nrd}_{D/K}(x) > 0$ in K_v, and hence $a \in N_{\mathbf{C}/\mathbf{R}}(C^*)$, from which the claim follows.

Proposition 3 is proved.

PROPOSITION 4. *Set* $U = \{x \in D^* \mid \mathrm{Nrd}_{D/K}(x) \in U_v,\ \forall v \in T\}$. *Suppose that* $x \in U$, $y \in D^*$ *and for any* $v \in T \cap V(\mathrm{Nrd}_{D/K}(y))$ *one has* $x \in U_v[D_v^{(1)}, D_v^{(1)}]$. *Then*

$$x^{-1}y^{-1}xy \in [D^{(1)}, D^{(1)}];$$

in particular, if $T = \varnothing$, *then* $D^{(1)} = [D^{(1)}, D^{(1)}]$.

PROOF. For $v \in T_f$ we construct maximal commutative K_v-subalgebras $H_v \subset D_v$ in the following way. For $v \in T$ we take for H_v the maximal totally unramified subfield of D_v. But if $v \in T_f \setminus T$, i.e., $D_v = M_{n_v}(\tilde{D}_v)$ for some division ring $\tilde{D}_v$ and integer $n_v > 1$, then we set $H_v = F_v \oplus R_v^{(nv-1)}$, where F_v and R_v are, respectively, the maximal unramified subfield and the maximal totally ramified subfield of the division ring $\tilde{D}_v$. We note that for $v \in T_f \setminus T$

$$\mathrm{Nrd}_{D_v/K_v}(H_v) = K_v, \tag{1}$$

since $\mathrm{Nrd}_{D_v/K_v}(F_v^*) = \mathrm{Nrd}_{D_v/K_v}(F_v^*) \supset U_v$, and $\mathrm{Nrd}_{D_v/K_v}(R_v^*) = \mathrm{Nrd}_{\tilde{D}_v/K_v}(R_v^*)$ contains a uniformizing element of the field K_v.

Using this property, we show that, under the conditions of the theorem, $[x, y] \approx [x_1, y_1]$ for some x_1, $y_1 \in U_f$, where the notation $a \approx b$ means that a and b are congruent modulo $[D^{(1)}, D^{(1)}]$, i.e., $b^{-1}(a) \in [D^{(1)}, D^{(1)}]$.

From (1) and the equality $[D_v^{(1)}, D_v^{(1)}] = D_v^{(1)}$ for $v \in T_f \setminus T$ it follows that for some $z_v \in [D_v^{(1)}, D_v^{(1)}]$ we have $yz_v \in H_v$ $(v \in T_f \setminus T)$. Hence, from Lemma 7 we find that there exists $z \in [D^{(1)}, D^{(1)}]$ such that $K_v(yz) \simeq H_v$ for all $v \in T_f \setminus T$. If we apply the weak approximation theorem to the field $K(yz)$ and take into account the fact that $v(\mathrm{Nrd}_{D_v/K_v}(H_v^*)) = \mathbf{Z}$ for $v \in T_f$, we find $d \in K(yz)$ with the property

$$v\big(\mathrm{Nrd}_{D/K}(d)\big) = v\big(\mathrm{Nrd}_{D/K}(x)\big) \quad \forall v \in T_f$$

(recall that $v(\mathrm{Nrd}_{D/K}(x)) = 0$ for $v \in T$). We then have

$$[x, y] \approx [x, yz] = \big[d^{-1}x, yz\big],$$

where $d^{-1}x \in U_f$ and $\mathrm{Nrd}_{D/K}(yz) = \mathrm{Nrd}_{D/K}(y)$. Thus, we may suppose that $x \in U_f$.

In a similar way we find that for any $v \in T_f \cap V(\mathrm{Nrd}_{D/K}(y))$ there exists $z_v \in [D_v^{(1)}, D_v^{(1)}]$ with the property: $xz_v \in H_v$. For $v \in T \cap V(\mathrm{Nrd}_{D/K}(y))$ this follows from the condition $x \in U_v[D_v^{(1)}, D_v^{(1)}]$, and for $v \in (T_f \setminus T) \cap V(\mathrm{Nrd}_{D/K}(y))$ it follows from (1) and the equality $[D_v^{(1)}, D_v^{(1)}] = D_v^{(1)}$. As before, according to Lemma 7 there exists $z \in [D^{(1)}, D^{(1)}]$ such that

$$K_v[xz] \simeq H_v \quad \forall v \in T_f \cap V\big(\mathrm{Nrd}_{D/K}(y)\big).$$

Since $v(\mathrm{Nrd}_{D_v/K_v}(H_v^*)) = \mathbf{Z}$ for $v \in T_f$, it follows from the weak approximation theorem that there exists $d \in K(yz)$ such that $\forall v \in T_f$

$$v\big(\mathrm{Nrd}_{D/K}(d)\big) = v\big(\mathrm{Nrd}_{D/K}(y)\big).$$

Then

$$[x, y] \approx [xz, y] = \big[xz, d^{-1}y\big],$$

where xz, $d^{-1}y \in U_f$. Thus, it suffices to show that for x, $y \in U_f$ we have $[x, y] \in [D^{(1)}, D^{(1)}]$.

According to Proposition 3, there exist z_1, $z_2 \in [D^{(1)}, D^{(1)}]$ and $d_i \in K(yz_i)$ $(i = 1, 2)$ such that $\mathrm{Nrd}_{D/K}(x) = \mathrm{Nrd}_{D/K}(d_1 d_2)$. Here we obviously have the following congruences:

$$[x, y] \approx [x, yz_1] = \left[d_1^{-1}x, yz_1\right] \approx \left[d_1^{-1}x, yz_2\right] = \left[d_2^{-1}d_1^{-1}x, yz_2\right]. \quad (2)$$

Thus, $[x, y]$ determines the same class modulo $[D^{(1)}, D^{(1)}]$ as is determined by some commutator $[s, t]$, where $s \in D^{(1)}$ and $t \in U_f$. Again by Proposition 3 we can find z_3, $z_4 \in [D^{(1)}, D^{(1)}]$ and $d_i \in K(sz_i)$ $(i = 3, 4)$ with the property that

$$\mathrm{Nrd}_{D/K}(t) = \mathrm{Nrd}_{D/K}(d_3 d_4).$$

Making computations analogous to (2), we can now easily establish that $[s, t] \in [D^{(1)}, D^{(1)}]$. But then $[x, y] \in [D^{(1)}, D^{(1)}]$, as was to be proved.

Then claim that when $T = \varnothing$ we have $D^{(1)} = [D^{(1)}, D^{(1)}]$, which now follows immediately from the fact that $D^{(1)}$ coincides with the commutator of D^*, since in this case $U = D^*$.

The proof of Proposition 4 is complete.

§5. Proof of Theorem 4

We first prove two auxiliary results.

PROPOSITION 5. *Suppose that* x, $y \in D^*$ *and* $x \in 1 + \mathfrak{P}_v$ *for all* $v \in T$. *Then* $[x, y] \in [D^*, D^{(1)}]$.

PROOF. We set $N = [D^*, D^{(1)}]$ and consider the bimultiplicative map $\psi\colon D^* \times D^* \to D^{(1)}/N$ which is given by the formula $\psi(x, y) = [x, y]N$; note that $\psi(x, y)$ depends only on $\mathrm{Nrd}_{D/K}(x)$, $\mathrm{Nrd}_{D/K}(y)$. We introduce the open subgroups $B_v = (1 + \mathfrak{p}_v)[D_v^{(1)}, D_v^{(1)}] \subset D_v^*$. Then the weak approximation theorem implies the existence of a factorization $x = x_1 \cdots x_d$, where $d = [T]$ and $T = \{v_1, \ldots, v_d\}$, in which the elements x_i satisfy the conditions

$$x_i \in xB_{v_i}, \quad x_i \in B_{v_j} \qquad (i, j = 1, \ldots, d;\, i \neq j).$$

Since $\psi(x, y) = \psi(x_1, y) \cdots \psi(x_d, y)$, it suffices to show that the commutators $[x, y]$ belong to the group N, where $x \in 1 + \mathfrak{P}_{v_0}$ and $x \in B_v$ for $v \in T \setminus \{v_0\}$, $v_0 \in T$. If $v_0(n) = 0$, then $x \in U_{v_0}[D_{v_0}^{(1)}, D_{v_0}^{(1)}]$ and, according to Proposition 4, $[x, y] \in [D^{(1)}, G^{(1)}] \subset N$. Otherwise, n can be represented in the form $n = p^t n_0$, where p is the prime number corresponding to v_0 and $(p, n_0) = 1$. Then $x^{p^t} \in U_v[D_v^{(1)}, D_v^{(1)}]$ for all $v \in T$, so that $\psi(x, y)^{p^t} = \psi(x^{p^t}, y) = e$ (e is the identity of the group $D^{(1)}/N$). Since $(p, n_0) = 1$, to prove the required equality $\psi(x, y) = e$ it suffices to show that $\psi(x, y)^{n_0} = \psi(x, y^{n_0}) = e$.

For $\alpha \in K_{v_0}^*$ we let $N_{v_0}(\alpha)$ denote the set of all elements $\beta \in K_{v_0}^*$ such that there exists a degree n extension P of K_{v_0} with the property that α, $\beta \in N_{P/K_{v_0}}(P^*)$. Let $\tilde{N}_{v_0}(\alpha)$ be the subgroup generated by $N_{v_0}(\alpha)$.

LEMMA 12. *If* $\theta = (\mathrm{Nrd}_{D/K}(y))^{n_0} \in \tilde{N}_{v_0}(\mathrm{Nrd}_{D/K}(x))$, *then* $\psi(x, y^{n_0}) = e$.

PROOF. Since any degree n extension to K_{v_0} can be imbedded in D_{v_0}, there exist $z_1, \ldots, z_d \in D_{v_0}^{(1)}$ and $a_i \in K_{v_0}(xz_i)$ such that

$$\theta = \mathrm{Nrd}_{D_{v_0}/K_{v_0}}(a_1) \cdots \mathrm{Nrd}_{D_{v_0}/K_{v_0}}(a_d).$$

From Lemma 6 and the weak approximation theorem for $D^{(1)}$ it follows that z_i can be assumed to belong to $D^{(1)}$. If we now use the weak approximation theorem applied to the fields $K(xz_i)$, we find that there exist $b_i \in K(xz_i)$ such that

$$v_0(\theta) = v_0\big(\mathrm{Nrd}_{D/K}(b_1) \cdots \mathrm{Nrd}_{D/K}(b_d)\big).$$

Then

$$\begin{aligned}\psi(x, y^{n_0}) &= \psi(xz_1, y^{n_0}) = \psi(xz_1, y^{n_0}b_1^{-1}) = \psi(xz_2, y^{n_0}b_1^{-1}) \\ &= \psi(xz_2, y^{n_0}b_1^{-1}b_2^{-1}) = \cdots = \psi(x, y^{n_0}b_1^{-1} \cdots b_d^{-1}) = e\end{aligned}$$

in accordance with Proposition 4.

Lemma 12 is proved.

It remains to show that

$$\big(\mathrm{Nrd}_{D/K}(y)\big)^{n_0} \in \tilde{N}_{v_0}\big(\mathrm{Nrd}_{D/K}(x)\big).$$

Since $x \in 1 + \mathfrak{P}_{v_0}$, it follows that $\mathrm{Nrd}_{D/K}(x) \in 1 + \mathfrak{p}_{v_0}$, and so the claim is a consequence of the following fact, which is applicable because $v_0(n) \neq 0$ and $v_0((n,2)) = 0$ imply that $v_0(2) = 0$.

LEMMA 13. *Suppose that $v_0(2) = 0$. If $\alpha \in 1 + \mathfrak{p}_{v_0}$, then $\tilde{N}_{v_0}(\alpha)$ has prime to p index in $K_{v_0}^*$.*

PROOF. It is well known that the multiplicative group $K_{v_0}^*$ of the field K_{v_0} has the structure $K_{v_0}^* \simeq \mathbf{Z} \times F \times \mathbf{Z}_p^l$, where F is the group of all roots of unity in K_{v_0}, p is the prime number corresponding to v_0, and $l = [K_{v_0} : \mathbf{Q}_p]$. Thus,

$$K_{v_0}^*/K_{v_0}^{*n} \simeq \mathbf{Z}/n\mathbf{Z} \times (\mathbf{Z}/p^t\mathbf{Z})^l \simeq \mathbf{Z}/n_0\mathbf{Z} \times \tilde{F} \times (\mathbf{Z}/p^t\mathbf{Z})^{l+1},$$

where $\tilde{F} = F/F^n$. The element α has p-primary order in the quotient group $K_{v_0}^*/K_{v_0}^{*n}$. From this one easily finds that for $l > 1$ the subgroups of index n in $K_{v_0}^*$ which contain α generate a subgroup of index prime to p. On the other hand, according to local class field theory any open subgroup of index n is a norm subgroup, i.e., it has the form $N_{P/K_{v_0}}(P^*)$, where P is a degree n extensions of K_{v_0}, and so is contained in $\tilde{N}_{v_0}(\alpha)$.

It remains to consider the case $l = 1$, i.e., $K_{v_0} = \mathbf{Q}_p$, $p \neq 2$. Let P be an unramified extension of degree n over K_{v_0}, and let $R = P(\sqrt[p]{p})$. Then $[R : \mathbf{Q}_p] = n$ and $[K_{v_0}^* \colon N_{R/K_{v_0}}(R^*)] = n_0$. In fact, since $\mathbf{Q}_p$ has no nontrivial pth roots of unity and P is unramified over $\mathbf{Q}_p$, it follows that the same is true for P. Therefore R does not contain any nontrivial abelian subextensions, so that $N_{R/P}(R^*) = P^*$ ([12], Chapter VI, §2.6, Proposition 2.4). Hence, $N_{R/K_{v_0}}(R^*) = N_{P/K_{v_0}}(P^*) = U_{v_0}K_{v_0}^{*n_0}$ and $[K_{v_0} : N_{R/K_{v_0}}(R^*)] = n_0$.In addition, $\alpha \in N_{R/K_{v_0}}(R^*)$, so that $N_{R/K_{v_0}}(R^*) \subset \tilde{N}_{v_0}(\alpha)$.

Lemma 13, and hence also Proposition 5, are proved.

PROPOSITION 6. *Let L be a maximal cyclic subfield of D such that L_v/K_v is unramified for all $v \in T$. Then $[D^*, D^{(1)}] = L'[D^{(1)}, D^{(1)}]$ where $L' = \{\sigma\alpha/\alpha \mid \alpha \in L^{(1)}\}$, $L^{(1)} = L \cap D^{(1)}$, and σ is a generator of the group* Gal(L/K).

PROOF. It suffices to show that for $x \in D^*$ and $y \in D^{(1)}$ one has $[x, y] \in L'[D^{(1)}, D^{(1)}]$. We keep the notation in the proof of Proposition 4. We represent y in the form $y = y_1 \cdots y_d$, where $y_i \in y(B_{v_i} \cap D_{v_i}^{(1)})$ and $y_i \in (B_{v_j} \cap D_{v_j}^{(1)})$ $(i, j = 1, \dots, d; i \neq j)$. Then, if we let $a \approx b$ denote the relation $b^{-1}a \in [D^{(1)}, D^{(1)}]$ we obtain

$$[x, y] = x^{-1}(y_1 \cdots y_d)^{-1}x(y_1 \cdots y_d)$$
$$\approx [x, y_1][x, y_2 \cdots y_d] \approx \cdots \approx \prod_{i=1}^{d}[x, y_i]. \tag{1}$$

By the Skolem-Noether theorem, there exists $a \in D^*$ with the property that $\sigma(l) = a^{-1}la$ for any $l \in L$. For $v \in T$ the algebra D_v is a division ring, so that $v(\mathrm{Nrd}_{D/K}(a)) \pmod n$ is a generator of the group $\mathbf{Z}/n\mathbf{Z}$. Hence, for any $i = 1, \dots, d$ there exists a power n_i such that $x = \alpha a^{n_i}x_i$, where $\alpha \in K^*$, $\mathrm{Nrd}_{D/K}(x_i) \in U_{v_i}$. Furthermore, $D_v^{(1)} = L_v^{(1)}[D_v^{(1)}, D_v^{(1)}]$, so that $D^{(1)} \subset L^{(1)}\prod_{i=1}^{1} B_{v_i}$. Thus, for any $i = 1, \dots, d$ we have $y_i = z_i t_i$, where $z_i \in L^{(1)}$, $t_i \in B_{v_j}$ for all $j = 1, \dots, d$. We then have

$$[x, y_i] = [a^{n_i}x_i, y_i] = x_i^{-1}[a^{n_i}, y_i]\,y_i^{-1}x_i y_i \approx [a^{n_i}, y_i]$$

by Proposition 4, since $T \cap V(\mathrm{Nrd}_{D/K}(x_i)) \subset T \setminus \{v_i\}$ and $y_i \in B_{v_j}$ for $j \neq i$.

Continuing the computations, we obtain

$$[a^{n_i}, y_i] = [a^{n_i}, z_i t_i] = a^{-n_i}t_i^{-1}a^{n_i}[a^{n_i}, z_i]t_i \approx [a^{n_i}, z_i] \in L'.$$

We have shown that $[x, y_i] \in L'[D^{(1)}, D^{(1)}]$ for any $i = 1, \dots, d$; taking (1) into account, we obtain the required conclusion.

Proposition 6 is proved.

PROOF OF THEOREM 4. According to Proposition 5 we have $[x, y] \in [D^*, D^{(1)}]$, and so, by Proposition 6, $[x, y] = zw$, where $z = (\sigma l)^{-l} l$, $l \in L^{(1)}$, $w \in [D^{(1)}, D^{(1)}]$. Since $[D^{(1)}, D^{(1)}] \subset 1 + \mathfrak{P}_v$, it follows that $z \in 1 + \mathfrak{P}_{L_v}$ for all $v \in T$ $(\mathfrak{P}_{L_v} = \mathfrak{P}_v \cap L_v)$. Then $\sigma l \equiv l \pmod{\mathfrak{P}_{L_v}}$, so that, since L_v/K_v is unramified, $l = cd$, $c \in U_v$, $d \in 1 + \mathfrak{P}_{L_v}$. We have the equality of the reduced norm $\mathrm{Nrd}_{D/K}(l) = c^n\,\mathrm{Nrd}_{D/K}(d) = 1$, which, because $\mathrm{Nrd}_{D/K}(d) \in 1 + \mathfrak{p}_v$, implies that $c^n \equiv 1 \pmod{\mathfrak{p}_v}$. By Hensel's lemma, there exists an nth root of unity $f \in K_v$ such that $c \equiv f \pmod{\mathfrak{p}_v}$. Then

$$l = f(cf^{-1}d) \in U_v[D_v^{(1)}, D_v^{(1)}],$$

since $cf^{-1}d \in D_v^{(1)} \cap (1 + \mathfrak{P}_v) = [D_v^{(1)}, D_v^{(91)}]$ (see [20]). Now, since $z = a^{-1}l^{-1}al$ for some $a \in D^*$, it follows that $z \in [D^{(1)}, D^{(1)}]$ by Proposition 4.

Theorem 4 is proved.

§6. Proof of Theorems 2 and 3

THEOREM 2. *For a maximal cyclic subfield L of a division ring D the relation $D^{(1)} = L^{(1)}[D^{(1)}, D^{(1)}]$ holds if and only if the field L_v is a maximal unramified subfield of the division ring D_v for any $v \in T$. If the algebra $D_{\bar{2}}$ is not a division ring for the $\bar{2}$-adic field $K_{\bar{2}}$, where $\bar{2}/2$, or if D has odd index, then such a field L always exists.*

The proof consists of two parts. We first establish sufficiency of our conditions. We shall need

LEMMA 12. *The group $D^{(1)}$ is generated by the set of commutators*

$$\{[x, y] = x^{-1}y^{-1}xy \mid x \in U, y \in D^*\}.$$

PROOF. It is well known that $D^{(1)} = [D^*, D^*]$, i.e., this group is generated by all commutators $[x, y]$, $x, y \in D^*$ [1]. Thus, it suffices to show that an arbitrary commutator $[x, y]$ can be represented as a product of commutators of the required type. Let F be a maximal subfield in D such that $F_{\bar{v}}/K_v$ is a totally ramified extension of degree n for all $v \in T$. By the weak approximation theorem, there exists $\alpha \in F^*$ with the property that

$$v\big(\mathrm{Nrd}_{D/K}(x)\big) = v\big(\mathrm{Nrd}_{D/K}(\alpha)\big)$$

for $v \in T$. Then $x = \alpha x_0$, where $x_0 = \alpha^{-1}x \in U$. Analogously, $y = \beta y_0$, $\beta \in F^*$, $y_0 \in U$. A direct calculation shows that

$$[x, y] = [x_0, \alpha][x_0\alpha, y_0][\alpha y_0, x_0][x_0, y],$$

and this gives us the lemma.

Lemma 12 is proved.

From Lemma 12 it follows that it suffices to show that the group $L^{(1)}[D^{(1)}, D^{(1)}]$ contains the commutators $[x, y]$ for $x \in U$ and $y \in D^*$. This is achieved using an argument which is an exact repetition of the proof of Proposition 6 of [22].

Proof of necessity. Suppose that for some $v \in V_f$ the algebra D_v remains a division ring and the extension L_v/K_v is not unramified. We choose a maximal unramified subfield $M \subset D_v$, and we let $\alpha \in M^{(1)}$ be an element such that the residue $\bar{\alpha}$ is primitive for the corresponding residue field extension $\overline{M}/\overline{K}_v$. This element can be constructed as follows. Let $\bar{\beta}$ be a primitive element for $\overline{M}/\overline{K}_v$ with norm 1 [19]. By Hilbert's Theorem 90, $\bar{\beta} = \bar{\gamma}^{-1}\bar{\sigma}(\bar{\gamma})$, where $\bar{\gamma} \in \overline{M}^*$ and $\bar{\sigma}$ is a generator of $\mathrm{Gal}(\overline{M}/\overline{K}_v)$; then $\alpha = \gamma^{-1}\sigma(\gamma)$, where $\gamma \in M^*$ and $\sigma \in \mathrm{Gal}(M/K_v)$ naturally correspond to $\bar{\gamma}$ and $\bar{\sigma}$, is the desired element. We claim that $\alpha \notin L_v^{(1)}[D_v^{(1)}, D_v^{(1)}]$. In fact, if $\alpha = ld$, $l \in L_v^{(1)}$, $d \in [D_v^{(1)}, D_v^{(1)}]$, then, passing to residues modulo the maximal ideal P_v and taking into account that $[D_v^{(1)}, D_v^{(1)}] \subset 1 + \mathfrak{P}_v$, we find that $\bar{\alpha} = \bar{l}$. Because $\bar{\alpha}$ is primitive, this would imply equality of residue fields $\overline{L}_v = \overline{M}$, and the extension L_v/K_v would be unramified. Thus, $D_v^{(1)} \neq L_v^{(1)}[D_v^{(1)}, D_v^{(1)}]$. But then also $D^{(1)} \neq L^{(1)}[D^{(1)}, D^{(1)}]$, by the weak approximation theorem.

The remaining claim, that there exists a cyclic subfield L/K with the required properties, follows from Wang [10].

Theorem 2 is proved.

We note that in the second part of the proof we did not use cyclicity of L/K, and so necessity of the unramified condition holds for any maximal subfield.

Theorem 2 is also connected with the problem at the end of [19], which asked whether the group $D^{(1)}$ coincides with the normal subgroup $\langle L^{(1)} \rangle$ generated by $L^{(1)}$ for a suitable maximal subfield L. If we restrict ourselves to cyclic subfields, which is completely natural to do, then, as noted above, for certain division rings D we have $\langle L^{(1)} \rangle \neq D^{(1)}$ for any L.

THEOREM 3. *Suppose that the division ring D satisfies the conditions of Theorem* 1, *i.e., $v((n, 2)) = 0$ for $v \in T$, and L is a maximal cyclic subfield for which $D^{(1)} = L^{(1)}[D^{(1)}, D^{(1)}]$. Then*

$$L^{(1)} \cap \prod_{v \in T} \left[D_v^{(1)}, D_v^{(1)}\right] \subset [D^{(1)}, D^{(1)}].$$

PROOF. Let

$$x \in L^{(1)} \cap \prod_{v \in T} [D_v^{(1)}, D_v^{(1)}].$$

By Hilbert's Theorem 90, $x = \sigma(l)l^{-1}$ for some $l \in L^*$, where σ is a generator of the Galois group $\mathrm{Gal}(L/K)$. Here, since the extension L_v/K_v is unramified for all $v \in T$, we may suppose that $l \in U$, i.e., it belongs to the unit group $U(L_v)$. We consider the reduction $\bar{l}$ modulo the ideal $\mathfrak{P}_{L_v} \subset O_{L_v}$. Since $x \in [D_v^{(1)}, D_v^{(1)}]$, it follows that $x \equiv 1 \pmod{\mathfrak{P}_{L_v}}$, and hence $\bar{\sigma}(\bar{l}) = \bar{l}$, where $\bar{\sigma}$ is the automorphism of the residue field $\bar{L}_v$ which is induced by σ. Since $\bar{\sigma}$ generates $\mathrm{Gal}(\bar{L}_v/\bar{K}_v)$, we have $\bar{l} \in \bar{K}$, i.e. $l = ab$, $a \in U_v$, $b \in 1 + \mathfrak{P}_{L_v}$.

It remains to note that, by the Skolem-Noether theorem, $\sigma(l) = y^{-1}ly$ for some $y \in D^*$, and hence $x = \sigma(l)l^{-1} = l^{-1}y^{-1}ly \in [D^{(1)}, D^{(1)}]$, in accordance with Theorem 4.

Theorem 3 is proved.

BIBLIOGRAPHY

1. V. P. Platonov, *The Tannaka-Artin problem and reduced K-theory*, Izv. Akad. Nauk SSSR Ser. Mat. **40** (1976), 227–261; English transl. in Math. USSR Izv. **10** (1976).

2. ______, *Algebraic groups and reduced K-theory*, Proc. Internat. Congr. Math. (Helsinki, 1978), Vol. 1, Acad. Sci. Fenn., Helsinki, 1980, pp. 311–317.

3. Jacques Tits, *Groupes Whitehead de groupes algébriques simples sur un corps* (*d'après V. P. Platonov et al.*), Sém. Bourbaki 1976/77, Expose 505, Lecture Notes in Math., vol. 677, Springer-Verlag, 1978, pp. 218–236.

4. V. P Platonov, *Arithmetic and structural problems in linear algebraic groups*, Proc. Internat. Congr. Math. (Vancouver, 1974), Vol. 1, Canad. Math. Congr., Montréal, 1975, pp. 471–478; English transl. in Amer. Math. Soc. Transl. (2) **109** (1977).

5. ______, *The arithmetic theory of algebraic groups*, Uspekhi Mat. Nauk **37** (1982), no. 3 (225), 3–54; English transl. in Russian Math. Surveys **37** (1982).

6. G. A. Margulis, *Finiteness of quotient groups of discrete subgroups*, Funktsional. Anal. i Prilozhen. **13** (1979), no. 3, 28–39; English transl. in Functional Anal. Appl. **13** (1979).

7. Martin Kneser, *Orthogonale Gruppen über algebraischen Zahlkörpern*, J. Reine Angew. Math. **196** (1956), 213–220.

8. V. P. Platonov and A. S. Rapinchuk, *On the group of rational points of three-dimensional groups*, Dokl. Akad. Nauk SSSR **247** (1979), 279–282; English transl. in Soviet Math. Dokl. **20** (1979).

9. G. A. Margulis, *On the multiplicative group of a quaternion algebra over a global field*, Dokl. Akad. Nauk SSSR **252** (1980), 542–546; English transl. in Soviet Math. Dokl. **21** (1980).

10. Shianghao Wang, *On Grunwald's theorem*, Ann. of Math. (2) **51** (1950), 471–484.

11. Helmut Hasse, *Neue Begrundung und Verallgemeinerung der Theorie des Normrestsymbols*, J. Reine Angew. Math. **162** (1930), 134–144.

12. J. W. S. Cassels and A. Fröhlich, editors, *Algebraic number theory* (Proc. Instructional Conf., Brighton, 1965), Academic Press, London, and Thompson, Washington, D. C., 1967.

13. J. [Issai] Schur, *Über die Darstellung der symmetrischen und der alternierenden Gruppen durch gebrochene lineare Substitutionen*, J. Reine Angew. Math. **139** (1911), 155–250.

14. Serge Lang, *Algebraic numbers*, Addision-Wesley, 1964.

15. B. L. van der Waerden, *Algebra*. Vol. I, 8th ed., and Vol. II, 5th ed., Springer-Verlag, 1971, 1967; English transl. of 7th and 5th eds., Ungar, New York, 1970.

16. S. Gurak, *The Hasse norm principle in non-abelian extensions*, J. Reine Angew. Math. **303 / 304** (1978), 314–318.

17. J. [Issai] Schur, *Über die Darstellung der endlichen Gruppen durch gebrochenelineare Substitutionen*, J. Reine Angew. Math. **127** (1904), 20–50.

18. V. P. Platonov, *The problem of strong approximation and the Kneser-Tits conjecture for algebraic groups*, Izv. Akad. Nauk SSSR Ser. Mat. **33** (1969), 1211–1219; English transl. in Math. USSR Izv. **3** (1969).

19. V. P. Platonov and V. I. Yanchevskiĭ, *On Harder's conjecture*, Dokl. Akad. Nauk SSSR **221** (1975), 784–787; English transl. in Soviet Math. Dokl. **16** (1975).

20. Carl Riehm, *The norm* 1 *group of* $\mathfrak{p}$*-adic division algebra*, Amer. J. Math. **92** (1970), 499–523.

21. Shianghaw Wang, *On the commutator group of a simple algebra*, Amer. J. Math. **72** (1950), 323–334.

22. V. P. Platonov and A. S. Rapinchuk, *The multiplicative structure of division rings over number fields, and the Hasse principle*, Dokl. Akad. Nauk SSSR **266** (1982), 560–564; English transl. in Soviet Math. Dokl. **26** (1982).

Translated by N. KOBLITZ

Труды Матем. Инст. Стеклов.
165(1984)

Proceedings of the Steklov
Institute of Mathematics
1985, Issue 3

HOMOLOGY OF GL_n, CHARACTERISTIC CLASSES AND MILNOR'S K-THEORY

UDC 514.7

A. A. SUSLIN

ABSTRACT. Several interconnected problems of algebraic K-theory are investigated: a description of the image of a characteristic class $c_{n,n}$, a connection between the Milnor and Quillen K-groups, and stability properties in homology of the group GL_n. The main result asserts that for an infinite field F the embeddings $\mathrm{GL}_n(F) \hookrightarrow \mathrm{GL}(F)$ induce isomorphisms $H_n(\mathrm{GL}_n(F)) \xrightarrow{\sim} H_n(\mathrm{GL}(F))$, and the homological product

$$F^* \times \cdots \times F^* = H_1(\mathrm{GL}_1(F)) \times \cdots \times H_1(\mathrm{GL}_1(F)) \to H_n(\mathrm{GL}_n(F))$$

induces an isomorphism $K_n^M(F) \xrightarrow{\sim} H_n(\mathrm{GL}_n(F))/H_n(\mathrm{GL}_{n-1}(F))$ for Milnor's K-groups K_n^M. It follows from the theorem that there is a homomorphism $\varphi\colon K_n(F) \to K_n^M(F)$ whose composition with the natural mapping $K_n^M(F) \to K_n(F)$ coincides with the image of the class. Moreover, it is shown that $\varphi(K_n(F)) \supset (n-1)!K_n^M)(F)$.

Bibliography: 21 titles.

The main purpose of this paper is to study the homomorphism $c_{n,n}\colon K_n(F) \to H^0(\operatorname{Spec} F, \mathscr{K}_n) = K_n(F)$ occuring in the theory of characteristic classes for higher K-groups. It is easy to see that $c_{1,1}$ is the identity map. It follows from this and the product formula that the restriction of $c_{n,n}$ to the symbolic part of $K_n(F)$ is the multiplication by $(-1)^{n-1}(n-1)!$. This fact allowed mathematicians to hope that $c_{n,n}$ itself coincided (at least up to torsion) with multiplication by $(-1)^{n-1}(n-1)!$. However, recently Soulé, using computations of K-groups of number fields, showed that this is not the case, and he conjectured that in the general case the image of $c_{n,n}$ is (up to factorials) the symbolic part of $K_n(F)$. We prove below (Theorem 5.2) that this is really the case. The main result of the paper (Theorem 3.4) states that for an infinite field F the embedding $\mathrm{GL}_n(F) \hookrightarrow \mathrm{GL}(F)$ of groups induces an isomorphism $H_n(\mathrm{GL}_n(F)) \xrightarrow{\sim} H_n(\mathrm{GL}(F))$ and the homological product

$$\underbrace{F^* \times F^* \times \cdots \times F^*}_{n} = H_1(\mathrm{GL}_1(F)) \times \cdots \times H_1(\mathrm{GL}_1(F)) \to H_n(\mathrm{GL}_n(F))$$

induces an isomorphism $K_n^M(F) \xrightarrow{\sim} H_n(\mathrm{GL}_n(F))/H_n(\mathrm{GL}_{n-1}(F))$ (here $K_n^M(F)$ are Milnor's K-groups of F; see [1] and [6]). This theorem gives rise to a natural

---

1980 *Mathematics Subject Classification*. Primary 18F25.

0081-5438/85 $1.00 + $.25 per page

homomorphism

$$\varphi\colon K_n(F) = \pi_n(\mathrm{BGL}(F)^+) \to H_n(\mathrm{BGL}(F)^+) = H_n(\mathrm{GL}(F))$$
$$= H_n(\mathrm{GL}_n(F)) \to H_n(\mathrm{GL}_n(F))/H_n(\mathrm{GL}_{n-1}(F)) = K_n^M(F),$$

and we prove in §5 that $c_{n,n}$ coincides with the composition $K_n(F) \xrightarrow{\varphi} K_n^M(F) \to K_n(F)$, where the second homomorphism is induced by the product in K-theory. We prove also that, if $l(a) \in K_1(F)$ denotes the element corresponding to $a \in F^*$, then

$$\varphi(l(a_1)\cdots l(a_n)) = (-1)^{n-1}(n-1)!\{a_1,\ldots,a_n\}$$

and in particular

$$\varphi(K_n(F)) \supset (n-1)!K_n^M(F).$$

The homomorphism φ is also of interest from other points of view. First of all the previous formula shows that the composition $K_n^M(F) \to K_n(F) \to K_n^M(F)$ coincides with multiplication by $(-1)^{n-1}(n-1)!$, and hence the kernel of the natural homomorphism $K_n^M(F) \to K_n(F)$ is anihilated by $(n-1)!$ (cf. [18]). Moreover, we prove that the image of $\varphi\colon K_3(F) \to K_3^M(F)$ coincides with the kernel of Milnor's homomorphism $K_3^M(F) \to I^3(F)/I^4(F)$ (see [6]). Thus, Milnor's conjecture that $K_3^M(F)/2K_3^m(F) \xrightarrow{\sim} I^3(F)/I^4(F)$ turns out to be equivalent to $\varphi(K_3(F)) = 2K_3^M(F)$. It is likely that $\varphi(K_n(F)) = (n-1)!K_n^M(F)$ in general.

One of the useful technical results proved in §1 is that for any infinite field F, the natural group embedding

$$\mathrm{GL}_n(F) \hookrightarrow \begin{Vmatrix} 1 & * \\ 0 & \mathrm{GL}_n(F) \end{Vmatrix}$$

is a homology isomorphism. The statement is trivial and well known for the fields of characteristic zero, but requires more delicate arguments in the case of positive characteristic.

After the preprint [19] of this paper appeared, the author learned that the last result was independently obtained by A. J. Berrick. The statement about the image of $c_{n,n}$ in the case of algebraically closed fields was proved by Soule [20], independently from the author, using results of [18].

Everywhere in this paper, unless we specify otherwise, the word "field" means an infinite field.

§1. Homology of affine groups

LEMMA 1.1. *Suppose that $\varphi_1,\ldots,\varphi_n\colon F \hookrightarrow \Omega$ are field embeddings, and $\prod_1^n\varphi_i(x) = 1$ for every $x \in F^*$. Then F is a finite field of p^m elements with $n \geqslant (p-1)m$.*

PROOF. Suppose first that F is finite and consists of p^m elements. In this case we can assume that $\Omega = F$ and $\varphi_i = \sigma^{t_i}$, where σ is the Frobenius automorphism ($\sigma(x) = x^p$). For any $x \in F^*$ we have

$$x^{\Sigma p^{t_i}} = \prod x^{p^{t_i}} = \prod \varphi_i(x) = 1;$$

hence $\Sigma_1^n p^{t_i} \equiv 0 \pmod{p^m - 1}$. Denote by n_j the number of t_i congruent to j modulo m. Then $n = \Sigma_0^{m-1} n_j$ and $\Sigma_0^{m-1} n_j p^j \equiv 0 \pmod{p^m - 1}$.

(1.1.1) SUBLEMMA. *If n_j $(0 \leqslant j \leqslant m-1)$ are nonnegative integers, not all of which are zero, such that*

$$\sum_0^{m-1} n_j p^j \equiv 0 \pmod{p^m - 1}, \quad \text{then } \sum_0^{m-1} n_j \geqslant (p-1)m.$$

PROOF. If $n_j \geqslant p$ for some j, then replacing n_j by $n_j - p$ and n_{j+1} by $n_{j+1}+1$ we obtain a set of n_j's with the same properties but with a smaller sum. Therefore we can assume that $n_j < p$ for all j. In this case

$$0 < \sum n_j p^j \leqslant (p-1)(1 + p + \cdots + p^{m-1}) = p^m - 1;$$

hence $n_j = p-1$ for all j and $\Sigma_0^{m-1} n_j = (p-1)m$.

(1.1.2) The above arguments prove that if F is finite and consists of p^m elements, then $n \geqslant (p-1)m$. This shows that F cannot be an infinite algebraic extension of a finite field. It is clear also that F cannot be a field of characteristic zero. Suppose finally that F is transcendental over its prime subfield F_p. Let $t \in F$ be transcendental over F_p and $t_i = \varphi_i(t)$. For any nonzero polynomial $f \in F_p[T]$ we have $\prod_1^n f(t_i) = \prod_1^n \varphi_i(f(t)) = 1$. Thus, $(t_1, \ldots, t_n)$ is an Ω-valued point of the variety $V \subset A^n_{F_p}$ given by the system of equations $\prod_1^n f(x_i) = 1$ for any nonzero $f \in F_p[T]$.

(1.1.3) SUBLEMMA. *The variety V is empty.*

Otherwise V would contain an algebraic point $x = (x_1, \ldots, x_n)$ and, choosing $f \in F_p[T]$ such that $f(x_1) = 0$, we obtain the contradiction $0 = \prod_1^n f(x_i) = 1$.

COROLLARY 1.2. *Consider the natural diagonal action of the group F^* on $F^{\otimes n}$ (the tensor power is taken over $\mathbf{Z}$ or the prime subfield, which makes no difference). Then $H_0(F^*, F^{\otimes n}) = 0$.*

PROOF. $H_0(F^*, F^{\otimes n}) = F^{\otimes n}/I$, where I is the ideal generated by elements of the form $f \otimes \cdots \otimes f - 1$ $(f \in F^*)$. According to Lemma 1.1, I is contained in no maximal ideal; hence $I = F^{\otimes n}$.

Let V_n denote the subring of symmetric tensors in $F^{\otimes n}$:

$$V_n = (F^{\otimes n})^{S_n}.$$

PROPOSITION 1.3. $H_0(F^*, V_n) = 0$.

PROOF. If $k_1, \ldots, k_r$ are natural numbers and $\Sigma_1^r k_i = n$, let $V_n^{k_1, \ldots, k_r}$ denote the sugroup of V_n generated by all tensors of the form

$$\sum_{\sigma \in S_n / S_{k_1} \times \cdots \times S_{k_r}} (\overbrace{f_1 \otimes \cdots \otimes f_1}^{k_1} \otimes \cdots \otimes \overbrace{f_r \otimes \cdots \otimes f_r}^{k_r})^{\sigma}.$$

Evidently, the subgroups $V_n^{k_1, \ldots, k_r}$ are F^*-invariant and $V_n = \Sigma V_n^{k_1, \ldots, k_r}$. Set

$$V_n^{(j)} = \sum_{\Sigma_j (k_j - 1) \leqslant i} V_n^{k_1, \ldots, k_r}.$$

Then the $V_n^{(j)}$ define an F^*-invariant filtration on V_n. It suffices to check that $H_0(F^*, V_n^{(j)}/V_n^{(j-1)}) = 0$. It is easy to see that when $\Sigma_j (k_j - 1) = i$ the mapping $F^r \to V_n^{(i)}/V_n^{(i-1)}$ given by

$$f_1 \times \cdots \times f_r \to \sum_{\sigma \in S_n / S_{k_1} \times \cdots \times S_{k_r}} (f_1 \otimes \cdots \otimes f_1 \otimes \cdots \otimes f_r \otimes \cdots \otimes f_r)^{\sigma}$$

is multilinear and, hence, defines an F^*-equivariant homomorphism $F^{\otimes r} \to V_n^{(i)}/V_n^{(i-1)}$. Thus, we obtain a surjective F^*-equivariant homomorphism

$$\coprod_{k_1, \ldots, k_r : \Sigma(k_j - 1) = i} F^{\otimes r} \to V_n^{(i)}/V_n^{(i-1)},$$

and our statement follows from Corollary 1.2.

LEMMA 1.4. *Let $n_1, \ldots, n_r$ be natural numbers and for each i let $T^{n_i}(F)$ denote either the n_ith symmetric or the n_ith external power of F. Set $n = \Sigma n_i$ and consider the natural projection $F^{\otimes n} \to T^{n_1}(F) \otimes \cdots \otimes T^{n_r}(F)$. The kernel K of this projection is a V_n-submodule of $F^{\otimes n}$, and hence $T^{n_1}(F) \otimes \cdots \otimes T^{n_r}(F)$ has a natural V_n-module structure.*

PROOF. Denote by K_i the kernel of the homomorphism $F^{\otimes n_i} \to T^{n_i}(F)$. Then

$$K = \sum_{i=1}^{r} F^{\otimes n_1} \otimes \cdots \otimes K_i \otimes \cdots \otimes F^{\otimes n_r}.$$

It suffices to check that every summand is a V_n-submodule of $F^{\otimes n}$. We restrict ourselves to the case $i = 1$. If T^{n_1} is the symmetric power, then $K_1 \otimes \cdots \otimes F^{\otimes n_r}$ is generated by the tensors of the form $v^{(i,j)} - v$, where $1 \leqslant i < j \leqslant n_1$, and for each $w \in V_n$ we have $w(v^{(i,j)} - v) = (uw)^{(i,j)} - uw \in K_1 \otimes \cdots \otimes F^{\otimes n_r}$. If $T^{n_1}(F)$ is the exterior power, then $K_1 \otimes \cdots \otimes F^{\otimes n_r}$ is generated by tensors v such that $v^{(i,j)} = v$ for some $1 \leqslant i < j \leqslant n_1$, and for any $w \in V_n$ we have $(wv)^{(i,j)} = wv$, so $wv \in K_1 \otimes \cdots \otimes F^{\otimes n_r}$.

LEMMA 1.5. *Suppose that A is a commutative ring, G an abelian group, and φ: $G \to A^*$ a homomorphism. Let M be an A-module. Consider the induced G-module structure on M. If $H_0(G, A) = 0$, then $H_p(G, M) = 0$ for all $p \geqslant 0$.*

PROOF. $H_0(G, A) = A/I$, where I is the ideal of A generated by elements $\varphi(g) - 1$ $(g \in G)$. Since M is an A-module and the action of G on M is A-linear, the groups $H_p(G, M)$ have natural A-module structures. Since the group G is abelian, it is well known that for any $g \in G$ the homomorphism g: $M \to M$ induces the identity maps on the homology groups. Thus, A-modules $H_p(G, M)$ are annihilated by I and hence are 0.

COROLLARY 1.6. *Under the conditions of Lemma* 1.4,

$$H_p\big(F^*, T^{n_1}(F) \otimes \cdots \otimes T^{n_r}(F)\big) = 0.$$

PROPOSITION 1.7. *Suppose that F is a field of characteristic $p > 0$ and V a vector space over F. Then $H_i(F^*, H_j(V, F_p)) = 0$ for any i and any $j > 0$.*

PROOF. It is well known that for any vector space V over F_p the graded abelian group $H_*(V, F_p)$ is naturally isomorphic to $S(V) \otimes \Lambda(V)$ when $p \neq 2$ (here generators of the symmetric algebra have degree two and generators of the exterior algebra have degree one) and to $S(V)$ when $p = 2$ (here generators have degrees one). So $H_j(V, F_p)$ is isomorphic as F^*-module to a direct sum of modules of the kind considered in the previous corollary.

COROLLARY 1.8. *Let F be an arbitrary field, V vector space over F, and k a prime field. Then $H_i(F^*, H_j(V,k)) = 0$ for any i and any $j > 0$.*

PROOF. If $\operatorname{char} k \neq \operatorname{char} F$, then $H_j(V,k) = 0$. If $\operatorname{char} k = \operatorname{char} F > 0$, then our statement was proved above. If $\operatorname{char} k = \operatorname{char} F = 0$, then $H_j(V,k) = \Lambda^j(V)$. For any natural number n the action of $n \in F^*$ on $H_j(V,k)$ coincides with multiplication by n^j. Since F^* is abelian, we conclude that $H_i(F^*, H_j(V,k))$ is annihilated by $n^j - 1$ and hence is 0 (because it is a vector space over $\mathbf{Q}$).

THEOREM 1.9. *Let $G_1 \subset GL_n(F)$ and $G_2 \subset GL_m(F)$ be two groups such that at least one of them contains the group of diagonal matrices. Let $M \subset M_{n,m}(F)$ be a vector subspace such that $G_1M = M = MG_2$. Then the natural embedding*

$$\lambda: \begin{Vmatrix} G_1 & 0 \\ 0 & G_2 \end{Vmatrix} \hookrightarrow \begin{Vmatrix} G_1 & M \\ 0 & G_2 \end{Vmatrix}$$

is a homology isomorphism.

PROOF. It suffices to check that λ induces an isomorphism on the homology groups with coefficients from any prime field. In view of the Hochschild-Serre spectral sequence it suffices to check that $H_i(G_1 \times G_2, H_j(M,k)) = 0$ for $j > 0$. Suppose, for example, that $G_1 \supset F^*$. Then our statement follows from Corollary 1.8 and the spectral sequence

$$H_p\big((G_1/F^*) \times G_2,\ H_q\big(F^*, H_j(M,k)\big)\big) \Rightarrow H_{p+q}\big(G_1 \times G_2, H_j(M,k)\big).$$

§2. Spaces of unimodular vectors

If F is a field and V and W are vector spaces over F, then $\operatorname{Um}(V,W)$ denotes the simplicial set whose nondegenerate p-simplices are sequences

$$\left(\begin{pmatrix} v_0 \\ w_0 \end{pmatrix}, \dots, \begin{pmatrix} v_p \\ w_p \end{pmatrix}\right),$$

where $v_i \in V$, $w_i \in W$ and $v_0, \dots, v_p$ are linearly independent, the ith face of such a simplex being obtained by omitting $\binom{v_i}{w_i}$. The following statement is a particular case of van der Kallen's acyclicity theorem [3]; the proof given below is taken from [3].

LEMMA 2.1. *The space* $\operatorname{Um}(V,W)$ *is* $(\dim V - 2)$*-acyclic.*

We will need a somewhat more precise statement. In the same notation as above denote by $\operatorname{GP}(V,W)$ the simplicial set whose nondegenerate p-simplices are sequences

$$\left(\begin{pmatrix} v_0 \\ w_0 \end{pmatrix}, \dots, \begin{pmatrix} v_p \\ w_p \end{pmatrix}\right)$$

such that the vectors $v_0, \dots, v_p$ are in general position (i.e. any n of them are linearly independent for $n \leqslant \dim V$). Evidently, $\operatorname{Um}(V,W)$ is the $(\dim V - 1)$-skeleton of $\operatorname{GP}(V,W)$, and (2.1) follows from the following lemma:

LEMMA 2.2. *The space* $\operatorname{GP}(V,W)$ *is acyclic.*

PROOF. Suppose that

$$x = \sum_i n_i \left(\begin{pmatrix} v_0^i \\ w_0^i \end{pmatrix}, \dots, \begin{pmatrix} v_p^i \\ w_p^i \end{pmatrix}\right)$$

is a p-cyclic. Since the field F is infinite, we can find a vector $v \in V$ such that v, $v_0^i, \ldots, v_p^i$ are in general position for every i. Choose any w in W and set

$$y = \sum_i n_i \left(\begin{pmatrix} v \\ w \end{pmatrix}, \begin{pmatrix} v_0^i \\ w_0^i \end{pmatrix}, \ldots, \begin{pmatrix} v_p^i \\ w_p^i \end{pmatrix} \right).$$

Then $dy = x$.

We will denote by $C_*(V, W)$ the augmented simplicial complex of $\operatorname{Um}(V, W)$ with dimensions shifted by one, i.e. $C_0(V, W) = \mathbf{Z}$ and $C_i(V, W) = C_{i-1}(\operatorname{Um}(V, W))$, the free abelian group with basis

$$\left(\begin{pmatrix} v_1 \\ w_1 \end{pmatrix}, \ldots, \begin{pmatrix} v_i \\ w_i \end{pmatrix} \right),$$

where $v_1, \ldots, v_i$ are linearly independent. According to Lemma 2.1, all the homology groups of $C_*(V, W)$ are zero except for H_n, where $n = \dim V$. This group $H_n(C_*(V, W))$ will be denoted by $H(V, W)$. Let $\operatorname{Aff}(V, W)$ denote the affine group

$$\left\| \begin{matrix} \operatorname{Aut}(V) & 0 \\ \operatorname{Hom}(V, W) & 1 \end{matrix} \right\|.$$

There is a natural action of $\operatorname{Aff}(V, W)$ on the complex $C_*(V, W)$. This action induces an action of $\operatorname{Aff}(V, W)$ on $H(V, W)$. We will denote the group

$$H_0(\operatorname{Aff}(V, W), H(V, W)) = H(V, W)_{\operatorname{Aff}(V, W)}$$

by $S(V, W)$. In the case when the space W is zero we will omit it in the notation. We will also use the following abbreviated notation:

$$C_*^{n,m} = C_*(F^n, F^m), \quad C_*^n(F) = C_*(F^n),$$
$$H_{n,m}(F) = H(F^n, F^m), \quad A_{n,m}(F) = \operatorname{Aff}(F^n, F^m).$$

2.3. As the space $\operatorname{GP}(F^n)$ is acyclic, we have an exact sequence

$$C_{n+1}(\operatorname{GP}(F^n)) \xrightarrow{d} C_n(\operatorname{GP}(F^n)) \xrightarrow{d} H_n(F) \to 0.$$

Denote by $e_1, \ldots, e_n$ the standard basis of F^n. If $a_i \in F^*$, then the vectors $e_1, \ldots, e_n, \Sigma_1^n a_i e_i$ are in general position, and by $\langle a_1, \ldots, a_n \rangle$ we will denote the image of $d(e_1, \ldots, e_n, \Sigma_1^n a_i e_i)$ in the group $S_n(F)$.

(2.3.1) Proposition. *The group $S_n(F)$ is generated by all $\langle a_1, \ldots, a_n \rangle$ with the following defining relations for the generators: if $\lambda_1, \ldots, \lambda_n \in F^*$ are all distinct, then*

$$\begin{aligned} &\langle \lambda_1 a_1, \ldots, \lambda_n a_n \rangle - \langle a_1, \ldots, a_n \rangle \\ &\quad = \sum_{i=1}^n (-1)^{i+n} \langle a_1(\lambda_1 - \lambda_i), \ldots, \widehat{a_i(\lambda_i - \lambda_i)}, \ldots, a_n(\lambda_n - \lambda_i), \lambda_i \rangle. \end{aligned}$$

Proof. From the exact sequence (2.3), factoring out the action of $\operatorname{GL}_n(F)$, we obtain

$$0 \leftarrow S_n(F) \xleftarrow{\bar{d}} C_n(\operatorname{GP}(F^n))_{\operatorname{GL}_n(F)} \xleftarrow{\bar{d}} C_{n+1}(\operatorname{GP}(F^n))_{\operatorname{GL}_n(F)}.$$

The action of $\operatorname{GL}_n(F)$ on $C_n(\operatorname{GP}(F^n))$ and $C_{n+1}(\operatorname{GP}(F^n))$ permutes the basis elements, and each element is equivalent under this action to a unique element of the form $(e_1, \ldots, e_n, a)$ (respectively, $(e_1, \ldots, e_n, a, b)$), where the corresponding vectors are in general position. The last condition is evidently equivalent to the following one: $a = \sum_1^n a_i e_i$ with $a_i \in F^*$ (respectively $a = \sum_1^n a_i e_i$ and $b = \sum_1^n \lambda_i a_i e_i$ with a_i, $\lambda_i \in F^*$ and $\lambda_i \neq \lambda_j$ for $i \neq j$). Thus, the above exact sequence takes the form

$$0 \leftarrow S_n(F) \overset{\bar{d}}{\leftarrow} \coprod_{(a_1,\dots,a_n),a_i \in F^*} Z \leftarrow \coprod_{\substack{(a_1,\dots,a_n),a_i \in F^* \\ (\lambda_1,\dots,\lambda_n),\lambda_i \in F^*, \lambda_i \neq \lambda_j \text{ if } i \neq j}} Z.$$

Denote the generator corresponding to (a) (to (a, λ)) by e_a (by $e_{a,\lambda}$). Then $\bar{d}(e_a) = \langle a \rangle$ and

$$\begin{aligned} \bar{d}(e_a) = \langle a \rangle,\ \bar{d}(e_{a,\lambda}) &= d(e_1, \dots, e_n, a, a\lambda) \mod GL_n(F) \\ &= \sum_{i=1}^{n} (-1)^i (e_1, \dots, \hat{e}_i, \dots, e_n, a, a\lambda + (-1)^n (e_1, \dots, e_n, a\lambda) \\ &\quad + (-1)^{n+1} (e_1, \dots, e_n, a) (\mod GL_n(F)). \end{aligned}$$

We have the following equivalences modulo the action of $GL_n(F)$:

$$((e_1, \dots, \hat{e}_i, \dots, e_n, a, a\lambda)) \sim \left(e_1, \dots, e_{n-1}, \begin{pmatrix} a_1 \\ \vdots \\ a_{i-1} \\ a_{i+1} \\ \vdots \\ a_n \\ a_i \end{pmatrix}, \begin{pmatrix} \lambda_1 a_1 \\ \vdots \\ \lambda_{i-1} a_{i-1} \\ \lambda_{i+1} a_{i+1} \\ \vdots \\ \lambda_n a_n \\ \lambda_i a_i \end{pmatrix} \right)$$

$$\sim \left(e_1, \dots, e_n, \begin{pmatrix} a_1(\lambda_1 - \lambda_i) \\ \cdots\ \cdots. \\ a_{i-1}(\lambda_{i-1} - \lambda_i) \\ a_{i+1}(\lambda_{i+1} - \lambda_i) \\ \cdots\ \cdots. \\ a_n(\lambda_n - \lambda_i) \\ \lambda_i \end{pmatrix} \right),$$

and hence

$$\bar{d}(e_a, \lambda) = \sum_{i=1}^{n} (-1)^i e_{(a_1(\lambda_1 - \lambda_i), \dots, \widehat{a_i(\lambda_i - \lambda_i)}, \dots, a_n(\lambda_n - \lambda_i), \lambda_i)} + (-1)^n e_{a\lambda} + (-1)^{n+1} e_a.$$

PROPOSITION 2.4. *Suppose that a_i, $\lambda_i \in F^*$ and $\lambda_i \neq \lambda_j$ if $i \neq j$. Then the following equality holds in $K_n^M(F)$:*

$$\begin{aligned} &\{\lambda_1 a_1, \dots, \lambda_n a_n\} - \{a_1, \dots, a_n\} \\ &\quad = \sum_{i=1}^{n} (-1)^{i+n} \{a_1(\lambda_1 - \lambda_i), \dots, \widehat{a_i(\lambda_i - \lambda_i)}, \dots, a_n(\lambda_n - \lambda_i), \lambda_i\}. \end{aligned}$$

PROOF. We use induction on n. When $n = 1$ everything is evident. In the general case we have

$$\sum_{i=1}^{n-1} (-1)^{i+n}\left\{a_1(\lambda_1-\lambda_i),\ldots,\widehat{a_i(\lambda_i-\lambda_i)},\ldots,a_n(\lambda_n-\lambda_i),\lambda_i\right\}$$

$$= \sum_{i=1}^{n-1} (-1)^{i+n}\left[\{a_1(\lambda_1-\lambda_i),\ldots,a_n\lambda_n,\lambda_i\} + \left\{a_1(\lambda_1-\lambda_i),\ldots,1-\frac{\lambda_i}{\lambda_n},\lambda_i\right\}\right]$$

$$= \sum_{i=1}^{n-1} (-1)^{i+n-1}\left\{a_1(\lambda_1-\lambda_i),\ldots,\widehat{a_i(\lambda_i-\lambda_i)},\ldots,a_{n-1}(\lambda_{n-1}-\lambda_i),\lambda_i,a_n\lambda_n\right\}$$

$$+ \sum_{i=1}^{n-1} (-1)^{i+n}\left\{a_1(\lambda_1-\lambda_i),\ldots,\widehat{a_i(\lambda_i-\lambda_i)},\ldots,a_{n-1}(\lambda_{n-1}-\lambda_n),\lambda_i-\lambda_n,\lambda_n\right\}.$$

Applying the induction hypothesis to $a_1,\ldots,a_{n-1}$, $\lambda_1,\ldots,\lambda_{n-1}$ and also to $a_1,\ldots,a_{n-1}$, $\lambda_1-\lambda_n,\ldots,\lambda_{n-1}-\lambda_n$, we see that the last expression is equal to

$$\{a_1\lambda_1,\ldots,a_{n-1}\lambda_{n-1},a_n\lambda_n\} - \{a_1,\ldots,a_{n-1},a_n\lambda_n\}$$
$$-\{a_1(\lambda_1-\lambda_n),\ldots,a_{n-1}(\lambda_{n-1}-\lambda_n),\lambda_n\} + \{a_1,\ldots,a_{n-1},\lambda_n\}$$
$$= \{a_1\lambda_1,\ldots,a_n\lambda_n\} - \{a_1,\ldots,a_n\}$$
$$-\{a_1(\lambda_1-\lambda_n),\ldots,a_{n-1}(\lambda_{n-1}-\lambda_n),\lambda_n\}.$$

(2.4.1) COROLLARY. *There is a unique homomorphism $S_n(F)\to K_n^M(F)$ taking $\langle a_1,\ldots,a_n\rangle$ to $\{a_1,\ldots,a_n\}$.*

2.5. *The pairings.* Let V be a vector space over F. We will say that $x = (v_0,\ldots,v_{p-1}) \in C_{p-1}(\mathrm{GP}(V))$ and $y = (w_0,\ldots,w_{q-1}) \in C_{q-1}(\mathrm{GP}(V))$ are in *general position* if the vectors $v_0,\ldots,v_{p-1}$, $w_0,\ldots,w_{q-1}$ are in general position. If this is the case, we define $xy \in C_{p+q-1}(\mathrm{GP}(V))$ to be the simplex $(v_0,\ldots,v_{p-1},w_0,\ldots,w_{q-1})$. We extend this definition by linearity to arbitrary chains (two chains are in general position if every simplex of the first chain is in general position with every simplex of the second chain). We take C_{-1} to be $\mathbf{Z}$ and extend the previous definition in the obvious way to (–1)-dimensional chains (a(–1)-dimensional chain is in general position with each chain). It is easy to see that if x and y are in general position then dx and y as well as x and dy are in general position, and $d(xy) = dx\cdot y + (-1)^{\deg x+1}x\cdot dy$ (compare with our proof of Lemma 2.2).

(2.5.1) Suppose that $V = V_1\oplus V_2$. We have evidently embeddings of complexes $C_*(V_1,V_2)\hookrightarrow C_*(V)\hookleftarrow C_*(V_2)$. Moreover, if $x\in C_p(V_1,V_2) = C_{p+1}(\mathrm{Um}(V_1,V_2))$ and $y\in C_q(V_2) = C_{q+1}(\mathrm{Um}(V_2))$, then x and y are in general position and $xy\in C_{p+q}(V)$. Thus, we obtain a pairing of complexes $C_*(V_1,V_2)\otimes C_*(V_2)\to C_*(V)$ and the induced pairing $H(V_1,V_2)\otimes H(V_2)\to H(V)$ of homology groups. To distinguish these everywhere defined pairings from the partially defined ones above, we will use the sign $\times$ to denote the latter pairings. If $x\in C_p(V_1,V_2)$, $y\in C_q(V_2)$, and $\alpha\in\mathrm{Aff}(V_1,V_2)$, then $\alpha x\times y = \alpha(x\times y)$ and hence a pairing $S(V_1,V_2)\otimes H(V_2)\to S(V)$ arises. Finally, if $V = V_1'\oplus V_2'$ and $\dim V_1' = \dim V_1, \dim V_2' = \dim V_2$, then images of the pairings $S(V_1,V_2)\otimes H(V_2)\to S(V)$ and $S(V_1',V_2')\otimes H(V_2')\to S(V)$ evidently coincide.

(2.5.2) In the above notation we have pairings

$$C_*(V_1)\otimes C_*(V_2)\to C_*(V),\quad H(V_1)\otimes H(V_2)\to H(V).$$

If $x \in C_p(V_1)$, $y \in C_q(V_2)$, $\alpha \in \mathrm{Aut}(V_1)$ and $\beta \in \mathrm{Aut}(V_2)$, then $\alpha x \times \beta y = \mathrm{diag}(\alpha, \beta) \times (x \times y)$, and hence we get the induced pairings $S(V_1) \otimes S(V_2) \to S(V)$. These pairings are associative in the obvious sense, so

$$H_*(F) = \coprod H_n(F) = \coprod H(F^n),\ S_*(F) = \coprod S_n(F) = \coprod S(F^n)$$

are naturally graded associative rings.

(2.5.3) LEMMA. *For any* $a_1, \ldots, a_p \in F^*$

$$\langle a_1 \rangle \times \cdots \times \langle a_p \rangle = \sum_{1 \leqslant i_1 < \cdots < i_k \leqslant p} (-1)^k \langle a_1, \ldots, \underset{i_1}{1}, \ldots, \underset{i_k}{1}, \ldots, a_p \rangle.$$

PROOF. $\langle a \rangle$ is the class modulo the action of $\mathrm{GL}_1(F)$ of the cycle $(a) - (1)$, and hence $\langle a_1 \rangle \times \cdots \times \langle a_p \rangle$ is the class modulo the action of $\mathrm{GL}_p(F)$ of the cycle

$$((a_1) - (1)) \times \cdots \times ((a_p) - (1))$$
$$= \sum_{1 \leqslant i_1 < \cdots < i_k \leqslant p} (-1)^{p-k} (e_1, \ldots, a_{i_1} e_{i_1}, \ldots, a_{i_k} e_{i_k}, \ldots, e_p)$$
$$= d\left(\sum_{1 \leqslant i_1 < \cdots < i_k \leqslant p} (-1)^k (e_1, \ldots, a_{i_1} e_{i_1}, \ldots, a_{i_k} e_{i_k}, \ldots, e_p, a_1 e_1 + \cdots + a_p e_p \right),$$

but the last cycle is evidently congruent modulo the action of $\mathrm{GL}_p(F)$ to the cycle

$$d\left(\sum_{1 \leqslant i_1 < \cdots < i_k \leqslant p} (-1)^k (e_1, \ldots, e_p, a_1 e_1 + \cdots + e_{i_1} + \cdots + e_{i_k} + \cdots + a_p e_p \right).$$

2.6. The long exact sequence

$$0 \leftarrow \mathbf{Z} = C_0(F^n) \leftarrow C_1(F^n) \leftarrow \cdots \leftarrow C_n(F^n) \leftarrow H_n(F) \leftarrow 0$$

of $\mathrm{GL}_n(F)$-modules gives the iterated connecting homomorphism

$$\delta_n \colon H_n(\mathrm{GL}_n(F), \mathbf{Z}) \to H_0(\mathrm{GL}_n(F), H_n(F)) = S_n(F).$$

(2.6.1) LEMMA. a) *The restriction of* δ_n *to* $H_n(\mathrm{GL}_{n-1}(F))$ *is zero.*

b) *For any n and m the following diagram commutes*:

$$\begin{array}{ccc} H_n(\mathrm{GL}_n(F)) \times H_m(\mathrm{GL}_m(F)) & \to & H_{n+m}(\mathrm{GL}_{n+m}(F)) \\ \downarrow & & \downarrow \\ S_n(F) \times S_m(F) & \to & S_{n+m}(F). \end{array}$$

PROOF. Part a) follows from the fact that the homomorphism $C_1(F^n) \to C_0(F^n)$ has $\mathrm{GL}_{n-1}(F)$-linear splitting given by $r \to r \cdot e_n$. Part b) follows immediately from comparison of the definitions of the operations on $H_n(\mathrm{GL}_n(f))$ and $S_n(F)$.

2.7. For any n we have the following natural mapping, induced by homological product:

$$\theta \colon F^* \times \cdots \times F^* = H_1(\mathrm{GL}_1(F)) \times \cdots \times H_1(\mathrm{GL}_1(F))$$
$$\to H_n(\mathrm{GL}_n(F)) \to H_n(\mathrm{GL}_n(F))/H_n(\mathrm{GL}_{n-1}(F)).$$

(2.7.1) LEMMA. *The mapping* θ *is multilinear. Moreover, if* $a_i + a_j = 1$ *for some* $i \neq j$, *then* $\theta(a_1, \ldots, a_n) = 0$.

PROOF. The first statement is evident. To prove the second, it suffices to check that $a \times (1 - a) \in H_2(\mathrm{GL}_1(F))$ for any $a \neq 0, 1$. If α and β are commuting elements of a group G, we will temporarily denote by $(\alpha, \beta) \in H_2(G, \mathbf{Z})$ the homology class of the 2-dimensional cycle $[\alpha, \beta] - [\beta, \alpha]$. With this notation

$$a \times (1-a) = (\operatorname{diag}(a,1), \operatorname{diag}(1,1-a))$$
$$\equiv (\operatorname{diag}(a,a^{-1}), \operatorname{diag}(1,1-a)) \mod H_2(\mathrm{GL}_1(F)).$$

From the Hoschschild-Serre spectral sequence corresponding to the group extension $1 \to \mathrm{SL}_n(F) \to \mathrm{GL}_n(F) \to F^* \to 1$ we obtain the exact sequence

$$0 \to H_2(\mathrm{SL}_n(F))_{F*} \to H_2(\mathrm{GL}_n(F)) \to H_2(F^*) \to 0.$$

If $n \geqslant 3$, then (see [7] and [8]) the group $H_2(\mathrm{SL}_n(F))$ coincides with $K_2(F)$ and the action of F^* is trivial. On the other hand, the group $H_2(\mathrm{SL}_2(F)) = K_2\,\mathrm{Sp}(F)$ is generated by symbols $\{a, b\}$, $a, b \in F^*$, which are subject to well-known relations [8]. The action of F^* on $K_2\,\mathrm{Sp}(F)$ is given by $c\{a, b\} = \{a, bc\} - \{a, c\}$. This shows that $H_2(\mathrm{SL}_2(F))_{F*}$ is also $K_2(F)$, and hence

$$H_2(\mathrm{GL}_2(F)) = H_2(\mathrm{GL}(F)).$$

The image of $(\operatorname{diag}(a, a^{-1}), \operatorname{diag}(1, 1-a))$ in $H_2(\mathrm{GL}(F))$ equals

$$(\operatorname{diag}(a, a^{-1}, 1), \operatorname{diag}(1, 1-a, (1-a)^{-1}) = \{a, 1-a\} = 0.$$

(2.7.2) Corollary. *θ defines a homomorphism (which will be denoted by the same letter θ)*

$$K_n^M(F) \to H_n(\mathrm{GL}_n(F))/H_n(\mathrm{GL}_{n-1}(F)).$$

From Lemma (2.5.3) we obtain immediately

(2.7.3) Corollary. *The composition*

$$K_n^M(F) \xrightarrow{\theta} H_n(\mathrm{GL}_n(F))/H_n(\mathrm{GL}_{n-1}(F)) \xrightarrow{\delta_n} S_n(F)$$

sends $\{a_1, \dots, a_n\}$ *to*

$$\sum_{1 \leqslant i_1 < \cdots < i_k \leqslant n} (-1)^k \langle a_1, \dots, \underset{i_1}{1}, \dots, \underset{i_k}{1}, \dots, a_n \rangle.$$

(2.7.4) Corollary. *The composition*

$$K_n^M(F) \to H_n(\mathrm{GL}_n(F))/H_{n-1}(\mathrm{GL}_{n-1}(F)) \to S_n(F) \to K_n^M(F)$$

is the identity map.

§3. The main theorem

The natural action of the group $A_{n,m}(F)$ on the complex $C_*^{n,m}(F)$ gives rise to a spectral sequence whose E^1 term has the form $E_{p,q}^1 = H_p(A_{n,m}(F), H_{n,m}(F))$, and its limit is the graded object $H_*(A_{n,m}(F), H_{n,m}(F))$. For any $0 \leqslant q \leqslant n$ the action of $A_{n,m}(F)$ permutes transitively the basis elements of $C_q^{n,m}(F)$, and the stabilizer of $(e_{n-q+1}, \dots, e_n)$ coincides with $A_{n-q,m+q}(F)$. This shows that

$$E_{p,q}^1 = H_p(A_{n-q,m+q}(F)) = H_p(\mathrm{GL}_{n-q}(F), \mathbf{Z}).$$

We will denote this spectral sequence by $\mathscr{E}^{n,m}$.

Proposition 3.1. *The natural embedding $H_n(F) \hookrightarrow H_{n,m}(F)$ induces isomorphisms*

$$H_i(\mathrm{GL}_n(F), H_n(F)) \xrightarrow{\sim} H_i(A_{n,m}(F), H_{n,m}(F)).$$

In particular, $S_n(F) \xrightarrow{\sim} S_{n,m}(F)$.

PROOF. The natural embedding $C^n_*(F) \hookrightarrow C^{n,m}_*(F)$ of complexes respects the actions of the corresponding groups and hence gives rise to a homomorphism $\mathcal{E}^n \to \mathcal{E}^{n,m}$ of spectral sequences. According to the above remarks, this homomorphism induces an isomorphism of E^1 terms and hence induces an isomorphism of the limits.

(3.1.1) COROLLARY. *If* $V = V_1 \oplus V_2$, *then the image of the pairing* $H(V_1, V_2) \otimes H(V_2) \to H(V) \to S(V)$ *coincides with the image of* $S(V_1) \otimes S(V_2) \to S(V)$.

(3.2.1) It is easy to see that the differential

$$d^1_{p,q}\colon E^1_{p,q} = H_p\big(\mathrm{GL}_{n-q}(F), \mathbf{Z}\big) \to E^1_{p,q-1} = H_p\big(\mathrm{GL}_{n-q+1}(F), \mathbf{Z}\big)$$

is induced by the embedding $\mathrm{GL}_{n-q}(F) \hookrightarrow \mathrm{GL}_{n-q+1}(F)$ when q is odd, and is zero when q is even (cf. [3]).

(3.2.2) The edge homomorphism

$$H_n(\mathrm{GL}_n(F), \mathbf{Z}) = E^1_{n,0} \to E^\infty_{n,0} \hookrightarrow S_n(F)$$

coincides evidently with the homomorphism δ_n from (2.6).

3.3. Let $S^0_n(F)$ denote the subgroup of decomposable elements in $S_n(F)$, i.e. the image of the homomorphism $S_1(F) \otimes S_{n-1}(F) \oplus \cdots \oplus S_{n-1}(F) \otimes S_1(F) \to S_n(F)$. By 3.1, for any decomposition $F^n = V_1 \oplus V_2$ the image of the homomorphism $H(V_1, V_2) \otimes H(V_2) \to S_n(F)$ is contained in $S^0_n(F)$.

(3.3.1) PROPOSITION. *For any* $\lambda, a_1, \ldots, a_n \in F^*$ *and any* i *the element*

$$\langle a_1, \ldots, \lambda a_i, \ldots, a_n\rangle - \langle a_1, \ldots, a_n\rangle$$

lies in $S^0_n(F)$.

PROOF. We assume that $i \neq 1, n$ (in these cases the proof below needs to be modified, and it becomes simpler). We have

$$\begin{aligned}
&\langle a_1, \ldots, \lambda a_i, \ldots, a_n\rangle - \langle a_1, \ldots, a_n\rangle \\
&\quad = \bar{d}\big((e_1, \ldots, \lambda^{-1}e_i, \ldots, e_n, a) - (e_1, \ldots, e_i, \ldots, e_n, a)\big) \\
&\quad = \bar{d}\big((e_1, \ldots, e_{i-1}) \cdot \big((\lambda^{-1}e_i) - (e_i)\big)(e_{i+1}, \ldots, e_n, a)\big) \\
&\quad = \text{the class modulo the action of } \mathrm{GL}_n(F) \text{ of the cycle} \\
&d(e_1, \ldots, e_{i-1}) \cdot \big((\lambda^{-1}e_i) - (e_i)\big) \cdot (e_{i+1}, \ldots, e_n, a) \\
&\qquad + (-1)^i (e_1, \ldots, e_{i-1}) \cdot \big((\lambda^{-1}e_i) - (e_i)\big) d(e_{i+1}, \ldots, e_n, a).
\end{aligned}$$

Set $u = a_1 e_1 + \cdots + a_i e_i = a - a_{i+1}e_{i+1} - \cdots - a_n e_n$. It is easy to see that for all $1 \leqslant k \leqslant i-1$ and $i+1 \leqslant l \leqslant n$ the following vectors are linearly independent:

$$e_1, \ldots, \hat{e}_k, \ldots, \hat{e}_l, \ldots, e_n,\ a,\ u;\ e_1, \ldots, \hat{e}_k, \ldots, e_n,\ u.$$

It follows that all the chains below are in general position, and hence the following computation makes sense. We have

$$(e_{i+1}, \ldots, e_n, a) = d(u, e_{i+1}, \ldots, e_n, a) + u \cdot d(e_{i+1}, \ldots, e_n, a),$$
$$(e_1, \ldots, e_{i-1}) = (-1)^{i-1}\big(d(e_1, \ldots, u) - d\big((e_1, \ldots, e_{i-1}) \cdot u\big)\big);$$

hence the above cycle is equal to

$$\begin{aligned} & d(e_1,\dots,e_{i-1})\cdot((\lambda^{-1}e_i)) - (e_i)\cdot d(u,e_{i+1},\dots,e_n,a) \\ & -d(e_1,\dots,e_{i-1},u)\cdot((\lambda^{-1}e_i)-(e_i))\cdot d(e_{i+1},\dots,e_n,a) \\ & +d(e_1,\dots,e_{i-1})\cdot[((\lambda^{-1}e_i)-(e_i))\cdot u + u\cdot((\lambda^{-1}e_i)-(e_i))] \\ & \cdot d(e_{i+1},\dots,e_n,a). \end{aligned}$$

It suffices now to check that every summand is decomposable. Denote by V_2 the subspace spanned by the vectors u, $e_{i+1},\dots,e_n$ and by V_2 the subspace spanned by $e_1,\dots,e_i$. The cycles

$$d(e_1,\dots,e_{i-1},u)\cdot((\lambda^{-1}e_i)-(e_i))$$

and

$$d(e_1,\dots,e_{i-1})\cdot[((\lambda^{-1}e_i)-(e_i))u + u((\lambda^{-1}e_i)-(e_i))]$$

are in $H_2(V_2)$. Since the cycle $d(e_{i+1},\dots,e_n,a)$ is in general position with V_2', it lies in $H(V_1',V_2')$, where V_1' is a subspace complementary to V_2'. Similarly, the first summand is in the image of $H(V_1,V_2)\otimes H(V_2)\to H_n(F)$ with V_1 such that $V_1\oplus V_2 = F^n$, so it is decomposable.

(3.3.2) Corollary. *For any $a_1,\dots,a_n\in F^*$ the element $\langle a_1,\dots,a_n\rangle - \langle 1,\dots,1\rangle$ is decomposable.*

(3.3.3) Proposition. a) *$S_0(F)=\mathbf{Z}$, and $S_1(F)=F^*$.*
b) *$S_2(F)=K_2(F)\oplus\mathbf{Z}$.*
c) *If $n\geqslant 3$, then $S_n(F)$ coincides with $S_n^0(F)$.*

Proof. Part a) is evident. It follows from (2.7.4) that $K_2(F)$ is a direct summand of $S_2(F)$. On the other hand, there is a surjective mapping $S_2(F)\to\mathbf{Z}$ ($\langle a,b\rangle\to 1$) whose kernel contains $K_2(F)$. By the last corollary the kernel is exactly $K_2(F)$, which proves part b). To prove the last part, it suffices, by Corollary 3.3.2, to check that when $n\geqslant 3$ the cycle $\langle 1,\dots,1\rangle$ is decomposable. If n is odd, we can argue as follows. Consider the main relations for the generators $\langle a_1,\dots,a_n\rangle$ (see 2.3.1). According to Corollary 3.3.2, its left-hand side is congruent to zero modulo $S_n^0(F)$ while the right-hand side is congruent to $\langle 1,\dots,1\rangle$. If n is even, we consider the decomposable element

$$\underbrace{\langle 1,\dots,1\rangle}_{n-2}\times\langle 1,1\rangle.$$

Using Corollary 3.3.2, we see that it is congruent to $\langle 1,\dots,1\rangle$ modulo $S_n^0(F)$.

(3.3.4) Corollary. *For $n\geqslant 2$*

$$S_n(F)=K_n^M(F)\oplus\langle 1,1\rangle\times S_{n-2}(F).$$

Proof. $K_n^M(F)$ is a direct summand of $S_n(F)$ by (2.7.4). It is easy to see that $\langle 1,1\rangle\times S_{n-2}(F)$ lies in the kernel of the homomorphism $S_n(F)\to K_n^M(F)$, so $S_n(F)=K_n^M(F)\oplus\langle 1,1\rangle\times S_{n-2}(F)\oplus ?$. To complete our proof, taking Proposition 3.3.3 into account and proceeding by induction on n, it suffices to establish the following fact:

(3.3.5) Lemma. *For any $a\in F^*$*

$$\langle a\rangle\times\langle 1,1\rangle = \langle 1,1\rangle\times\langle a\rangle.$$

PROOF.

$$\langle a\rangle \times \langle 1,1\rangle - \langle 1,1\rangle \times \langle a\rangle = [(a)\times\langle 1,1\rangle - \langle \bar{1},1\rangle \times (a)] \\ -[(1)\times\langle 1,1\rangle - \langle 1,1\rangle \times (1)].$$

The first cycle is obtained from the second one by the action of the matrix diag$(a,1,1)$, and hence their difference is zero in $S_3(F)$.

THEOREM 3.4. a) *The differentials $d^2, d^3, \ldots$ in the spectral sequence $\mathcal{E}^n$ are zero.*

b) *The homomorphism $S_{n-2}(F) \to S_n(F)$ given by $x \to \langle 1,1\rangle \times x$ is injective.*

c) *The homomorphism $H_i(\mathrm{GL}_{n-1}(F)) \to H_i(\mathrm{GL}(F))$ is an isomorphism when $i \leqslant n - 1$.*

d) *θ: $K_n^M(F) \to H_n(\mathrm{GL}_n(F))/H_n(\mathrm{GL}_{n-1}(F))$ is an isomorphism.*

PROOF. We proceed by induction on n. If $n = 1$ or 2, all the statements are evident. In the general case, pick a cycle $\sigma \in H_2(F)$ whose image in $S_2(F)$ is $\langle 1,1\rangle$, and consider the homomorphism $C_*^{n-2}(F)(2) \to C_*^n(F)$ of complexes given by multiplication by σ ($x \to \sigma \times x$). This homomorphism respects the actions of the groups $\mathrm{GL}_{n-2}(F)$ and $\mathrm{GL}_n(F)$, and hence induces a homomorphism $\tilde{\mathcal{E}}^{n-2} \to \mathcal{E}^n$ of spectral sequences, where $\tilde{\mathcal{E}}^{n-2}$ is obtained from $\mathcal{E}^{n-2}$ shifting by 2 the index q.

The resulting homomorphisms $\tilde{E}^1_{p,q} \to E^1_{p,q}$ are isomorphisms for $q \geqslant 2$; on the other hand, $\tilde{E}^1_{p,q} = 0$ for $q < 2$. These observations together with the induction hypothesis prove the part a). The homomorphism induced on the limits $S_{n-2}(F) \to S_n(F)$, evidently, coincides with the multiplication by $\langle 1,1\rangle$. The filtration of $S_{n-2}(F)$ arising from the spectral sequence $\tilde{\mathcal{E}}$ has the following properties: $S_{n-2}(F)^0 = S_{n-2}(F)^1 = 0$ and the induced homomorphisms $S_{n-2}(F)^{i+1/i} \to S_n(F)^{i+1/i}$ are isomorphisms for $i \geqslant 1$. This shows that $S_{n-2}(F) \to S_n(F) \to S_n(F)/S_n(F)^1$ is an isomorphism, and hence the multiplication by $\langle 1,1\rangle$: $S_{n-2}(F) \to S_n(F)$ is injective. Finally, we see from the same observations that $S_n(F)^1 \cap \langle 1,1\rangle \times S_{n-2}(F) = 0$ and, since

$$S_n(F)^0 = \delta_n\big(H_n(\mathrm{GL}_n(F))/H_n(\mathrm{GL}_{n-1}(F)\big) \supset K_n^M(F),$$

Corollary 3.3.4 implies that the group

$$H_n(\mathrm{GL}_n(F))/H_n(\mathrm{GL}_{n-1}(F)) \approx E^2_{n,0} = E^\infty_{n,0} = S_n(F)^0$$

coincides with $K_n^M(F)$ and

$$\mathrm{Ker}\big(H_{n-1}(\mathrm{GL}_{n-1}(F)) \to H_{n-1}(\mathrm{GL}_n(F))\big) = E^2_{n-1,1} = E^\infty_{n-1,1} = S_n(F)^{1/0} = 0.$$

Part c) follows from what we have proved and the fact that $E^2_{p,q} = E^\infty_{p,q} = 0$ for $p + q \leqslant n - 1$.

(3.4.1) COROLLARY. $S_n(F) = K_n^M(F) \oplus K_{n-2}^M(F) \oplus \cdots \oplus K_{n-2}^M(F)$, *where r is the integral part of $n/2$.*

§4. Homomorphism from Quillen K-theory to Milnor K-theory

Using Theorem 3.4 we can define a homomorphism $H_*(\mathrm{GL}(F)) \to K_*^M(F)$ as the composition

$$H_n(\mathrm{GL}(F)) = H_n(\mathrm{GL}_n(F)) \to H_n(\mathrm{GL}_n(F))/H_n(\mathrm{GL}_{n-1}(F)) \approx K_n^M(F).$$

Lemma 2.6.1 shows that this homomorphism is a graded ring homomorphism. Combining the homomorphism with the Hurewicz homomorphism

$$K_n(F) = \pi_n(\mathrm{BGL}(F)^+) \to H_n(\mathrm{BGL}(F)^+) = H_n(\mathrm{GL}(F)),$$

we obtain a map φ: $K_n(F) \to K_n^M(F)$. Probably this homomorphism coincides with the homomorphism constructed in [18] via Mennicke symbols, but this problem stays open. The map φ: $K_*(F) \to K_*^M(F)$ is not a ring homomorphism. As far as one can see, it satisfies the following product formula which is usual in theory of characteristic classes:

(4.1). *If* $x \in K_n(F)$ *and* $y \in K_m(F)$, *then*

$$\varphi(x \cdot y) = -\frac{(m+n-1)!}{(m-1)!(n-1)!}\varphi(x)\varphi(y).$$

Below we prove this formula in the case $m = 1$ (the case $m = 2$ follows automatically, since $K_2(F)$ is generated by products of elements of $K_1(F)$). Moreover, it follows from results of the next section and the product formula [13] for characteristic classes that (4.1) holds up to torsion.

The product in K-theory is defined as follows (see [4]). The matrix tensor product defines a homomorphism

$$\psi_{n,m}\colon \mathrm{GL}_n(F) \times \mathrm{GL}_m(F) \to \mathrm{GL}_{mn}(F)$$

(defined up to conjugation by a permutation matrix) and hence gives a well-defined (up to homotopy) mapping

$$f_{n,m}\colon \mathrm{BGL}_n(F)^+ \times \mathrm{BGL}_m(F)^+ \to \mathrm{BGL}_{mn}(F)^+ \to \mathrm{BGL}(F)^+.$$

Using the H-group structure on $\mathrm{BGL}(F)^+$ we define a mapping $\gamma_{n,m}$ by $\gamma_{n,m} = f_{n,m} - f'_{n,m} - f''_{n,m}$, where

$$f'_{n,m}(x, y) = f_{n,m}(x_0, y),\ f''_{n,m}(x, y) = f_{n,m}(x, y_0)$$

and x_0 and y_0 are the base points in the corresponding spaces. The mapping $\gamma_{n,m}$ is homotopically trivial on the join $\mathrm{BGL}_n(F)^+ \vee \mathrm{BGL}_m(F)^+$ and hence defines (again since $\mathrm{BGL}(F)^+$ is an H-group) a unique (up to homotopy) mapping

$$\hat{\gamma}_{n,m}\colon \mathrm{BGL}_n(F)^+ \wedge \mathrm{BGL}_m(F)^+ \to \mathrm{BGL}(F)^+$$

such that the diagram

$$\begin{array}{ccc} \mathrm{BGL}_n(F)^+ \times \mathrm{BGL}_m(F)^+ & \xrightarrow{\gamma_{n,m}} & \\ \downarrow & & \mathrm{BGL}(F)^+ \\ \mathrm{BGL}_n(F)^+ \wedge \mathrm{BGL}_m(F)^+ & \xrightarrow{\hat{\gamma}_{n,m}} & \end{array}$$

commutes up to homotopy. The maps $\hat{\gamma}_{n,m}$ agree with each other, and taking limits we obtain a uniquely (up to weak homotopy) defined mapping

$$\hat{\gamma}\colon \mathrm{BGL}(F)^+ \wedge \mathrm{BGL}(F)^+ \to \mathrm{BGL}(F)^+.$$

The map $\hat{\gamma}$ defines pairings of the homotopy groups of $\mathrm{BGL}(F)^+$, i.e. of the K-groups of F.

LEMMA 4.2. *Suppose that* $x \in K_n(F)$ *and* $y \in K_m(F)$. *Denote by* λ: $K_*(F) \to H_*(\mathrm{GL}(F))$ *the Hurewicz homomorphism. Then*

$$\lambda(x \cdot y) = \psi_*(\lambda(x) \times \lambda(y)) - mn\lambda(x) \cdot \lambda(y).$$

PROOF. Set $\sigma = \lambda(x) \in H_n(\mathrm{GL}(F))$ and $\tau = \lambda(y) \in H_m(\mathrm{GL}(F))$. The elements σ and τ are primitive homology classes, i.e. if $\Delta\colon \mathrm{GL} \to \mathrm{GL} \times \cdots \times \mathrm{GL}$ denotes the diagonal map then

$$\Delta_*(\sigma) = \sigma \times 1 \times \cdots \times 1 + \cdots + 1 \times 1 \times \cdots \times \sigma$$

and the same is true for τ. We can assume that $\sigma \in H_n(\mathrm{GL}_n(F))$ and $\tau \in H_m(\mathrm{GL}_m(F))$. From the description of products in K-theory given above it follows that $\lambda(xy) = (\gamma_{n,m})_*(\sigma \times \tau)$. The map $\gamma_{m,n}$ can be decomposed as follows:

$$\mathrm{BGL}_n^+ \times \mathrm{BGL}_m^+ \xrightarrow{\Delta} (\mathrm{BGL}_n^+ \times \mathrm{BGL}_m^+) \times (\mathrm{BGL}_n^+ \times \mathrm{BGL}_m^+) \times (\mathrm{BGL}_n^+ \times \mathrm{BGL}_m^+)$$

$$\xrightarrow{f_{n,m} \times (-f'_{n,m}) \times (-f''_{n,m})} \mathrm{BGL}^+ \times \mathrm{BGL}^+ \times \mathrm{BGL}^+ \xrightarrow{\Sigma} \mathrm{BGL}^+.$$

Since σ and τ are primitive, we obtain

$$\begin{aligned} \Delta_*(\sigma \times \tau) & \\ = (\sigma \times \tau) &\times (1 \times 1) \times (1 \times 1) + (\sigma \times 1) \times (1 \times \tau) \times (1 \times 1) \\ &+ (\sigma \times 1) \times (1 \times 1) \times (1 \times \tau) + (-1)^{mn}(1 \times \tau) \times (\sigma \times 1) \times (1 \times 1) \\ &+ (1 \times 1) \times (\sigma \times \tau) \times (1 \times 1) + (1 \times 1) \times (\sigma \times 1) \times (1 \times \tau) \\ &+ (-1)^{mn}(1 \times \tau) \times (1 \times 1) \times (\sigma \times 1) + (-1)^{mn}(1 \times 1) \times (1 \times \tau) \\ &\times (\sigma \times 1) + (1 \times 1) \times (1 \times 1) \times (\sigma \times \tau). \end{aligned}$$

Since the restriction of ψ to $\mathrm{GL}_n \times 1$ coincides with

$$\Delta\colon \mathrm{GL}_n \to \underbrace{\mathrm{GL}_n \times \cdots \times \mathrm{GL}_n}_{m} \hookrightarrow \mathrm{GL}_{n+m}$$

and since σ is primitive, we conclude that $f_*(\sigma \times 1) = f''_*(\sigma \times 1) = m\sigma$ and $f'_*(\sigma \times 1) = 0$. Similarly, $f_*(1 \times \tau) = f'_*(1 \times \tau) = n\tau$, $f''_*(1 \times \tau) = 0$ and $f'_*(\sigma \otimes \tau) = f''_*(\sigma \times \tau) = 0$. Finally, the mapping $(\text{—})_*$ is the change of sign on primitive elements, so applying $(f \times (-f') \times (-f''))_*$ to the above sum we obtain

$$\begin{aligned} f_*(\sigma \times \tau) \times 1 \times 1 &+ m\sigma \times (-n\tau) \times 1 \\ &+ (-1)^{mn} n\tau \times 1 \times (-m\sigma) + (-1)^{mn}(1 \times (-n\tau) \times (-m\sigma)). \end{aligned}$$

Now, applying Σ_*, we get

$$f_*(\sigma \times \tau) - mn\sigma\tau - (-1)^{mn} mn\tau\sigma + (-1)^{mn} mn\tau\sigma = f_*(\sigma \times \tau) - mn\sigma\tau.$$

COROLLARY 4.3. *If $m = 1$ or $n = 1$, then*

$$\varphi(xy) = -mn\varphi(x)\varphi(y).$$

PROOF. In the case under consideration $\psi_*(\lambda(x) \times \lambda(y)) \in H_{mn}(\mathrm{GL}_{mn}(F))$ goes to zero under the homomorphism $H_{m+n}(\mathrm{GL}(F)) \to K^M_{m+n}(F)$.

COROLLARY 4.4. *If $a_1, \dots, a_n \in F^*$ and $l(a_i) \in K_1(F) = F^*$ are the corresponding elements, then*

$$\varphi(l(a_1) \cdot \cdots \cdot l(a_n)) = (-1)^{n-1}(n-1)!\{a_1, \dots, a_n\}.$$

This corollary shows that the image of $K_n(F) \xrightarrow{\varphi} K_n^M(F)$ contains $(n-1)!K_n^M(F)$. It may be conjectured that the image is always exactly $(n-1)!K_n^M(F)$, but the problem seems to be rather complicated, as for example the following result shows:

PROPOSITION 4.5. *The image of the homomorphism $K_3(F) \to K_3^M(F)$ is the kernel of the Milnor* [6] *homomorphism $K_3^M(F) \to I^3(F)/I^4(F)$.*

This proposition shows that for $n = 3$ the above conjecture is equivalent to Milnor's conjecture [6] that

$$\ker\left(K_n^M(F) \to I^n(F)/I^{n+1}(F)\right) = 2K_n^M(F).$$

PROOF OF THE PROPOSITION. The Hurewicz theorem applied to the simply connected space $\mathrm{BSL}(F)^+$ shows that the natural mapping $K_3(F) = \pi_3(\mathrm{BSL}(F)^+) \to H_3(\mathrm{SL}(F))$ is surjective; hence

$$K_3^M(F)/\operatorname{Im} K_3(F) = H_3(\mathrm{GL}(F))/\left(H_3(\mathrm{GL}_2(F)) + H_3(\mathrm{SL}(F))\right).$$

Consider the Hochschild-Serre spectral sequence associated with the group extension $1 \to \mathrm{SL}_n(F) \to \mathrm{GL}_n(F) \to F^* \to 1$. It looks as follows:

$$\begin{array}{llll}
H_3(F^*) & 0 & * & * \\
H_2(F^*) & 0 & * & * \\
H_1(F^*) & 0 & H_1(F^*, H_2(\mathrm{GL}_n(F))) & * \\
H_0(F^*) & 0 & H_0(F^*, H_2(\mathrm{SL}_n(F))) & H_3(\mathrm{SL}_n(F))_{F^*}
\end{array}$$

Since the homomorphism $\mathrm{GL}_n(F) \to F^*$ splits, all the differentials starting in the first column are zero, and hence

$$H_3(\mathrm{GL}_n(F))/H_3(\mathrm{SL}_n(F)) + H_3(\mathrm{GL}_1(F)) = H_1\left(F^*, H_2(\mathrm{SL}_n(F)\right).$$

This shows that

$$K_3^M(F)/\operatorname{Im}(K_3(F)) = \operatorname{Coker}\left(H_1\left(F^*, H_2(\mathrm{SL}_2(F))\right) \to H_1\left(F^*, H_2(\mathrm{SL}(F))\right)\right).$$

Consider the exact sequence of F^*-modules

$$0 \to R(F) \to H_2(\mathrm{SL}_2(F)) \to H_2(\mathrm{SL}(F)) \to 0$$

(where $R(F)$ is defined by the sequence). The corresponding long exact homology sequence has the form

$$0 \to K_3^M(F)/\operatorname{Im}(K_3(F)) \to R(F)_{F^*} \to H_2(\mathrm{SL}_2(F))_{F^*} \to H_2(\mathrm{SL}(F)) \to 0.$$

As we mentioned in our proof of Lemma 2.7.1, $H_2(\mathrm{SL}_2(F)) \xrightarrow{\sim} H_2(\mathrm{SL}(F))$; so $K_3^M(F)/\operatorname{Im}(K_3(F)) = R(F)_{F^*}$. It remains to use the following fact (cf. [16] and [17]).

PROPOSITION 4.6 [5]. *The F^*-module $R(F)$ is canonically isomorphic to $I^3(F) \subset W(F)$, where the action of F^* on $I^3(F)$ is induced by the natural action of F^* on $W(F)$ and has the form*

$$a \cdot \langle\langle x, y, z\rangle\rangle = \langle\langle x, y, z\rangle\rangle + \langle\langle x, y, z, a\rangle\rangle.$$

§5. The characteristic class $c_{n,n}$: $K_n(F) \to K_n(F)$

We briefly recall the construction of the characteristic classes $c_{n,n}$: $K_n(X) \to H^i(X, \mathcal{K}_{n+i})$ (see [11]–[13] and [21]).

(5.1.1) For any scheme X we denote by $\mathcal{K}_n$ the sheaf (in the Zariski topology) associated with the presheaf $U \to K_n(\Gamma(U, \mathcal{O}_X))$. It is evident from the definition that for any point $x \in X$ the stalk $(\mathcal{K}_n)_x$ of the sheaf is $K_n(\mathcal{O}_{X,x})$.

(5.1.2) If G is a discrete group acting on a topological space X and $\mathcal{F}$ an abelian sheaf on X endowed with a linear C-action, then let $H^n(X, G, \mathcal{F})$ denote the corresponding equivariant cohomology groups (see [2]). The group $H^1(X, G, \mathcal{O}_X^*)$ classifies the line G-bundles over X.

(5.1.3) Suppose that X is a smooth variety over a field F and G a discrete group acting on X. Let E/X be a vector G-bundle of constant rank n. Consider the projectivization $P = P(E)$ of E and the canonical line bundle $L = \mathcal{O}_P(-1)$ on P. There is a natural action of G on P and L. Let $\mathcal{E} \in H^1(P, G, \mathcal{O}_P^*) = H^1(P, G, \mathcal{K}_1)$ denote the cohomology class corresponding to the G-bundle L. The following result is the principal tool for construction of characteristic classes with values in K-cohomology groups. A simple proof of a particular case (of trivial group G) can be found in [14] and [15].

(5.1.4) THEOREM [1], [21]. *The bigraded abelian group $H^*(P, G, \mathcal{K}_*)$ is isomorphic to*

$$H^*(X, G, \mathcal{K}_*) \oplus \xi H^*(X, G, \mathcal{K}_*) \oplus \cdots \oplus \xi^{n-1} H^*(X, G, \mathcal{K}_*),$$

i.e.

$$H^i(P, G, \mathcal{K}_j) = H^i(X, G, \mathcal{K}_j) \oplus \xi H^{i-1}(X, G, \mathcal{K}_{j-1}) \oplus \cdots \oplus \xi^{n-1} H^{i-n+1}(X, G, \mathcal{K}_{j-n+1}).$$

(5.1.5) Now we can act in the standard way and construct equivariant classes $c_i(E) \in H^i(X, G, \mathcal{K}_i)$. Namely, we write ξ^n in the form $-c_1\xi^{n-1} - \cdots - c_n$ and set $c_i(E) = c_i$. By definition, $c_0(E) = 1 \in H^0(X, G, \mathcal{K}_0)$. The total characteristic class $c(E) = \Sigma_1^n c_i(E)$ has the usual additive property: if $0 \to E' \to E \to E'' \to 0$ is an exact sequence of vector G-bundles, then $c(E) = c(E') \cdot c(E'')$.

(5.1.6) In the case when the action of G on X is trivial, the equivariant cohomology can be computed as follows. Let $\mathcal{F}$ be a trivial G-sheaf and I^* its injective resolution. Denote by $C_*(G)$ a G-projective resolution of the trivial G-module $\mathbf{Z}$ (the standard resolution, for example). Then the total complex associated with the bicomplex $\operatorname{Hom}(C_*(G), I^*)$ is a G-injective resolution of $\mathcal{F}$, so $H^*(X, G, \mathcal{F})$ coincides with the homology of the complex

$$\Gamma^G(\operatorname{Hom}(C_*(G), I^*)) = \operatorname{Hom}(C_*(G), \Gamma(I^*))^G = \operatorname{Hom}(C_*(G)_G, \Gamma(I^*)).$$

The natural mapping

$$H_*(\operatorname{Hom}(C_*(G)_G, \Gamma(I^*))) \to \operatorname{Hom}(H_*(C_*(G)_G), H_*(\Gamma(I^*)))$$

defines homomorphisms

$$H^p(X, G, \mathcal{F}) \to \coprod_{i=0}^{p} \operatorname{Hom}(H_{p-i}(G, \mathbf{Z}), H^i(X, \mathcal{F})).$$

(5.1.7) Suppose that G acts trivially on X and E is a vector G-bundle of rank n. By (5.1.6) the equivariant classes $c_p(E)$ define homomorphisms

$$c_{k,p}(E)\colon H_k(G,\mathbf{Z}) \to H^{p-k}(X,\mathcal{K}_p),$$

and, evidently, $c_{k,p}(E) = 0$ for $p > n$.

(5.1.8) Suppose that $X = \operatorname{Spec} A$ is a smooth affine variety. Set $G = \mathrm{GL}_n(A)$. There is a natural action of G on the trivial bundle A^n, and hence the corresponding homomorphisms

$$c_{k,p}(A^n)\colon H_k(\mathrm{GL}_n(A)) \to H^{p-k}(X,\mathcal{K}_p).$$

It is easy to see that the restriction of $c_{k,p}(A^{n+1})$ to $H_k(\mathrm{GL}_n(A))$ is $c_{k,p}(A^n)$, so passing to limit we obtain a natural homomorphism

$$c_{k,p}(A^\infty)\colon H_k(\mathrm{GL}(A)) \to H^{p-k}(X,\mathcal{K}_p).$$

Combining these homomorphisms with the Hurewicz homomorphisms $K_k(A) \to H_k(\mathrm{GL}(A))$, we obtain finally the desired characteristic classes

$$c_{k,p}\colon K_k(A) = K_k(X) \to H^{p-k}(X,\mathcal{K}_p).$$

(5.1.9) Suppose that X is smooth and irreducible. Denote by F the field of rational function on X. The group $H^0(X,\mathcal{K}_p)$ can be identified with a subgroup of $K_p(F)$ (see [9]). By virtue of the functorial character of the characteristic classes we have the commutative diagram

$$\begin{array}{ccc} K_p(X) & \xrightarrow{c_{p,p}} & H^0(X,\mathcal{K}_p) \\ \downarrow & & \cap \\ K_p(F) & \xrightarrow{c_{p,p}} & H^0(\operatorname{Spec} F,\mathcal{K}_p) = K_p(F). \end{array}$$

This shows that studying $c_{p,p}$ for an arbitrary X reduces to studying this homomorphism in the case of a field. In the latter case the homomorphism $c_{p,p}$ maps $K_p(F)$ into $H^0(\operatorname{Spec} F,\mathcal{K}_p) = K_p(F)$, and the other characteristic classes are, evidently, zero. It is easy to see that $c_{1,1}$ is the identity map. This and the product formula [13], [21], which in the given case has the form

$$c_{m+n,m+n}(xy) = -\frac{(m+n-1)!}{(m-1)!(n-1)!} c_{mn}(x)\cdot c_{mn}(y)$$

(where $x \in K_m(F)$ and $y \in K_n(F)$), imply that, on the decomposable elements, $c_{p,p}$ is the multiplication by $(-1)^{p-1}(p-1)!$. The following result is the main goal of the paper:

THEOREM 5.2. *The homomorphism $c_{p,p}\colon K_p(F) \to K_p(F)$ coincides with the composition $K_p(F) \to K_p^M(F) \to K_p(F)$.*

PROOF. The homomorphism $c_{p,p}(F^{p-1})\colon H_p(\mathrm{GL}_{p-1}(F)) \to K_p(F)$ is trivial according to (5.1.7), and it follows from Theorem 3.4 that $c_{p,p}(F)\colon H_p(\mathrm{GL}(F)) \to K_p(F)$ decomposes into the composition

$$\begin{aligned} H_p(\mathrm{GL}(F)) &= H_p(\mathrm{GL}_p(F)) \to H_p(\mathrm{GL}_p(F))/H_p(\mathrm{GL}_{p-1}(F)) \\ &= K_p^M(F) \xrightarrow{\lambda} K_p(F). \end{aligned}$$

To compute the homomorphism λ, since the isomorphism

$$K_p^M(F) \tilde{\to} H_p(\mathrm{GL}_p(F))/H_p(\mathrm{GL}_{p-1}(F))$$

is induced by $F^* \times \cdots \times F^* \to H_p(\mathrm{GL}_1 \times \cdots \times \mathrm{GL}_1) \to H_p(\mathrm{GL}_p)$, it suffices to compute the equivariant class

$$c_p(F^p) \in H^p(\operatorname{Spec} F, G, \mathcal{K}_p) = H^p(G, K_p(F)),$$

where $G = \mathrm{GL}_1(F) \times \cdots \times \mathrm{GL}_1(F)$ is the group of all diagonal matrices. The G-module F^p is the direct sum of one-dimensional modules $F_{(i)}$, where $F_{(i)}$ is F with the action of G defined by the ith projection $G \to \mathrm{GL}_1(F)$. By virtue of the multiplicativity of the total characteristic class we have

$$c_p(F^p) = c_1(F) \times \cdots \times c_1(F),$$

where $c_1(F) \in H^1(\mathrm{GL}_1(F), K_1(F)) = \mathrm{Hom}(F^*, F^*)$ is the identity map. Therefore

$$\langle a_1 \times \cdots \times a_p, c_p(F^p)\rangle = l(a_1) \cdot \cdots \cdot l(a_p),$$

where $l(a) \in K_1(F)$ is the element corresponding to a under the canonical isomorphism $F^* \simeq K_1(F)$. Thus, the homomorphism c_p takes $\{a_1, \ldots, a_p\}$ to $l(a_1) \cdot \cdots \cdot l(a_p)$, i.e. is a homogeneous component of the canonical ring homomorphism $K_*^M(F) \to K_*(F)$.

COROLLARY 5.3. *If F is an algebraic extension of a finite field, then $c_{p,p} = 0$ for $p \geqslant 2$.*

PROOF. If F is infinite, then the statement follows from Theorem 5.2 and the fact that $K_p^M(F) = 0$. If F is finite, then we have to use that $c_{p,p}$ is functorial and the result of Quillen [10] that the mapping $K_p(F) \to K_p(\bar{F})$ (where $\bar{F}$ is an algebraic closure of F) is injective.

COROLLARY 5.4. *If F is a global field without real places, then $c_{p,p} = 0$ for $p \geqslant 3$. For an arbitrary global field F the homomorphism $c_{3,3}$ is zero.*

PROOF. If F has no real places, then $K_p^M(F) = 0$ for $p \geqslant 3$ (see [1]). For an arbitrary global field F the group $K_p^M(F)$ with $p \geqslant 3$ equals $\coprod \mathbf{Z}/2\mathbf{Z}$ (the summation is over the real places of F), so $2K_p^M(F) = 0$. Moreover, the homomorphism $K_p^M(F)/2K_p^M(F) \to I^p(F)/I^{p+1}(F)$ is, evidently, an isomorphism, and the statement about $c_{3,3}$ follows from Proposition 4.5.

BIBLIOGRAPHY

1. H. Bass and J. Tate, *The Milnor ring of a global field*, Algebraic K-theory. II (Proc. Conf., Seattle, 1972), Lecture Notes in Math., vol. 342, Springer-Verlag, 1973, pp. 349–446.

2. Alexandre Grothendieck, *Sur quelques points d'algèbre homologique*, Tôhoku Math. J. (2) **9** (1957), 119–221.

3. Wilferd van der Kallen, *Homology stability for linear groups*, Invent. Math. **60** (1980), 269–295.

4. Jean-Louis Loday, *K-théorie algébrique et représentations de groupes*, Ann. Sci. École Norm. Sup. (4) **9** (1976), 309–377.

5. A. A. Suslin, *Torsion in K_2 of fields*, Preprint E-2-82, Leningrad Branch, Steklov Inst. Math. Acad. Sci. USSR, Leningrad, 1982. (English)

6. John Milnor, *Algebraic K-theory and quadratic forms*, Invent. Math. **9** (1969/70), 318–344.

7. ______, *Introduction to algebraic K-theory*, Princeton Univ. Press, Princeton, N.J., and Univ. of Tokyo Press, Tokyo, 1971.

8. Calvin C. Moore, *Group extensions of p-adic and adelic linear groups*, Inst. Hautes Études Sci. Publ. Math. No. 35 (1968), 157–222.

9. Daniel Quillen, *Higher algebraic K-theory*. I, Algebraic K-theory. I (Proc. Conf., Seattle, 1972), Lecture Notes in Math., vol. 341, Springer-Verlag, 1973, pp. 85–147.

10. ______, *On the cohomology and K-theory of the general linear groups over a finite field*, Ann. of Math. (2) **96** (1972), 552–586.

11. C. Soulé, *K-théorie des anneaux d'entiers de corps de nombres et cohomologie étale*, Invent. Math. **55** (1979), 251–295.

12. V. V. Shekhtman, *Algebraic K-theory and characteristic classes*, Uspekhi Mat. Nauk **33** (1978), no. 6(204), 239–240; English transl. in Russian Math. Surveys **33** (1978).

13. ______, *The Riemann-Roch theorem and the Atiyah-Hirzebruch spectral sequence*, Uspekhi Mat. Nauk **35** (1980), no. 6(216), 179–180; English transl. in Russian Math. Surveys **35** (1980).

14. Clayton C. Sherman, *Some theorems on the K-theory of coherent sheaves*, Comm. Algebra **7** (1979), 1489–1508.

15. ______, *K-cohomology of regular schemes*, Comm. Algebra **7** (1979), 999–1027.

16. A. A. Suslin, *Cancellation problem for projective modules and related topics*, Preprint R-2-77, Leningrad Branch, Steklov Inst. Math. Acad. Sci. USSR, Leningrad, 1977; English transl., [17].

17. ______, *The cancellation problem for projective modules and related topics*, Ring Theory (Proc. Conf., Waterloo, Ont., 1978), Lecture Notes in Math., vol. 734, Springer-Verlag, 1979, pp. 323–338.

18. ______, *Mennicke symbols and their applications in the K-theory of fields*, Algebraic K-theory. Part I (Oberwolfach, 1980), Lecture Notes in Math., vol. 966, Springer-Verlag, 1982, pp. 334–356.*

19, ______, *Homology of* GL_n, *characteristic classes and Milnor K-theory*, Preprint E-4-82, Leningrad Branch, Steklov Inst. Math. Acad. Sci. USSR, Leningrad, 1982. (English)**

20. C. Soulé, *Operations in algebraic K-theory*, Preprint, 1982.

21. Henri Gillet, *Riemann-Roch theorems for higher algebraic K-theory*, Advances in Math. **40** (1981), 203–289.

Translated by L. N. VASERSTEIN

**Editor's note.* The Russian cites a preprint of this article.

***Editor's note.* Also published in Algebraic K-theory, Number Theory, Geometry and Analysis (Bielefeld, 1982), Lecture Notes in Math., vol. 1046, Springer-Verlag, 1984, pp. 357–375.

Труды Матем. Инст. Стеклов.
165(1984)

Proceedings of the Steklov
Institute of Mathematics
1985, Issue 3

LOCAL AND GLOBAL INVARIANTS OF A FOUR-DIMENSIONAL PSEUDO-RIEMANNIAN MANIFOLD

UDC 512.7

A. N. TYURIN

ABSTRACT. A global invariant is constructed for a four-dimensional Riemannian manifold, and both local and global invariants are computed for a Kähler surface. In §1 the local invariant of a pseudo-Riemannian manifold is described, in §2 the indicatrix is computed for a pseudo-Riemannian manifold and its physical meaning is discussed in the context of general relativity, in §3 the local invariant is computed for a Kähler manifold, and in §4 the construction of the global invariant and a description of its geometrical structure are given for a Riemannian four-dimensional manifold.

Bibliography: 12 titles.

Introduction

Twenty years ago in the seminar on algebraic geometry at Moscow State University, while discussing the contents of the monograph *Algebraic surfaces* [12], I. R. Shafarevich pointed out that comparatively little was known about the topology and metric properties of complex Kähler and algebraic surfaces.

It is easy to enumerate the increase in our knowledge on this question over the last twenty years: the theory of the decomposition of a surface into a connected sum $r\boldsymbol{P}^2 \# k\overline{\boldsymbol{P}}^2$ (Moishezon [4]), the Bogomolov-Miyaoka inequality $c_1^2 \leqslant 3c^2$ [3] and, following that, the Calabi-Yau existence theorem for metrics [11] and theorems following from this.

However, in the last five years, thanks to the physicists, interest has grown considerably in the metric theory of four-dimensional manifolds.

In this paper we study the local and global invariants of a four-dimensional pseudo-Riemannian manifold. The motive for this investigation was the desire to apply, as did Lefschetz (see the Introduction to [2]) the powerful methods of differential geometry to algebraic geometry. However, as sometimes happens, the opposite effect was obtained: the algebraic geometry began to work on the differential geometry.

Throughout the paper we will use a special notation to distinguish just the vector spaces over $\boldsymbol{R}$ (or fiber bundles with fiber a vector space over $\boldsymbol{R}$). For example, the symbol $\boldsymbol{R}T$ denotes a four-dimensional $\boldsymbol{R}$-vector space, and the symbol T will denote a four-dimensional $\boldsymbol{C}$-vector space, its complexification. The symbol $\boldsymbol{R}TX$

1980 *Mathematics Subject Classification*. Primary 53C20, 53C80; Secondary 53C55, 83E15.

denotes the tangent bundle to a four-dimensional differentiable manifold X, and the symbol TX denotes its complexification; the symbol $\boldsymbol{R}g$ denotes an $\boldsymbol{R}$-isomorphism of $\boldsymbol{R}$-vector spaces and g denotes its complexification; and so on.

§1. A local invariant

Let X be a four-dimensional differentiable (C^∞) manifold, and let $\boldsymbol{R}TX$ be its tangent bundle. A pseudo-Riemannian structure on X (or a pseudo-Riemannian metric) is any selfadjoint isomorphism

$$\boldsymbol{R}g\colon \boldsymbol{R}TX \to \boldsymbol{R}T^*X, \qquad \boldsymbol{R}g^* = \boldsymbol{R}g, \tag{1.1}$$

where $\boldsymbol{R}T^*X$ is the cotangent bundle over X.

At each point $x \in X$ the isomorphism (1.1) defines a quadratic form $\langle\ ,\ \rangle_x$ on the four-dimensional vector space $\boldsymbol{R}T_x$: $\langle t_1, t_2\rangle_x = g_x(t_1)(t_2)$.

In the projectivization $\boldsymbol{PR}T_x$ of the space $\boldsymbol{R}T_x$ a quadratic surface is defined by $\boldsymbol{PR}g_x = \{t \in \boldsymbol{R}T \mid \langle t, t\rangle_x = 0\}$ which can be:

1) the empty set $\varnothing$ — the Riemannian case, or

2) a sphere S^2 — the Lorentz case, or (1.2)

3) a hyperbola — the (2.2)-case.

There is a fiber-preserving antilinear involution θ on the complexification of the tangent bundle TX for which the set of fixed points is the real tangent bundle $\mathrm{Re}_\theta TX = \boldsymbol{R}TX$.

The isomorphism (1.1) defines a fiber-preserving $\boldsymbol{C}$-linear selfadjoint isomorphism:

$$g\colon TX \to T^*X, \qquad g^* = g, \tag{1.1$'$}$$

the projectivization of which at the point $x \in X$ defines a nonsingular quadratic surface $\boldsymbol{P}g_x$ in $\boldsymbol{P}^3 = \boldsymbol{P}T_x$.

This $\boldsymbol{P}g_x$ is isomorphic, as a complex-analytic manifold, to the direct product of two complex projective lines

$$\boldsymbol{P}g_x = q_+ \times q_-, \qquad q_\pm = \boldsymbol{P}^1. \tag{1.3}$$

Each projective line $q_\pm$ parametrizes the straight-line generators of the quadric $\boldsymbol{P}g_x \subset \boldsymbol{P}T_x = \boldsymbol{P}^3$. Two straight-line generators from different families intersect at a single point which gives the decomposition (1.3).

The antilinear involution θ giving the real structure on T_x commutes with the g of (1.1$'$) and acts on the quadric $\boldsymbol{P}(g_x) = q_+ \times q_-$.

This action is distinguished by the type of pseudo-Riemannian metric (1.2):

1) In the Riemannian case we have $\theta(q_+) = q_+$ and $\theta(q_-) = q_-$; that is, θ defines a real structure on the projective lines $q_\pm = \boldsymbol{P}^1$ with the set of real points $\mathrm{Re}_\theta\, q_\pm = \varnothing$.

2) In the Lorentz case we have $\theta(q_+) = q_-$ and $\theta(q_-) = q_+$; and

$$\boldsymbol{PR}g_x = S^2 = \bigcup_{L \in q_+} (L \cap \theta(L)). \tag{1.4}$$

3) In the (2.2)-case we have $\theta(q_+) = q_+$ and $\theta(q_-) = q_-$; that is, θ defines a real structure on the projective lines q_+ and q_-, where both must be standard structures, that is to say, $\mathrm{Re}_\theta(q_\pm) = S^1_\pm$ is a circle,

$$\boldsymbol{PR}g_x = S^1_+ \times S^1_- \quad \text{(a torus)}, \qquad S^1_\pm = \mathrm{Re}_\theta(q_\pm). \tag{1.5}$$

Choosing a volume form $\omega \in \Gamma\Lambda^4 \boldsymbol{R}TX$ on X, we obtain three real quadratic forms on the six-dimensional vector bundle $\Lambda^2 \boldsymbol{R}TX$,

$$\boldsymbol{R}v\colon \Lambda^2\boldsymbol{R}TX \to \Lambda^2\boldsymbol{R}T^*X, \qquad t_1 \wedge t_2 = v(t_1, t_2)\cdot\omega, \tag{1.6}$$

$$\Lambda^2 g\colon \Lambda^2\boldsymbol{R}TX \to \Lambda^2\boldsymbol{R}T^*X, \tag{$\Lambda^2(1.1)$}$$

$$R\colon \Lambda^2\boldsymbol{R}TX \to \Lambda^2\boldsymbol{R}T^*X. \tag{1.7}$$

The last form is the tensor curvature of the Riemannian connection defined by the Riemannian structure (1.1).

There are three quadratic forms v, $\Lambda^2 G$ and R on the complexification $\Lambda^2 TX$ which commute with the involution θ giving the real structure on $\Lambda^2 TX$.

We consider the projectivizations at the point $x \in X$ and the intersection of these three quadratic forms:

$$S_x = \boldsymbol{P}(v_x) \cap \boldsymbol{P}(\Lambda^2 g_x) \cap \boldsymbol{P}(R_x). \tag{1.8}$$

(The projectivization $\boldsymbol{P}(v_2)$ does not depend on the choice of volume form ω.)

There is the following alternative:

PROPOSITION 1. *Either S_x is an algebraic surface (possibly reducible) or the manifold X at the point x has constant curvature in any two-dimensional direction (see the introduction to* [8]).

A structure (1.1) at the point $x \in X$ is said to be *thick* if S_x is an irreducible surface having only simple double points (see [8], §2, Definition 4).

If the pseudo-Riemannian structure (1.1) is thick at a point $x \in X$, then S_x is a surface of type K3 (see [8] and [10]).

The quadric $\boldsymbol{P}(v_x) \in \boldsymbol{P}\Lambda^2 T_x$ is a Plücker embedding of a complex Grassmann manifold $G(2, T_x)$. Each of its points represents a projective line L in a three-dimensional projective space $\boldsymbol{P}T_x$.

The intersection

$$\boldsymbol{P}(v_x) \cap \boldsymbol{P}(\Lambda^2 g_x) = K_x \tag{1.9}$$

represents the manifold of lines in $\boldsymbol{P}T_x$ which are tangent to the quadric $\boldsymbol{P}(g_x) \subset \boldsymbol{P}T_x$ (1.3) (see [8], (1.7)).

The straight-line generators $L \in q_{+}$, $L \subset \boldsymbol{P}T_x$, of the quadric (1.3) run over a conic $q_{\pm} \subset \boldsymbol{P}(v_x) \subset \boldsymbol{P}\Lambda^2 T_x$ lying in K_x and forming the singular point set of this manifold:

$$\operatorname{Sing} K_x = q_{+} \cup q_{-}. \tag{1.10}$$

Therefore the surface S_x must have singularities

$$\operatorname{Sing} S_x \supset \operatorname{Sing} K_x \cap \boldsymbol{P}(R) = (q_{+} \cap \boldsymbol{P}(R_x)) \cup (q_{-} \cap \boldsymbol{P}(R_x)). \tag{1.11}$$

The pseudo-Riemannian structure (1.1) is thick at the point $x \in X$ if the conic $q_{\pm}$ intersects the quadric $\boldsymbol{P}(R)$ at four distinct points and so the surfaces S_x has eight simple double points

$$s_1^{\pm} \cup s_2^{\pm} \cup s_3^{\pm} \cup s_4^{\pm} = q_{\pm} \cap \boldsymbol{P}(R). \tag{1.12}$$

It is possible to resolve these double points and obtain a surface $\tilde{S}_x$ of type K3. In the two-dimensional cohomology lattice $H^2(\tilde{S}_x, \boldsymbol{Z})$ of this surface there are a fixed sublattice of algebraic cycles A spanned by h (the class of hyperplane sections of the embedding $S_x \subset \boldsymbol{P}^5 = \boldsymbol{P}\Lambda^2 T_x)$ and eight rational nonsingular curves $L_i^{\pm}$, $i = 1, \ldots, 4$, where $L_i^{\pm}$ is obtained by blowing up the singular point s_i (1.12).

The correspondence between the tangent to the quadric $\boldsymbol{P}(g_x)$ and the points of contact on the quadric defines a rational transformation $\pi\colon K_x \to q_+ \times q_-$, which extends to a regular double covering

$$\pi\colon \tilde{S}_x \to q_+ \times q_- \tag{1.13}$$

(see [8], (1.22)).

The double coverings of the straight-line generators of the quadric $q_+ \times q_-$ define two elliptical pencils $|E_{\pm}|$ on S_x whose cohomology classes are

$$E_+ = \frac{1}{2}\left(h - \sum_{i=1}^{4} L_i^-\right), \qquad E_- = \frac{1}{2}\left(h - \sum_{i=1}^{4} L_i^+\right) \tag{1.14}$$

(see [8], (1.35)).

These divisibility conditions have to be added to the relations defining the lattice A.

Thus a thick point $x \in X$ of the pseudo-Riemannian structure (1.1) defines

1) a surface S_x of type K3, and

2) a sublattice of algebraic cycles $A \subset \operatorname{Pic} \tilde{S}_x \subset H^2(\tilde{S}_x, \boldsymbol{Z}) = L$ spanned by the generators

$$h,\ L_i^{\pm},\ i = 1, \ldots, 4, \qquad E_{\pm} = \frac{1}{2}\left(h - \sum_{i=1}^{4} L_i^{\pm}\right)$$

with intersection matrix

$$\begin{gathered} h^2 = 8,\ \left(L_i^{\pm}\right)^2 = -2, \forall i \neq j \qquad \left(L_i^{\pm}, L_j^{\pm}\right) = 0, \forall i, j \\ \left(L_i^{\pm}, L_j^{\mp}\right) = 0, \qquad \left(h, L_i^{\pm}\right) = 0. \end{gathered} \tag{1.15}$$

It is necessary to add to the above the real structure on $\tilde{S}_x$ given by the involution θ, since the quadrics v, $\Lambda^2 g$ and R commute with θ and, hence, θ acts on S_x, on $(q_+ \cup q_-) \subset \boldsymbol{P}(R)$ and on $\tilde{S}_x$ (and the generators (1.13) commute with this action).

By the global Torelli theorem for a K3 surface [10] the action of θ on $\tilde{S}_x$ gives a unique action of θ on the lattice $L = H^2(\tilde{S}_x, \boldsymbol{Z})$. Moreover the sublattice A (1.15) is invariant with respect to this action and the action of θ on A is distinguished by the pseudo-Riemannian structure (1.2):

1) in the Riemannian case:

$$\theta(h) = -h, \quad \theta\left(L_1^{\pm}\right) = -L_2^{\pm}, \quad \theta\left(L_3^{\pm}\right) = -L_4^{\pm}, \quad \operatorname{Re}_\theta \tilde{S}_x = \varnothing; \tag{1.16}$$

2) in the Lorentz case:

$$\begin{gathered} \theta(h) = -h, \quad \theta\left(L_i^{\pm}\right) = -L_i^{+}, \\ \operatorname{Re}_\theta(\tilde{S}_x) = \begin{cases} \text{either } \varnothing, \\ \text{or } 2T_1, \\ \text{or } T_g \cup kT_0, \end{cases} \end{gathered} \tag{1.17}$$

where kT_g is the disjoint union of k copies of a surface of genus g;

3) in the (2.2)-case there are six types of action of θ on a lattice A which are described in detail in [5].

DEFINITION 1. For a thick point $x \in X$ of the pseudo-Riemannian structure (1.1) the collection (S_x, A, θ) consisting of a K3 surface S_x, a sublattice A (1.15) and an involution θ acting on A, is called the *local invariant* of the pseudo-Riemannian structure at the point x.

In [8] one can find a definition of the local invariant of a Riemannian structure at any point $x \in X$ which is not flat (that is, at which the curvature is not the same in all two-dimensional directions).

REMARK 1. Among the simple metrics which are almost flat in the sense that one of their *-components is zero, only the Einstein metrics are thick (see [8], §5). Their local invariant is a special Kummer surface.

REMARK 2. It was proved in [6] that for any real quadratic form R in $\Lambda^2\boldsymbol{R}T$, $(R, v) = 0$, a Riemannian structure can be constructed in a neighborhood of the origin in $\boldsymbol{R}^4$, with curvature tensor which coincides with R in $\Lambda^2\boldsymbol{R}T_0 = \Lambda^2\boldsymbol{R}T$. From this it follows that any real K3 surface with the sublattice A (1.15) of algebraic cycles can be realized as the local invariant of a Riemannian structure.

REMARK 3. Observe that the definition of a local invariant (see [8], Definition 1) is analogous to defining the Gauss curvature of a surface as the projective invariant of the pencil of quadratic forms (spanned by the first and second quadratic form).

Although this has not yet been done, one would like very much to give an intuitive description of the "deformation" of four-dimensional space which preserves the local invariant (see [8], §3).

§2. Moduli and indicatrices

The set X_0 of thick points on a four-dimensional pseudo-Riemannian manifold X is an open submanifold separated into a finite number of connected components by local C^∞-membranes. Each point $x \in X_0$ has as local invariant a surface S_x of type K3 and thus defines a point in the moduli manifold of K3 surfaces. Although the surface S_x is algebraic, it is far from being a general algebraic surface, since $\operatorname{Pic} S_x \supset A$ is the lattice (1.15).

Let us denote by $M_{\mathrm{K3}}(A)$ the moduli manifold of such surfaces. It is not difficult to see that $M_{\mathrm{K3}}(A)$ is a nonsingular complex-analytic manifold of dimension 11.

In fact, for a K3 surface the choice of the sublattice A (1.15) in $\operatorname{Pic} S$ generalizes the choice of polarization (in the case when $A = \mathbf{Z}$). Therefore the proof of the existence of the moduli manifold $M_{\mathrm{K3}}(A)$ and the universal family over it $S \to^{\pi} M_{\mathrm{K3}}(A)$ is analogous to the proof of the existence of these objects for polarized K3 surfaces. This was precisely what was done by Nikulin in [5] and was based on the global Torelli theorem for polarized K3 surfaces, as proved by Shafarevich and Piatetski-Shapiro in [10].

The correspondence between a thick point $x_0 \in X$ and the local invariant defines a map

$$i\colon X_0 \to M_{\mathrm{K3}}(A), \qquad i(x) = (\tilde{S}_x, A \subset \operatorname{Pic} \tilde{S}_x), \tag{2.1}$$

which is called the *indicatrix* of the pseudo-Riemannian structure (1.1).

The construction of the universal family of K3 surfaces

$$\pi\colon S \to M_{\mathrm{K3}}(A) \tag{2.2}$$

defines an embedding $a\colon A \to \operatorname{Pic} S/M_{\mathrm{K3}}(A)$ of the lattice A into the corresponding Picard group of the map (2.2) and enables one to construct a cross-section s: $A \to \operatorname{Pic} S$.

Thus one defines a sequence of sheaves over the manifold $M_{\mathrm{K3}}(A)$

$$\{R^i\pi(s(a))\}, \tag{2.3}$$

where $R^i\pi$ is the ith direct image of the corresponding inverse sheaf over S.

The indicatrix i (2.1) enables one to carry the sequence of sheaves (2.3) over to X_0: $\{i^*R^i\pi(s(a))\}$. These are generalized spinor sheaves, since the sheaf

$$i^*R^0\pi\left(s\left(\frac{1}{2}\left(h - \sum_{i=1}^{4} L_i^{\pm}\right)\right)\right) \tag{2.4}$$

coincides with the spinor sheaf of the pseudo-Riemannian metric (1.1) (see [8], §3).

Furthermore,

$$T_X = i^*R^0\pi\left(s\left(h - \frac{1}{2}\sum_{i=1}^{4}(L_i^+ + L_i^-)\right)\right) \tag{2.5}$$

and the natural isomorphism

$$\begin{aligned} R^0\pi&\left(s\left(h - \frac{1}{2}\sum_{i=1}^{4}(L_i^+ + L_i^-)\right)\right) \\ &= R^0\pi\left(s\left(\frac{1}{2}\left(h - \sum_{i=1}^{4} L_i^*\right)\right)\right) \otimes R^0\pi\left(s\left(\frac{1}{2}\left(h - \sum_{i=1}^{4} L_i^-\right)\right)\right), \end{aligned}$$

induced by the isomorphism

$$s\left(h - \frac{1}{2}\sum_{i=1}^{4}(L_i^+ + L_i^-)\right) = s\left(1\left(h - \sum_{i=1}^{4} L_i^+\right)\right) \otimes s\left(\frac{1}{2}\left(h - \sum_{i=1}^{4} L_i^-\right)\right)$$

gives a conformal class of pseudo-Riemannian structures (1.1).

The real structure θ on $\tilde{S}_x$ makes it possible to make the image of the indicatrix more precise.

The triple $(\tilde{S}_x, A, \theta)$ is what Nikulin called the real algebraic K3 surface $(\tilde{S}_x)$ with a condition on the group of algebraic cycles (the sublattice $A \subset \operatorname{Pic}\tilde{S}_x$ which is the analogue of polarization) and an involution θ with a condition on the sublattice A (that is, a description of the action of θ on A, for example, in the Riemannian case (1.17), in the Lorentz case (1.17) and so on).

In [5] Nikulin proved the existence of the moduli manifold $M_{\mathrm{K3}}(A, \theta)$ of K3 surfaces with these strict conditions and studied its properties.

PROPOSITION 2 (NIKULIN [5]). 1) *$M_{\mathrm{K3}}(A,\theta)$ is a real analytic manifold of dimension* 11 *separated into a finite number $N(A,\theta)$ of connected components*

$$M_{\mathrm{K3}}(A,\theta) = \bigcup_{i=1}^{N(A,\theta)} M_{\mathrm{K3}}(A,\theta)_i \subset M_{\mathrm{K3}}(A). \tag{2.6}$$

2) *Each component $M_{\mathrm{K3}}(A,\theta)$ is the quotient of the direct product*

$$M_{\mathrm{K3}}(A,\theta)_i = \Delta^+ \times \Delta^- \tag{2.7}$$

of the interiors of two polyhedra in the Lobachevsky space

$$\Delta^+ \subset \mathscr{L}^+, \quad \Delta^- \subset \mathscr{L}^-, \quad \dim\mathscr{L}^+ + \dim\mathscr{L}^- = 11$$

with respect to the componentwise action of a discrete group. (*The interior of a polyhedron is a connected component of the complement of a union of hyperplanes in Lobachevsky space.*)

Enumerating the connected components of the manifold $M_{\mathrm{K3}}(A, \theta)$ (that is, a strict isotopy classification of K3 surfaces with an involution θ and with conditions on the sublattice of algebraic cycles (A, θ)) leads to a classification of integral "polarized" involutions $[L, \varphi, (S, \theta)]$, where L is a unimodular lattice (in our case $H^2(\tilde{S}_x, \boldsymbol{Z})$ with the Poincaré intersection); φ is an involution on L (in our case the action of the complex conjugate on two-dimensional cohomology of the surface); and S is a sublattice at L which is invariant with respect to φ, $\varphi|_S = \theta$ (in our case S is the sublattice A (1.15) with the action of θ (1.16) or (1.17), and so on).

In all the geometrically interesting cases this problem was solved by Nikulin in [5]. First of all he proved that the type (that is, the dimension) of the component itself, say $\mathscr{L}_i^+$, depends only on the topological type of the set of real points $\mathrm{Re}_\theta S_x$.

PROPOSITION 3 (NIKULIN [5]). 1) *If* $\mathrm{Re}_\theta \tilde{S}_x = \varnothing$ *or* $\mathrm{Re}_\theta \tilde{S}_x = 2T_1$, *then*

$$\dim \Delta^+ = \dim \mathscr{L}^+ = 5, \qquad \dim \Delta^- = \dim \mathscr{L}^- = 6.$$

2) *If* $\mathrm{Re}_\theta \tilde{S}_x = T_g \cup kT_0$, *then*

$$\dim \Delta^+ = \dim \mathscr{L}^+ = 6 + k - g,$$
$$\dim \Delta^- = \dim \mathscr{L}^- = 5 + g - k.$$

Thus in the majority of geometrically interesting cases, components of $M_{\mathrm{K3}}(A, \theta)$ have the local construction of the direct product of two Lobachevsky spaces of dimension 5 and 6.

PROPOSITION 4 (NIKULIN [5]). 1) *In the Riemannian case* (*when* θ *acts on the lattice* (1.15), *as in* (1.16)), $M_{\mathrm{K3}}(A, \theta)$ *has two components.*

2) *In the Lobachevsky case* (*when* θ *acts on the lattice* (1.15), *as in* (1.17)), *and when* $\tilde{S}_x = \varnothing$, *the manifold* $M_{\mathrm{K3}}(A, \theta)$ *has two components.*

In the general Lobachevsky case (when $\tilde{S}_x \neq \varnothing$), it has at least another 18 connected components (see [5]).

Finally, the indicatrix (2.1) can be represented as the direct product of its projections

$$i_+ = X_0 \to \Delta^+, \quad i_- = X_0 \to \Delta^-, \qquad i = i_+ \times i_-. \tag{2.8}$$

These results and constructions possibly have some physical interpretation.

Physical digression. Let X_0 be the universe in the general theory of relativity; that is, we have the Lorentz structure (1.1) and the Ricci quadric is positive definite at each point $x \in X_0$.

The concept of the totality of all types of mutual interaction (see [7], §3) presupposes the existence of a large universe Y which has $\mathrm{SU}(3) \times \mathrm{SU}(2) \times \mathrm{U}(1)$-symmetry and Lorentz group symmetry.

The minimal dimension of such a universe Y is 11 (U(1) acts on the circle S^1, SU(2) on the sphere S^2 and SU(3) on the complex projective plane $\boldsymbol{P}^2$).

The gravitational universe X_0 is embedded in the 11-dimensional manifold Y, and the normal bundle N_Y^X can be decomposed into one-dimensional (corresponding to electromagnetism U(1)), two-dimensional (corresponding to SU(2) of the weak interaction) and four-dimensional (corresponding to SU(3) of the strong interaction) subbundles.

It remains for us to note that the indicatrix (2.1) is an embedding of the gravitational universe X_0 into the 11-dimensional manifold $\Delta_5 \times \Delta_6$ (2.7) and the normal bundles of the components of the embedding of the indicatrix (2.8) define one-dimensional and two-dimensional subbundles in $N^X_{\Delta_5 \times \Delta_6}$.

If the indicatrix actually has a physical meaning, then Δ_5 is the Kaluza-Klein universe unifying gravitation and electomagnetic force and Δ_6 is the universe unifying gravitation and the weak interaction.

Moreover, this distinguishes the classical embedding of Einstein from the Kaluza-Klein unification: the spinor sheaf (2.4) and many of its generalizations $R^i\pi(s(a))$, $a \in A$ (2.3), are defined on the manifold $\Delta_5 \times \Delta_6 \subset M_{K3}(A, \theta)$. (We refer the reader to Salam's paper [7] for details of the geometrical formulation of the unification problem.)

§3. The local invariant of a Kähler surface

Let us suppose now that a complex structure has been introduced on the four-dimensional manifold X. This means, in particular, that $\mathbf{R}TX$ has the structure of a complex vector space, which is equivalent to having a complexification automorphism

$$J\colon TX \to TX, \qquad J^2 = -1, \tag{3.1}$$

which defines the decomposition

$$TX = T^{1,0} \oplus T^{0,1}, \quad J|_{T^{1,0}} = \sqrt{-1}\cdot \mathrm{id}, \quad J|_{T^{0,1}} = -\sqrt{-1}\cdot \mathrm{id}. \tag{3.2}$$

A pseudo-Riemannian structure (1.1′) is said to be *Hermitian with respect to* J if the isomorphism g is invariant with respect to J (that is, $JgJ^* = g$).

On the projectivized space $\boldsymbol{P}T_x$ the automorphism J (3.1) acts as an involution $\boldsymbol{P}(J)$ and defines two pointwise fixed lines $L^{1,0}_x = \boldsymbol{P}(T^{1,0}_x)$ and $L^{0,1}_x = \boldsymbol{P}(T^{0,1}_x)$.

The quadric $\boldsymbol{P}(g_x) = q_+ \times q_-$ (1.3) is invariant with respect to $\boldsymbol{P}(J)$.

It follows from (3.2) that the lines $L^{1,0}_x$ and $L^{0,1}_x$ on the quadric $\boldsymbol{P}(g_x)$ belong to a single system of line-generators, let us say q_+, and the involution $\boldsymbol{P}(J)$ acts on $\boldsymbol{P}(g_x)$ as follows:

$$\begin{aligned} &\boldsymbol{P}(J)(q_-) = q_-, \quad \boldsymbol{P}(J)|_{q_-} = \mathrm{id}, \\ &\boldsymbol{P}(J)(q_+) = q_+, \quad \boldsymbol{P}(J)|_{q_+} = \text{harmonic involution} \\ &\qquad\qquad\qquad\qquad\qquad \text{with fixed points } L^{1,0}_x,\ L^{0,1}_x. \end{aligned} \tag{3.3}$$

Further it follows from (3.2) that

$$J\cdot g = h\colon TX \to T^*X, \qquad h^* = -h \tag{3.4}$$

is a skew-symmetric correlation; that is, $h_x \in \Lambda^2 TX_x$.

Let us fix once and for all the point $x \in X$ and omit the suffix x. Then the polarity (1.1′), the automorphism J and everything else are defined on the vector space $T = (T_x)$.

The quadric G (the Grassman manifold of lines in $\boldsymbol{P}(T)$) lies in the projective space $\boldsymbol{P}\Lambda^2 T$. The line generators of the quadric $\boldsymbol{P}(g)$ map into two conics $q_+ \cup q_- \subset G$, and their linear envelopes, the planes

$$\boldsymbol{P}^2_\pm = \langle q_\pm \rangle \subset \boldsymbol{P}^5 = \boldsymbol{P}\Lambda^2 T, \qquad q_\pm = G \cap \boldsymbol{P}^2_\pm, \tag{3.5}$$

define a decomposition of Λ^2 into the eigenspaces of the operator * given by (1.1):

$$\Lambda^2 T = \Lambda_+ \oplus \Lambda_-, \quad *|_{\Lambda_+} = \mathrm{id}, \quad *|_{\Lambda_-} = -\mathrm{id}, \quad \boldsymbol{P}(\Lambda_\pm) = \boldsymbol{P}^2_\pm \tag{3.6}$$

(see [8], §1).

Then the lines $L^{0,1}$ and $L^{1,0}$ in $\boldsymbol{P}(T)$ define a pair of points $L^{0,1}$ and $L^{1,0}$ on the conic q_+, and it is not difficult to see that $\boldsymbol{P}(h) \in \boldsymbol{P}^2_+$ and

$$\boldsymbol{P}(h) = \bigcap_{L \in q_+} \langle L, \boldsymbol{P}(J)(L)\rangle. \tag{3.7}$$

Thus

$$\boldsymbol{P}(\Lambda_+) = \langle L^{0,1}, L^{1,0}, \boldsymbol{P}(h)\rangle. \tag{3.8}$$

The operator $\Lambda^2 J$ which is an involution on $\Lambda^2 T$ can be decomposed into operators on the eigenspaces

$$\begin{gathered} \Lambda^2 T = (H^{2,0} \oplus H^{0,2}) \oplus H^{1,1}, \qquad \Lambda^2 J|_{H^{2,0}+H^{0,2}} = -\mathrm{id}, \\ \Lambda^2 J|_{H^{1,1}} = \mathrm{id}. \end{gathered} \tag{3.9}$$

Here $\boldsymbol{P}(H^{2,0}) = L^{1,0}$ and $\boldsymbol{P}(H^{0,2}) = L^{0,1}$.

Put $\boldsymbol{P}^{1,1} = \boldsymbol{P}^3 = \boldsymbol{P}(H^{1,1})$. According to (3.4), $\boldsymbol{P}^2 \subset \boldsymbol{P}^{1,1}$ and $\boldsymbol{P}(h) \in \boldsymbol{P}^{1,1}$. Thus

$$\boldsymbol{P}^{1,1} = \langle \boldsymbol{P}(h), \boldsymbol{P}(\Lambda_-)\rangle.$$

Consider the projection of $\boldsymbol{P}^5 = \boldsymbol{P}(\Lambda^2 T)$ into $\boldsymbol{P}^{1,1}$ with center in the line $\langle L^{1,0}, L^{0,1}\rangle = \boldsymbol{P}(H^{2,0} \oplus H^{0,2})$. It is easy to see that the special quadratic complex K (1.9) projects into the quadratic cone

$$C = \bigcup_{l \in q_-} \langle L, \boldsymbol{P}(h)\rangle \tag{3.10}$$

with section the conic which is the projection of $\boldsymbol{P}^2_-$ onto q_+.

Let us now suppose that g is a Kähler metric. Then for the curvature quadric (1.7)

$$R\colon \Lambda^2 T \to \Lambda^2 T^*$$

we have $\Lambda^2 J \cdot R = R = R \cdot \Lambda^2 J$ (see [1]).

Geometrically this is equivalent to the condition

$$\operatorname{Sing} \boldsymbol{P}(R) \supset \langle L^{1,0}, L^{0,1}\rangle, \tag{3.11}$$

where $\operatorname{Sing} \boldsymbol{P}(R) = \boldsymbol{P}(\ker R)$; that is, $\boldsymbol{P}(R)$ is the cone with vertex $\langle L^{1,0}, L^{0,1}\rangle$ and base $\boldsymbol{P}(R^{1,1}) = \boldsymbol{P}^{1,1} \cap \boldsymbol{P}(R)$, which is a two-dimensional quadric in $\boldsymbol{P}^3 = \boldsymbol{P}^{1,1}$.

The intersection of the cone C (3.10) and the quadric $\boldsymbol{P}(R^{1,1})$ defines an elliptic curve

$$E = \boldsymbol{P}(R^{1,1}) \cap C, \tag{3.12}$$

which is a double covering of q under projection from the vertex of the cone $\boldsymbol{P}(h)$ by the mapping

$$\varphi\colon E \to q_-. \tag{3.13}$$

The surface (1.8)

$$S = \boldsymbol{P}(\Lambda^2 g) \cap G \cap \boldsymbol{P}(R) = K \cap \boldsymbol{P}(R)$$

is projected from the line $\langle L^{0,1}, L^{1,0}\rangle$ onto the curve E (3.13) with section the conic q_+.

From this it is easy to see that the surface $\tilde{S}$ (1.13) is obtained by the following construction:

Let i be the involution on the elliptic curve E giving the covering (3.13) and let $\boldsymbol{P}(J)$ be the involution (3.3) on q_+. Then $\tilde{S}' = q_+ \times E$, and the two-leafed covering (3.13) is the quotient under the involution $(\boldsymbol{P}(J) \times i)$.

Thus for a Kähler structure (g, J) the manifold of thick points X_0 is always empty.

If we identify $\Lambda^2 T$ with $\Lambda^2 T^*$ using $\Lambda^2 g$, then we can consider the quadrics G and R as symmetric operators on this space. The operator G then coincides with the operator $*$ and the operator R transforms the plane $\boldsymbol{P}(\Lambda_+)$ into the plane $\boldsymbol{P}^2$, which intersects the Grassman manifold G in the conic $q = G \cap R(\boldsymbol{P}(\Lambda_+))$.

If q is a nonsingular conic, then the lines in $\boldsymbol{P}(T)$ corresponding to the points of q sweep out the quadric $\boldsymbol{P}(\operatorname{Ric} P)$ in $\boldsymbol{P}(T)$.

It is easy to see that $\boldsymbol{P}(\operatorname{Ric} R)$ is the projectivized Ricci tensor.

In the Kähler case $R(\boldsymbol{R}(\Lambda_+))$ is defined as the image of a single point, say $\boldsymbol{P}(h)$, since $\boldsymbol{P}(\Lambda_+) \supset \langle L^{1,0}, L^{0,1} \rangle \subset \boldsymbol{P} \ker R$; that is,

$$R(\boldsymbol{P}(\Lambda_+)) = \langle L^{1,0}, L^{0,1}, R(\boldsymbol{P}(h)) \rangle. \tag{3.14}$$

Thus two quadrics $\boldsymbol{P}(g)$ and $\boldsymbol{P}(\operatorname{Ric} R)$ are defined in $\boldsymbol{P}(T)$ which, in the Kähler case, intersect in the pair of lines $L^{1,0}$ and $L^{0,1}$ and the pair of lines $L^{1,0}_-$ and $L^{0,1}_-$ which come from the second system of line generators q_-:

$$\boldsymbol{P}(g) \cap \boldsymbol{P}(\operatorname{Ric} R) = L^{0,1} \cup L^{1,0} \cup L^{1,0}_- \cup L^{0,1}_-. \tag{3.15}$$

If $\operatorname{Ric} R_x$ is nondegenerate at every point $x \in X$, then they define a second pseudo-Riemannian structure on X:

$$\operatorname{Ric} R \colon TX \to T^*X.$$

Three operations are possible:

1) Replacing the complex structure by

$$\boldsymbol{P}(T^{1,0}) = L^{1,0}_-, \qquad \boldsymbol{P}(T^{0,1}) = L^{0,1}_-. \tag{3.16}$$

In this case the pseudo-Riemannian metric g remains Hermitian but in the general case will be no longer Kähler ($\langle L^{0,1}, L^{1,0} \rangle \not\subset \operatorname{Sing} \boldsymbol{P}(R)$).

2) Retaining the complex structure but replacing the pseudo-Riemannian structure by $\operatorname{Ric} g$. The new structure will be Hermitian. What is its tensor curvature?

3) Replacing both the complex and the pseudo-Riemannian structures.

The local invariant of a thin pseudo-Riemannian structure is the projective invariant of the quadric pencil $\alpha = \langle \Lambda^2 g, G, R \rangle$ (see [8]).

In the Kähler case we can present the corresponding matrices in the form

$$\Lambda^2 g = \left\| \begin{matrix} 1 & & & & & \\ & 1 & & 0 & & \\ & & 1 & & & \\ & & & 1 & & \\ & 0 & & & 1 & \\ & & & & & 1 \end{matrix} \right\|, \quad G = \left\| \begin{matrix} 1 & & & & & \\ & 1 & & 0 & & \\ & & 1 & & & \\ & & & -1 & & \\ & 0 & & & -1 & \\ & & & & & -1 \end{matrix} \right\|, \quad R = \left\| \begin{array}{cc|ccc} 0 & & & 0 & \\ & \lambda & a & b & c \\ \hline & a & & & \\ 0 & b & & R_{--} & \\ & c & & & \end{array} \right\|; \tag{3.17}$$

that is, the decomposition (3.9) of the space $\Lambda^2(T)$ gives a decomposition of the pencil α into a direct sum

$$\begin{gathered}\alpha_{2,0} = a|_{H^{2,0}}, \quad \alpha_{0,2} = \alpha|_{H^{0,2}}, \quad \alpha_{1,1} = \alpha|_{H^{1,1}}, \\ \alpha = \alpha_{2,0} \oplus \alpha_{0,2} \oplus \alpha_{1,1}.\end{gathered} \tag{3.18}$$

The components $\alpha_{2,0}$ and $\alpha_{0,2}$ are constant and do not have any invariants, but

$$\begin{gathered}\alpha_{1,1} = \langle \Lambda^2 g^{1,1}, G^{1,1}, R^{1,1} \rangle, \\ \Lambda^2 g^{1,1} = \Lambda^2 g|_{H^{1,1}}, \quad G^{1,1} = G|_{H^{1,1}}, \quad R^{1,1} = R|_{H^{1,1}}.\end{gathered}$$

$$\begin{Vmatrix} 1 & & & 0 \\ & 1 & & \\ & & 1 & \\ 0 & & & 1 \end{Vmatrix}, \quad \begin{Vmatrix} 1 & & & 0 \\ & -1 & & \\ & & -1 & \\ 0 & & & -1 \end{Vmatrix}, \quad \begin{Vmatrix} \lambda & a & b & c \\ a & & & \\ b & & R_{--} & \\ c & & & \end{Vmatrix}.$$

The Hessian of the pencil $\alpha_{1,1}$ is the rational curve with four distinguished points defining the elliptic curve E (3.12).

Thus, in many special but important cases (self-dual metrics [8], §4, Kähler metrics) the local invariant of the structure has little significance, and we turn to the global invariant.

§4. The global invariant

Let X be an orientable four-dimensional compact Riemannian manifold. We define two operators on the space $\Gamma(\Lambda^2 T^*X)$ of the complex-valued C^∞-exterior 2-forms over X:

$$\begin{aligned} *&: \Gamma(\Lambda^2 T^*X) \to \Gamma(\Lambda^2 T^*X), \\ R&: \Gamma(\Lambda^2 T^*X) \to \Gamma(\Lambda^2 T^*X). \end{aligned} \tag{4.1}$$

These operators are obtained by transferring to cross-sections the operators (1.6) and (1.7) defined over each point. (Using $\Lambda^2 g$ we identify the polarities (1.7) and (1.8) with operators on the fibers of $\Lambda^2 T^*X$.)

We define three symmetric bilinear forms on $\Gamma(\Lambda^2 T^*X)$:

$$\begin{aligned} v(\omega_1, \omega_2) &= \int_X \omega_1 \wedge \omega_2, \\ w(\omega_1, \omega_2) &= \int_X \omega_1 \wedge {}^*\omega_2, \\ r(\omega_1, \omega_2) &= \int_X \omega_1 \wedge R\omega_2. \end{aligned} \tag{4.2}$$

Let Δ be the Laplace operator of the metric g. We restrict the quadratic forms (4.2) to the kernel $H^2(X, \boldsymbol{C}) = H$ of this operator; this is a $b_2(X)$-dimensional $\boldsymbol{C}$-vector space. Then

$$\begin{gathered} v|_{H^2(X,\boldsymbol{C})} \text{ is the Poincaré intersection form} \\ w|_{H^2(X,\boldsymbol{C})} \text{ is the Weil form} \\ r|_{H^2(X,\boldsymbol{C})}(\omega_1, \omega_2) = \int_X \omega_1 \wedge P_\Delta R(\omega_2), \end{gathered} \tag{4.3}$$

where ω_i is a harmonic 2-form over X and P_Δ is projected onto harmonic forms (see for example [9], §2).

Thus there is a pencil of quadratic forms $\alpha = \langle v, w, r\rangle$ in the space H ($= H^2(X, \boldsymbol{C})$). As usual, the group $\mathrm{GL}(H)$ acts on the pencil, specifically $\langle gvg^*, gwg^*, grg^*\rangle$ for $g \in \mathrm{GL}(H)$, and the group $\mathrm{GL}(3, \boldsymbol{C})$ acts by replacing the basis (v, w, r) of the pencil.

DEFINITION 2. The class of equivalent pencils $\langle v, w, r\rangle$ under the action of $\mathrm{GL}(H) \times \mathrm{GL}(3, \boldsymbol{C})$ is called the *global invariant* of the Riemannian structure (1.1).

A class of quadratic pencils is defined by either

1) an algebraic curve $H(\alpha)$, the Hessian of α; or

2) a two-dimensional linear sequence $|\eta|$ on $H(\alpha)$ giving the map φ: $H(\alpha) \to \boldsymbol{P}^2$ (the basis of the pencil); or

3) a nondegenerate θ-characteristic curve $H(\alpha)$ (θ is a sheaf over $H(\alpha)$, $\theta^2 = \Omega H(\alpha)$ and $h^0(H(\alpha), \theta) = 0$); or an algebraic variety

$$S = \boldsymbol{P}(v) \cap \boldsymbol{P}(w) \cap \boldsymbol{P}(r) \subset \boldsymbol{P}(H) \tag{4.4}$$

which is the intersection of three quadrics in $\boldsymbol{P}(H)$.

Both the Hessian $H(\alpha)$ and the manifold S (4.4) have special properties which stipulate the presence in the pencil of a special subpencil $\langle v, w\rangle$.

We decompose the space H ($= H^2(X, \boldsymbol{C})$) into the eigenspaces of the involution * (4.1):

$$\begin{gathered} H = H_+ \oplus H_-, \quad *|_{H_+} = \mathrm{id}, \quad *|_{H_-} = -\mathrm{id}, \\ \dim H_+ = b_+(X), \qquad \dim H_- = b_-(X), \\ b_+(x) + b_-(x) = b_2(X), \qquad b_+(X) - b_-(X) = \sigma(x), \end{gathered} \tag{4.5}$$

where $\sigma(x)$ is the signature of the manifold X.

The quadrics $\boldsymbol{P}(v)$ and $\boldsymbol{P}(w)$ (4.3) have the same restriction to $\boldsymbol{P}(H_+)$ and $\boldsymbol{P}(H_-)$:

$$\boldsymbol{P}(v) \cap \boldsymbol{P}(H_\pm) = \boldsymbol{P}(w) \cap \boldsymbol{P}(H_\pm) = Q_\pm \tag{4.6}$$

which is a nonsingular quadric in $\boldsymbol{P}(H_\pm)$.

There are only two singular quadrics in the quadric pencil $\langle v, w\rangle$; these are

$$G_\pm = \bigcup_{t \in Q_\pm} \langle t, \boldsymbol{P}(H_\mp)\rangle \tag{4.7}$$

of rank $b_\pm$. Their intersections and consequently the intersection of the quadrics $\boldsymbol{P}(v)$ and $\boldsymbol{P}(w)$ are

$$K = \bigcup_{\substack{t_+ \in Q_+ \\ t_- \in Q_-}} \langle t_+, t_-\rangle = G_+ \cap G_- = \boldsymbol{P}(v) \cap \boldsymbol{P}(w). \tag{4.8}$$

COROLLARY. 1) *The Hessian $H(\alpha)$ is a curve of genus $b_+(X)b_-(X) - b_2(X) + 1$.*

2) *There exist two one-dimensional linear sequences $|\eta_+| = \boldsymbol{P}^1_+$ and $|\eta_-| = \boldsymbol{P}^1_-$, where $\deg|\eta_\pm| = b_\mp$. These sequences give a map φ_0: $H(\alpha) \to \boldsymbol{P}^1_+ \times \boldsymbol{P}^1_-$ into a quadric. The map φ: $H(\alpha) \to \boldsymbol{P}^2$ onto the base of the pencil α is obtained by stereographic projection of the quadric $\boldsymbol{P}^1_+ \times \boldsymbol{P}^1_-$ onto the plane $\boldsymbol{P}^2$.*

All these assertions are proved in the same way as the corresponding assertions for the Hessian for the global local invariant in [8], §1.

Let us turn to the description of the manifold S (4.4). First of all the set of singular points of the manifold K (4.8) can be written as

$$\operatorname{Sing} K = Q_+ \cup Q_-. \tag{4.9}$$

Therefore the intersection $\boldsymbol{P}(v) \cap \boldsymbol{P}(w) \cap \boldsymbol{P}(r)$ has singularities

$$\operatorname{Sing} S \supset (Q_+ \cup Q_-) \cap \boldsymbol{P}(r). \tag{4.10}$$

The projection of $\boldsymbol{P}(H)$ onto $\boldsymbol{P}(H_\mp)$ with center $\boldsymbol{P}(H_\pm)$ defines a map of the manifold K (4.8) onto the quadric $Q_\mp$ and thus a map

$$\pi_\pm \colon S \to Q_\pm, \tag{4.11}$$

whose fiber is the intersection of two quadrics in $P^{b_\mp}$.

The direct project of these maps

$$\pi = (\pi_+ \times \pi_-) \colon S \to Q_+ \times Q_- \tag{4.12}$$

is a double covering branching at the hypersurface $D_S = (s)_0$ of zeros of the section $s \in H^0(Q_+ \times Q_-, \mathcal{O}_{Q_+ \times Q_-}$ (2.2)).

All the maps described above are rational. In order to make them regular we can use the canonical resolution of the singular points of the manifold K (4.8); we have the diagram

$$\begin{array}{c} \boldsymbol{P}\left(\mathcal{O}_{Q_+ \times Q_-}(-1,0) \oplus \mathcal{O}_{Q_+ \times Q_-}(0,-1)\right) \xrightarrow{\pi} Q_+ \times Q_- \\ \varphi \downarrow \\ \boldsymbol{P}(H) = \boldsymbol{P}\left(\left[H^0\left(\mathcal{O}_{Q_+ \times Q_-}(1,0)\right)\right]^* \oplus \left[H^0\left(\mathcal{O}_{Q_+ \times Q_-}(0,1)\right)\right]^*\right), \end{array} \tag{4.13}$$

where the map φ is given by the standard embedding

$$\begin{array}{l} 0 \to \mathcal{O}_{Q_+ \times Q_-}(-1,0) \oplus \mathcal{O}_{Q_+ \times Q_-}(0,-1) \\ \quad \xrightarrow{\varphi} \left[H^0\left(\mathcal{O}_{Q_+ \times Q_-}(1,0) \oplus \mathcal{O}_{Q_+ \times Q_-}(0,1)\right]^* \otimes \mathcal{O}_{Q_+ \times Q_-} \\ \qquad \| \\ \qquad H \end{array}$$

followed by projection into $\boldsymbol{P}(H)$.

The quadratic form r on H is defined by the cross-section $r \in H^0(\boldsymbol{P}(H), \mathcal{O}_{\boldsymbol{P}(H)}$ (2)). But then the direct image

$$R^0\pi\varphi^*(r) \in H^0\left(Q_+ \times Q_-, S^2\left(\mathcal{O}_{Q_+ \times Q_-}(1,0) \oplus \mathcal{O}_{Q_+ \times Q_-}(0,1)\right)\right), \tag{4.14}$$

that is, r induces the quadratic form

$$\begin{array}{c} s_r \colon \mathcal{O}_{Q_+ \times Q_-}(-1,0) \oplus \mathcal{O}_{Q_+ \times Q_-}(0,-1) \to \mathcal{O}_{Q_+ \times Q_-}(1,0) \oplus \mathcal{O}_{Q_+ \times Q_-}(0,1), \\ s_r = R^0\pi\varphi^*(r) \end{array} \tag{4.15}$$

on the two-dimensional sheaf $\mathcal{O}_{Q_+ \times Q_-}(-1,0) \oplus \mathcal{O}_{Q_+ \times Q_-}(0,-1)$.

The isotropy subspaces of the quadratic form give the manifold

$$S \subset \boldsymbol{P}\left(\mathcal{O}_{Q_+ \times Q_-}(-1,0) \oplus \mathcal{O}_{Q_+ \times Q_-}(0,-1)\right).$$

It is easy to see that $\tilde{S}$ is the canonical resolution of the singularity $(Q_+ \cup Q_-) \cap \boldsymbol{P}(r)$ on the manifold S (4.4).

The line bundle $\boldsymbol{P}(\mathcal{O}_{Q_+\times Q_-}(-1,0) \oplus \mathcal{O}_{Q_+\times Q_-}(0,-1))$ contains two nonintersecting cross-sections

$$S_+ = \boldsymbol{P}\big(\mathcal{O}_{Q_+\times Q_-}(-1,0)\big), \qquad S_- = \boldsymbol{P}\big(\mathcal{O}_{Q_+\times Q_-}(0,-1)\big). \tag{4.16}$$

In the space $\boldsymbol{P}(H)$ the quadrics Q_+ and Q_- intersect $\boldsymbol{P}(r)$ in manifolds

$$W_+ = Q_+ \cap \boldsymbol{P}(r), \qquad W_- = Q_- \cap \boldsymbol{P}(r). \tag{4.17}$$

Then the cross-sections $S_\pm$ (4.16) intersect $\tilde{S}$ in

$$W_+ \times Q_- = S_+ \cap \tilde{S}, \qquad Q_+ \times W_- = S_- \cap \tilde{S}. \tag{4.18}$$

The operator $*$ on H lifts to an involution $*$ on the two-dimensional sheaf $\mathcal{O}_{Q_+\times Q_-}(-1,0) \oplus \mathcal{O}_{Q_+\times Q_-}(0,-1)$:

$$*|\mathcal{O}_{Q_+\times Q_-(-1,0)} = \mathrm{id}, \qquad *|\mathcal{O}_{Q_+\times Q_-(0,-1)} = -\mathrm{id}. \tag{4.19}$$

The symmetric square (4.14) of the two-dimensional sheaf decomposes into a direct sum

$$S^2\big(\mathcal{O}_{Q_+\times Q_-}(1,0) \oplus \mathcal{O}_{Q_+\times Q_-}(0,1)\big) = \mathcal{O}_{Q_+\times Q_-}(2,0) \oplus \mathcal{O}_{Q_+\times Q_-}(1,1) \oplus \mathcal{O}_{Q_+\times Q_-}(0,2).$$

Therefore the cross-section s_r can also be decomposed into components:

$$s_r^+ \in H^0\big(\mathcal{O}_{Q_+\times Q_-}(2,0)\big), \quad s^- \in H^0\big(\mathcal{O}_{Q_+\times Q_-}(1,1)\big), \quad s_r^- \in H^0\big(\mathcal{O}_{Q_+\times Q_-}(0,2)\big),$$

$$s_r = s_r^+ + s_r^{+-} + s_r^- . \tag{4.20}$$

The zeros of the corresponding components are

$$(s_r^+)_0 = W_+ \times Q_-, \quad (s_r^-) = Q_+ \times W_-, \quad (s_r^{+-})_0 = E_r.$$

Since s_r is defined up to a multiplying factor, the invariant of the Riemannian structures is just the manifold of zeros in the corresponding components.

DEFINITION 3. 1) The divisor

$$W_+ \times Q_- \cup Q_+ \times W_- \subset Q_+ \times Q_- \tag{4.21}$$

is called the *global Weil invariant* of the Riemannian structure (1.1).

2) The manifolds W_+ and W_- are called the *$(\pm)$-components* of the Weil invariants.

3) The divisor $E_r \subset Q_+ \times Q_-$ is called the *global Ricci invariant* of the Riemannian structure (1.1).

Thus the global invariant of the Riemannian structure S is a two-leafed covering

$$\pi\colon \tilde{S} \to Q_+ \times Q_-, \tag{4.12'}$$

which branches on the divisor D_S of type (2.2) in $\mathrm{Pic}(Q_+ \times Q_-)$.

Thus, as in Lemma 5.1 in [8], one can prove

PROPOSITION 5. *The divisor*

$$D_S \in \langle (W_+ \times Q_- + Q_+ \times W_-), 2E_r \rangle$$

lies in the pencil of divisors spanned by the Weil invariant and the doubled Ricci invariant.

COROLLARY. *The dimension of the moduli manifold* M_G *of the global invariants is* $b_+(X) \cdot b_-(X) + b_2(X) - 4$.

DEFINITION 4. The Riemannian manifold X (1.1) is said to be
1) *Einstein in the large* if $s_r^{+-} = 0$,
2) *self-dual in the large* if $s_r^+ = 0$, and
3) *conformally flat in the large* if $s_r^+ = s_r^- = 0$.

It follows immediately from the definition that the structures (1.1) which are locally Einstein, locally self-dual or locally conformally flat have the corresponding global invariants.

If X is a complex analytic surface and (1.1) is the Kähler metric, then the space H has another decomposition besides (4.5):

$$\begin{gathered} H = H^{2,0} \oplus H^{0,2} \oplus H^{1,1}, \\ H_+ = \langle H^{2,0} \oplus H^{0,2}, h \rangle, \qquad H^{1,1} = \langle h, H_- \rangle, \end{gathered} \tag{4.22}$$

where h is the cohomology class of Kähler forms. In this case $W_+ = 2L$, where L is the hyperplane cross-section

$$L = Q_+ \cap \boldsymbol{P}(H^{2,0} \oplus H^{0,2}). \tag{4.23}$$

The Ricci invariant

$$E_r = (L \times Q_-) \cup (Q_+ \times E'),$$

where E' is separated from Q_- by a hyperplane in $\boldsymbol{P}(H_-)$.

Finally $\tilde{S} = Q_+ \times E$, where E is a two-leafed covering of Q_- branching at W_- (that is, E is the intersection of a pair of quadrics in $\boldsymbol{P}(H^{1,1})$) is a cone with vertex h and base Q_- in $\boldsymbol{P}(r)$.

The invariant of the intersection of a pair of quadrics at $\boldsymbol{P}(H^{1,1})$ is a collection of $h^{1,1} = \dim H^{1,1}$ points in the projective line $\boldsymbol{P}^1$ defined up to a bilinear transformation.

COROLLARY. *The dimension of the manifold* M_{GK} *of global invariants of Kähler structure is* $h^{1,1} - 3$.

Note that this dimension is complex without taking into account the real structure of the global invariant.

EXAMPLE. Let X be a K3 surface and let h be a Kähler metric which is not an Einstein metric. Then M_{GK} is the moduli manifold of hyperelliptic curves of genus 9. The surface (X, h) corresponds to a hyperelliptic curve of genus 9. Since the dimension of the variation space of a Kähler structure on X is 20, and the dimension of the space of global invariants is 17, then in this case the global invariant does not depend on the complex structure of X.

Finally, let us recall that the quadric (4.3) is real; therefore there is a real structure on the global invariant S given by the involution θ. In the Riemannian case

$$\theta(W_+) = W_+, \quad \theta(W_-) = W_-, \quad \theta(E_r) = E_r;$$

that is, θ defines a real structure on every component of the global invariant S.

The real structure defines a finite number of connected submanifolds $M_{GK}(\theta)_i$ in the moduli manifold of global structures M_{GK}. The description of these components is similar to a description of Nikulin's components $M_{K3}(A, \theta)$ which has not yet been obtained.

REMARK (Continuation of Remark 1 from §1). If the local invariant of a four-dimensional pseudo-Riemannian manifold is the generalization of Gauss curvature, then the global invariant is the generalization of its integral over the surface. By analogy with the Gauss-Bonnet formula we can conjecture that the components of the manifold $M_{\mathrm{GK}}(\theta)$ distinguish distinct classes of structures and can be called the genera of four-dimensional manifolds.

BIBLIOGRAPHY

1. Shoshichi Kobayashi and Katsumi Nomizu, *Foundations of differential geometry*, Vols. 1, 2, Interscience, 1963, 1969.

2. Solomon Lefschetz, *A page of mathematical autobiography*, Bull. Amer. Math. Soc. **74** (1968), 854–879.

3. Yoichi Miyaoka, *On the Chern numbers of surfaces of general type*, Invent. Math. **42** (1977), 225–237.

4. Boris Moishezon, *Complex surfaces and connected sums of complex projective planes*, Lecture Notes in Math., vol. 603, Springer-Verlag, 1977.

5. V. V. Nikulin, *Involutions of integral quadratic forms and their applications to real algebraic geometry*, Izv. Akad. Nauk SSSR Ser. Mat. **47** (1983), 109–188; English transl. in Math. USSR Izv. **22** (1984).

6. P. K. Rashevskiĭ, *Riemannian geometry and tensor analysis*, 4th ed., "Vysshaya Shkola", Moscow, 1978; German transl. of 1st ed., VEB Deutscher Verlag Wiss., Berlin, 1959.

7. Abdus Salam, *Gauge interactions, elementarity and superunification*, Philos. Trans. Roy. Soc. London Ser. A **304** (1982), 135–153.

8. A. N. Tyurin, *A local invariant of a Riemannian manifold*, Izv. Akad. Nauk SSSR Ser. Mat. **45** (1981), 824–851; English transl. in Math. USSR Izv. **19** (1982).

9. S. S. Chern, *Complex manifolds*, Inst. Fis. Mat. Univ. Recife, Recife, 1959.

10. I. I Pyatelskiĭ-Shapiro [Piatetski-Shapiro] and I. R. Shafarevich, *A Torelli theorem for algebraic surfaces of type* K3, Izv. Akad. Nauk SSSR Ser. Mat. **35** (1971), 530–572; English transl. in Math. USSR Izv. **5** (1971).

11. Shing Tung Yau, *Calabi's conjecture and some new results in algebraic geometry*, Proc. Nat. Acad. Sci. U.S.A. **74** (1977), 1798–1799.

12. I. R. Shafarevich et al., *Algebraic geometry*, Trudy Mat. Inst. Steklov. **75** (1965); English transl., Proc. Steklov Inst. Math. **75** (1965).

Translated by A. WEST

Труды Матем. Инст. Стеклов.
165(1984)

Proceedings of the Steklov
Institute of Mathematics
1985, Issue 3

REPRESENTATIONS OF LATTICES OVER DIVISION RINGS

UDC 517.98

A. V. YAKOVLEV

ABSTRACT. A representation of a lattice over a ring k is a homomorphism of the lattice into the lattice of all submodules of a free k-module of finite rank. A lattice is called algebraic over a field k if every representation over an arbitrary extension of k is induced by a representation over an algebraic extension of k. A lattice that is not algebraic over k is called transcendental over k. It is proved that for a lattice transcendental over an algebraically closed field k there exist infinitely many dimensions in which there are infinitely many nonisomorphic indecomposable representations of the lattice, and for an algebraic lattice of finite length there exist only finitely many nonisomorphic indecomposable representations. Representations of complemented lattices of finite length are also studied.

Bibliography: 6 titles.

In the past decade the scope of representation theory has been considerably extended; along with representations of groups and rings, the study of representations of quite different objects, such as quivers, partially ordered sets, etc. has been initiated. This paper makes the first steps towards a theory of representations of lattices over division rings. It seems to the author that this is not just another representation theory, but a quite general context into which many of the above-mentioned theories can be included. In §1 we give the definitions and explain how representations of partially ordered sets and quivers can be considered as representations of lattices. §2 gives some simple but typical examples of representations of lattices. The fundamental result is contained in §3, where we prove the existence, for finitely presented transcendental lattices, of an infinite number of dimensions, in each of which there are an infinite number of indecomposable representations, while for algebraic lattices there are only a finite number of indecomposable representations.

§1. Representations of lattices and problems which lead to them

By a representation of a lattice M over a ring k, we mean a homomorphism φ from M to the lattice $L(V)$ of submodules of some free k-module V of finite rank. In other words, the representation φ associates to every element m from M a submodule $V_m = \varphi(m)$ of V, while $V_{x+y} = V_x + V_y$ and $V_{xy} = V_x \cap V_y$. The rank of the free module V is called the dimension of the representation. One defines morphisms, direct sums and indecomposable representations in a natural way.

1980 *Mathematics Subject Classification*. Primary 06B15.

A congruence relation θ is introduced in the lattice M by saying that $x\theta y$ if $\varphi(x) = \varphi(y)$ for any homomorphism φ from M to the lattice $L(V)$ of submodules of a free k-module V, whose rank is not necessarily finite. If θ is the identity congruence, the lattice M is said to be faithfully representable over k. In the general case the quotient lattice M/θ is faithfully representable over k, and any representation of M is induced by some representation φ_1: $M/\theta \to L(V)$, i.e. has the form

$$M \to M/\theta \overset{\varphi_1}{\to} L(V)$$

(where the left arrow is the canonical epimorphism on a quotient structure). We may therefore confine ourselves to studying the representations of faithfully representable lattices. In such lattices all identities of submodules of free k-modules are fulfilled; in particular, independently of the choice of the ring k, they are Dedekind (modular) and Desarguesian (the latter term refers to a fairly complicated identity, which is the lattice-theoretic form of the well-known geometrical theorem of Desargues).

In what follows M will denote a lattice with 0 and 1, faithfully representable over a division ring T; we shall consider only those representations φ: $M \to L(V)$ for which $\varphi(0) = 0$ and $\varphi(1) = 1$. The class of all lattices faithfully representable over T will be denoted by $\mathfrak{M}_T$. It is clear that $\mathfrak{M}_T = \mathfrak{M}_k$, where k is the prime subfield of T.

To end this section, we indicate two problems of representation theory which can be reduced to the problem of describing the representations of a certain lattice. The first of these is the problem of describing the representations of a partially ordered set in the sense of Nazarova and Roĭter [4]. Let N be a partially ordered set and M the lattice freely generated by N; then a representation of N over the division ring T is exactly the same as a representation of M over T. One can replace M by the free Dedekind and Desarguesian lattice, or every by the free lattice from $\mathfrak{M}_T$ generated by N.

The second example is concerned with representations of quivers in the sense of Gabriel [6]. To each vertex i of a quiver, we associate a two-element chain $X_i = \{x_i < y_i\}$, and to each arrow α a four-element chain $\{a_\alpha < b_\alpha < c_\alpha < d_\alpha\} = A_\alpha$. Consider the free product of all lattices X_i, A_α and then form the quotient lattice M of this free product by associating to each arrow α: $i \to j$ the following relations:

$$x_i = a_\alpha, \quad y_i = c_\alpha, \quad x_j = b_\alpha, \quad y_j = d_\alpha,$$

and to the sublattice generated by all x_i and y_j the distributivity condition. Representations of the lattice M are exactly the same as representations of the original quiver.

The last example can be substantially generalized so as to include problems of classifying indecomposable objects in many important additive (not necessarily abelian) categories and even in some nonadditive categories. This will be carried out in a subsequent paper.

§2. Representations of complemented lattices of finite length

As an illustration of a typical situation arising in the theory of representations of lattices, let us study the representations of complemented lattices of finite length.

THEOREM 1. *A lattice of length 2 has a finite number of indecomposable representations over any division ring if and only if the number of its atoms does not exceed 3. If it has more than 3 atoms, then the number of pairwise nonisomorphic representations is infinite, their dimensions being unbounded. If, furthermore, the division ring T is infinite, then in every dimension there are infinitely many nonisomorphic irreducible representations.*

This theorem is well known. The lattices mentioned in it have the form

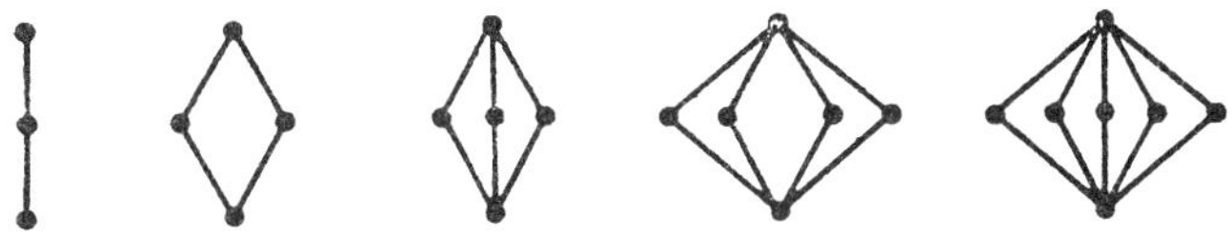

etc. The first two are not simple and each of them has two irreducible representations of dimension 1; the third has a unique irreducible representation of dimension 2 (for a basis e_1, e_2 of a two-dimensional space, the atoms are subspaces spanned by e_1, e_2, $e_1 + e_2$). Representations of the fourth lattice are quadruples of subspaces restricted by the following condition: any two have a zero intersection, while the sum of any two is the entire space. However, such quadruples of subspaces form an infinite series, which is typical for all "tame" problems of representation theory. For the remaining lattices, the problem of describing their representations is "wild".

The proof of the following theorem follows the usual methods of representation theory and therefore will only be sketched.

THEOREM 2. *Suppose that D and T are division rings. The category of representations of the lattice $L(D^n)$ $(n \geqslant 3)$ over T is equivalent to the category of representations of D by matrices over T.*

PROOF. We shall indicate the correspondence between a representation of the division ring D by matrices over T and a representation of the lattice $L(D^n)$, as well as the inverse correspondence. However, we shall omit the computational check that these correspondences define mutually inverse functors.

First suppose that $\varphi\colon D \to T_m$ is a homomorphism of the division ring D to the ring of $m \times m$ matrices over the division ring T. Then φ induces a homomorphism φ_n of the lattice of subspaces of the n-dimensional vector space D^n over D to the lattice of submodules of a free T_m-module of rank n; however, the latter lattice is canonically isomorphic to the lattice of subspaces of an mn-dimensional vector space over T. In this way, φ_n yields a representation of the lattice $L(D^n)$ over T, having dimension mn.

For the construction of the inverse correspondence, we remind the reader how one can recover the ring k from the lattice $L(k^n)$ $(n \geqslant 3)$ of submodules of a free n-dimensional k-module, if one is given 4 elements U, X, Y, Z of this lattice such that for some basis $e_1, \dots, e_n$ of the free module k^n

$$X = ke_1, \quad Y = ke_2, \quad Z = ke_3, \quad U = k(e_1 + e_2).$$

On the set A of complements of Y in $X + Y$ introduce two operations by the following method. Suppose V_1 and V_2 are two distinct complements of X in $X + Z$; for elements $P, Q \in A$, let

$$P \oplus Q = \left[(P + V_1)(Y + V_2) + (Y + V_1)(Q + V_2)\right](X + Y). \tag{1}$$

Suppose now that W_1 and W_2 are two distinct complements of Y in $Y + Z$; let

$$P \circ Q = \left[(U + W_1)(W_2 + Q) + \{X + (U + W_1)(W_2 + Q)\}(W_1 + P)\right](X + Y). \tag{2}$$

The Desarguesian identity easily implies that these definitions do not depend on the choice of the subspaces Z, V_1, V_2, W_1 and W_2. It is not difficult to check that, with these operations, A is a ring isomorphic to k, the isomorphism $k \to A$ being given by the correspondence $\alpha \to k(e_1 + \alpha e_2)$, so that X is the zero or the ring A and U its unit element (the details can be found, for example, in [1], [2] or [5]).

Suppose now that φ: $L(D^n) \to L(T^N)$ is a representation. Since all atoms of the lattice $L(D^n)$ are projectively equivalent, their images in $L(T^N)$ are also projectively equivalent and have the same dimension m; this shows that $N = mn$. The lattice $L(T^N)$ may be canonically identified with $L((T_m)^n)$. Considering the vector space D^n as the space of row vectors of length n, let x, y, z and u denote the subspaces generated by the corresponding vectors $(1, 0, 0, \dots, 0)$, $(0, 1, 0, \dots, 0)$, $(0, 0, 1, \dots, 0)$ and $(1, 1, 0, \dots, 0)$, and let a be the set of complements of x in $x + y$. Furthermore, let $X = \varphi(x)$, $Y = \varphi(y)$, $Z = \varphi(z)$ and $U = \varphi(u)$, and let A be the set of complements of X in $X + Y$. Introducing operations on a and A as indicated above, one can check from (1) and (2) that φ preserves these operations:

$$\varphi(p \oplus q) = \varphi(p) \oplus \varphi(q), \qquad \varphi(p \circ q) = \varphi(p) \circ \varphi(q),$$

i.e. that φ is a homomorphism of the ring a to the ring A. However, as remarked above, $a \cong D$ and $A \cong T_m$; in this way, we obtain a homomorphism $D \to T_m$.

THEOREM 3. *A complemented lattice M from the class $\mathfrak{M}_T$ has a finite number of indecomposable representations if and only if it is a direct sum of a finite number of simple lattices of the following types*:

(1) *the two-element lattice consisting of* 0 *and* 1;

(2) *the lattice* 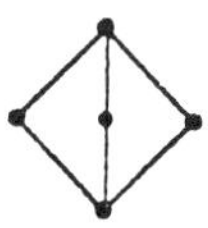 ;

(3) *the lattice $L(T_1^n)$, where $n \geqslant 3$ and the division ring T_1 has only a finite number of representations over T.*

Each indecomposable representation of the lattice M contracts into zero spaces all the direct summands except one, and therefore is in fact an irreducible representation of this simple summand (all such representations are described in Theorems 1 and 2).

PROOF. *Necessity.* If M has only a finite number of irreducible representations over T, then M can be imbedded into the direct sum of certain lattices of the form $L(T^n)$ (since M is exactly representable over T) and therefore has finite length. A complemented lattice of finite length splits into a direct sum of simple complemented lattices. Furthermore, a simple Dedekind and Desarguesian complemented lattice of finite length is either a lattice of length 2, or a lattice of the form $L(T_1^n)$, where T_1 is some division ring and $n \geqslant 3$ ([2], Theorem IV.15). Since a representation of each of the direct summands induces a representation of the entire lattice

M, these direct summands can only have a finite number of irreducible representations over T and therefore by Theorems 1 and 2 must belong to the types indicated in the statement of the theorem.

The sufficiency of the conditions is clear.

Note that this argument not only proves Theorem 3, but also allows one to reduce the description of representations of a complemented lattice of finite length to the problem of describing the representations of one division ring by matrices over another and to the representation of complemented lattices of length 2 with at most 5 elements.

§3. Representations of finitely presented lattices over fields

We say that a lattice is finitely presented if it has a finite number of generators on which are imposed only a finite number of relations. The lattices described at the end of §1 are finitely presented (if the partially ordered set and the quiver are finite). Furthermore, we say that a lattice is algebraic([1]) over k if every representation of it over a field $K \supset k$ is induced by some representation over an algebraic extension of k contained in K. Otherwise, the lattice is called transcendental.

THEOREM 4. *If M is a finitely presented transcendental lattice over an algebraically closed field k, then there exist an infinite number of dimensions, in each of which M has an infinite number of nonisomorphic indecomposable representations.*

PROOF. By the definition of transcendence, there exists some representations φ of M over a transcendental extension of K which is not equivalent to any representation over k. To specify the representation, it suffices to give a finite number of vectors, generating a finite number of subspaces which correspond to the finite system of generators of the lattice. The number of all components of these vectors is also finite. Replacing K by the subfield obtained by adjoining to k all these components, we obtain a finitely generated subfield $K_1 \subset K$ such that the representation φ is already defined over K_1. We may therefore assume that K is finitely generated, i.e. that K is a finite extension of a purely transcendental extension $k(x_1, \dots, x_n)$.

Let m be the dimension of the representation φ and S the system of generators of M. Then $\varphi(s)$ for $s \in S$ is a subspace of K^m; denote by $A_s(x_1, \dots, x_n)$ the matrix whose components are algebraic functions in $x_1, \dots, x_n$ and whose rows are elements of a basis of the subspace $\varphi(s)$. For any lattice polynomial f in elements of S, we can construct a matrix B_f whose components are rational functions of components of the matrix A_s in such a way that the rows of B_f generated the subspace $f(\varphi(s))$. A relation $f = g$ means that the ranks of the matrices B_f, B_g and $\binom{B_f}{B_g}$ are equal, which impose on the components of the matrices $A_s(x_1, \dots, x_n)$ certain algebraic relations.

Consider now the representation $\overline{\varphi}$ of the lattice M over a field k, having the same dimension as φ and such that $\dim_K \varphi(s) = \dim_k \overline{\varphi}(s)$ for any $s \in S$. Let A_s be a matrix with components from k, whose rows form a basis of $\overline{\varphi}(s)$. If $\overline{\varphi}$ and φ were equivalent, there would exist nonsingular matrices X_s and Y with components in K

([1]) The term "algebraic lattice" is usually employed in a quite different way in lattice theory (see [2]).

for which

$$X_s \bar{A}_s = A_s Y \tag{$*$}$$

for all $s \in S$. These equations form a homogeneous linear system with respect to the matrices X_s and Y; let $X_s^{(j)}$ and $Y^{(j)}$ ($j = 1, \ldots, N$) form a basis of the set of its solutions, so that any solution has the form $X_s = X_s^{(1)} t_1 + \cdots + X_s^{(N)} t_N$, $Y = Y^{(1)} t_1 + \cdots + Y^{(N)} t_N$. However, the representations are assumed to be inequivalent, i.e. for any $t_1, \ldots, t_N$ at least one of the matrices X_s or Y is singular. This means that the polynomial

$$\det\left(Y^{(1)} t_1 + \cdots + Y^{(N)} t_N\right) \cdot \prod_{s \in S} \det\left(X_s^{(1)} t_1 + \cdots + X_s^{(N)} t_N\right)$$

is identically zero.

For all algebraic functions constituting the matrices A_s, consider their specializations for $x_1 \to a_1, \ldots, x_n \to a_n$ ($a_i \in k$). We would like to construct infinitely many pairwise nonisomorphic specializations, which would yield representations of M over k. Suppose that a certain number $\varphi_1, \ldots, \varphi_r$ of such specializations have been constructed. The next specialization should be chosen in such a way that

1) the ranks of all the matrices A_s are not diminished;

2) for all defining relations $f = g$, the ranks of the matrices B_f and B_g are not diminished; and

3) for all φ_j, $j = 1, \ldots, r$, the dimension of the space of solutions of the system analogous to ($*$) is not increased, i.e. every such solution has the form $X_s = X_s^{(1)} t_1 + \cdots + X_s^{(N)} t_N$, $Y = Y^{(1)} t_1 + \cdots + Y^{(N)} t_N$, and therefore for any solution the determinant of at least one of the matrices X_s or Y is equal to 0.

All these conditions are equivalent to finite systems of algebraic inequalities. The first two show that the subspaces spanned by the rows of matrices $A_s(a_1, \ldots, a_n)$ generate the representation of M; a representation of this kind will be called a nonsingular specialization of φ. The last group of conditions guarantees that the representation we obtain is not equivalent to any of the representations $\varphi_1, \ldots, \varphi_r$.

The representations $\varphi_1, \varphi_2, \ldots$ need not be indecomposable. However, among their direct summands there are infinitely many pairwise nonisomorphic ones, and there will exist a dimension $l \leqslant m$ for which there exist infinitely many pairwise nonisomorphic irreducible direct summands of dimension l. Replacing the specialization

$$x_1 \to a_1, \quad x_2 \to a_2, \quad \ldots, \quad x_n \to a_n$$

by the specialization

$$x_1 \to \begin{Vmatrix} a_1 & 1 & 0 & \cdots & 0 & 0 \\ 0 & a_1 & 1 & \cdots & 0 & 0 \\ \cdots & \cdots & \cdots & \cdots & \cdots & \cdots \\ 0 & 0 & 0 & \cdots & a_1 & 1 \\ 0 & 0 & 0 & \cdots & 0 & a_1 \end{Vmatrix}, \quad \ldots, \quad x_n \to \begin{Vmatrix} a_n & 1 & 0 & \cdots & 0 & 0 \\ 0 & a_n & 1 & \cdots & 0 & 0 \\ \cdots & \cdots & \cdots & \cdots & \cdots & \cdots \\ 0 & 0 & 0 & \cdots & a_n & 1 \\ 0 & 0 & 0 & \cdots & 0 & a_n \end{Vmatrix}$$

(all are Jordan matrices of order p), we obtain infinitely many pairwise nonisomorphic irreducible representations of dimension lp.

The theorem is therefore proved.

THEOREM 5. *If M is an algebraic lattice over an algebraically closed field k which is also a finitely presented lattice of finite length, then M has only a finite number of pairwise nonisomorphic irreducible representations.*

For a representation φ of M and an interval $[a, b]$, denote by $v_\varphi[a, b]$ the number $\dim\varphi(b) - \dim\varphi(a)$. Let $p_1, \dots, p_l$ be a full system of pairwise incomparable simple intervals of the lattice M and associate to a representation φ the l-dimensional vector with real nonnegative components

$$V_\varphi = \big(v_\varphi(p_1), v_\varphi(p_2), \dots, v_\varphi(p_l)\big).$$

LEMMA 1. *Under the hypotheses of the theorem suppose that φ and ψ are representations of the lattice M over k such that $V_\varphi = V_\psi$. Then φ and ψ are isomorphic as representations.*

We first indicate how Lemma 1 implies Theorem 5. Let A be the set of vectors V_φ corresponding to all pairwise nonisomorphic irreducible representations of M. From Lemma 1 and the Krull-Schmidt theorem, which evidently holds for representations of lattices, it follows that no two different linear combinations of vectors from A with integral nonnegative coefficients can be equal; otherwise we would obtain two nonisomorphic representations with the same dimension vectors. This in turn means that the vectors from $\mathfrak{A}$ are linearly independent over the field of rational numbers, so that their number cannot exceed l.

PROOF OF LEMMA 1. In the lattice M, choose an arbitrary maximal chain $0 < m_1 < m_2 < \cdots < m_t = 1$; for an arbitrary representation φ: $M \to L(V)$, choose in the space V a basis compatible with the chain of subspaces $0 \subset \varphi(m_1) \subset \varphi(m_2) \subset \cdots \subset \varphi(m_t) = V$. For any $m \in M$ we can clearly choose a basis of $\varphi(m)$ such that its elements, written as the rows of a matrix, have the form

$$A_m = \left\|\begin{matrix} 0\cdots E & B^{(m)}_{1,i_1+1} & \cdots & 0 & B^{(m)}_{1,i_2+1} & \cdots & 0B^{(m)}_{1,i_r+1} & \cdots & B^{(m)}_{1,t} \\ 0\cdots 0 & 0 & \cdots & E & B^{(m)}_{2,i_2+1} & \cdots & 0B^{(m)}_{2,i_r+1} & \cdots & B^{(m)}_{2,t} \\ \cdots & \cdots & \cdots & \cdots & \cdots & \cdots & \cdots & \cdots & \cdots \\ 0\cdots 0 & 0 & \cdots & 0 & 0 & \cdots & EB^{(m)}_{r,i_r+1} & \cdots & B^{(m)}_{r,t} \end{matrix}\right\|$$

(here E denotes unit matrices of possibly different orders). The numbers $i_1, \dots, i_r$ for a given element are not uniquely defined; however, one can fix them for each m in advance knowing only the lattice and not the representation. We shall call a matrix of the form A_m a canonical form of a matrix describing the subspace $\varphi(m)$.

We now construct a field $K \supset k$ and a representation Φ such that any representation with vector of dimensions $V_\varphi = V_\psi$ (in particular, φ and ψ themselves) is a specialization of Φ. Note first that the equality $V_\varphi = V_\psi$ implies that $\dim\varphi(m) = \dim\psi(m)$ for any $m \in M$. Let A_s be the canonical matrices for all elements s of a finite system S of generators of M. The components of all A_s contained in nonzero and nonunit blocks $B^{(s)}_{i,j}$ will be considered as independent variables and adjoined to k in a transcendental manner. We shall now impose relations on these variables so that, for any lattice polynomial f in elements of S, the dimension of the subspace obtained by substituting for $s \in S$ the linear spans of the rows of matrices A_s is equal to $\dim \varphi(f(S))$. These relations will be added by induction on the length of f. Suppose, for example, that $f = g + h$ and that the required conditions have already been met for polynomials g and h. We now add the limitations which guarantee that the dimension of the sum of subspaces corresponding to g and h is not greater than $\dim \varphi(f(S))$; these are algebraic relations which can be expressed by saying that certain determinants are equal to zero. Let K' be the field obtained by adding all such relations and determine over K' the intersection of subspaces corresponding to g and h; impose on the indeterminates relations (again saying that certain determinants are equal to zero) guaranteeing that the dimension of the intersection is not greater than $\dim \varphi(g(S) \cdot h(S))$. Then the dimensions of the subspaces corresponding to $f = g + h$ and gh are equal, respectively, to $\dim \varphi(f(S))$ and $\dim \varphi(g(S) \cdot h(S))$. By applying this construction to all lattice polynomials, we obtain the required field K and representation Φ.

Suppose now that M is an algebraic lattice over the field k; then the representation Φ is equivalent to some representation φ_0 over k. We shall denote the canonical matrices describing the representation Φ and φ_0 by A_m and A_m^0 respectively. Equivalence of Φ and φ_0 indicates the existence of square matrices Y and Z_s ($s \in S$), with components in K, such that $Z_s A_s^0 = A_s Y$ for all $s \in S$. It is not hard to see that these matrices must be of echelon type:

$$Z_s = \begin{Vmatrix} E & * & \cdots & * \\ 0 & E & \cdots & * \\ \cdots & \cdots & \cdots & \cdots \\ 0 & 0 & \cdots & E \end{Vmatrix}, \qquad Y = \begin{Vmatrix} E & * & \cdots & * \\ 0 & B & \cdots & * \\ \cdots & \cdots & \cdots & \cdots \\ 0 & 0 & \cdots & E \end{Vmatrix}$$

(the number of steps in Y is equal to t, and in Z_s it is the number of steps in A_s), and that the matrices Z_s are completely determined by Y.

LEMMA 2. *It is possible to choose the matrices Y and Z_s in such a way that all their components are polynomials in the components of the matrices A_s with coefficients in k.*

The proof is by induction on t. We first determine matrices Y' and Z_s' such that the first $t-1$ vertical strips coincide for all matrices $A_s' = (Z_s')^{-1} A_s Y'$ and A_s^0, and then matrices Y'' and Z_s'' such that in all matrices $A_s' = (Z_s'')^{-1} A_s' Y_s''$ and A_s^0 the blocks coincide except, possibly, for the upper right ones in those matrices which have a unit matrix in the upper left corner; this is also possible by the inductive hypothesis applied to matrices obtained from A_s' and A_s^0 by discarding the first vertical and (for those matrices for which the discarded strip is nonzero) the first horizontal rows. Replacing A_s by A_s'', we see that it remains to prove the lemma only in the case when the elements which do not belong to k occur only in the upper right corners of the matrices A_s. But in this case the assertion is trivial, since for the determination of matrices Z_s and Y for nondiagonal blocks, we obtain (using the hypothesis that M is algebraic) a system of equations with coefficients in k, for which only certain free terms are not contained in k.

LEMMA 3. *Suppose that ρ is a representation of M over k such that $V_\rho = V_\varphi = V_\psi$. Then ρ is equivalent to φ_0.*

Indeed, ρ is a specialization of Φ. Applying the same specialization to the matrices Y and Z_s in Lemma 2, we obtain matrices establishing the equivalence of ρ and φ_0 over K and therefore also over the infinite field k.

In particular, the representations φ and ψ are equivalent to φ_0 and therefore to each other, which completes the proof of Lemma 1 and Theorem 5.

REMARK. The theorem ceases to be true if the lattice is generated by a finite set, but these generators are linked by an infinite system of relations. For constructing a suitable example, we shall need the following result (cf. [3]): there exists a number N such that the lattice of all subspaces of a vector space of dimension n over a finite field k is generated by N elements for all n. Let P be a free lattice with N generators $p_1, \dots, p_N$; according to the cited theorem, for every n we have an epimorphism φ_n: $P \to L(k^n)$. These epimorphisms induce a homomorphism $\Phi = \prod_n \varphi_n$ of the lattice P to the product of all lattices $L(k^n)$. Let M be the image of this homomorphism and $\bar{\varphi}_n$: $M \to L(k^n)$ the homomorphism which makes the diagram

$$\begin{array}{ccc} P & \xrightarrow{\Phi} & \prod_n L(k^n) \\ \downarrow & & \downarrow \\ M & \xrightarrow{\bar{\varphi}_n} & L(k^n) \end{array}$$

commutative, where the right arrow is the natural projection. The representations $\bar{\varphi}_n$: $M \to L(k^n)$ are indecomposable and are the only indecomposable representations of the finitely generated lattice M over the algebraic closure of the field k.

It seems likely that a finitely presented algebraic lattice has finite length, but the author has not yet succeeded in proving this.

BIBLIOGRAPHY

1. Garrett Birkhoff, *Lattice theory*, 2nd rev. ed., Amer. Math. Soc., Providence, R. I., 1948.

2. George Grätzer, *General lattice theory*, Birkhäuser, Basel, and Academic Press, New York, 1980.

3. A. A. Kravchenko, *The minimal number of generators of the lattice of subspaces of a finite-dimensional linear space*, Zap. Nauchn. Sem. Leningrad. Otdel. Mat. Inst. Steklov. (LOMI) **114** (1982), 148–149; English transl. in J. Soviet Math. **27** (1984), no. 4.

4. L. A. Nazarova and A. V. Roĭter, *Representations of partially ordered sets*, Zap. Nauchn. Sem. Leningrad. Otdel. Mat. Inst. Steklov. (LOMI) **28** (1972), 5–31; English transl. in J. Soviet Math. **3** (1975), no. 5.

5. L. A. Skornyakov, *Complemented modular lattices and regular rings*, Fizmatgiz, Moscow, 1961; English transl., Oliver and Boyd, London, 1964.

6. Peter Gabriel, *Unzerlegbare Darstellungen*. I, Manuscripta Math. **6** (1972), 71–103, 309.

Translated by G. MAXWELL

ABCDEFGHIJ–AMS–89876